ENERGY CONSERVATION IN HEATING, COOLING, AND VENTILATING BUILDINGS

SERIES IN THERMAL AND FLUIDS ENGINEERING

JAMES P. HARTNETT and THOMAS F. IRVINE, JR., Editors
JACK P. HOLMAN, Senior Consulting Editor

Cebeci and Bradshaw	• Momentum Transfer in Boundary Layers
Chang	• Control of Flow Separation: Energy Conservation, Operational Efficiency, and Safety
Chi	• Heat Pipe Theory and Practice: A Sourcebook
Eckert and Goldstein	• Measurements in Heat Transfer, 2d edition
Edwards, Denny, and Mills	• Transfer Processes: An Introduction to Diffusion, Convection, and Radiation
Fitch and Surjaatmadja	• Introduction to Fluid Logic
Ginoux	• Two-Phase Flows and Heat Transfer with Application to Nuclear Reactor Design Problems
Hsu and Graham	• Transport Processes in Boiling and Two-Phase Systems, Including Near-Critical Fluids
Kestin	• A Course in Thermodynamics, revised printing
Kreith and Kreider	• Principles of Solar Engineering
Lu	• Introduction to the Mechanics of Viscous Fluids
Moore and Sieverding	• Two-Phase Steam Flow in Turbines and Separators: Theory, Instrumentation, Engineering
Nogotov	• Applications of Numerical Heat Transfer
Richards	• Measurement of Unsteady Fluid Dynamic Phenomena
Sparrow and Cess	• Radiation Heat Transfer, augmented edition
Tien and Lienhard	• Statistical Thermodynamics, revised printing
Wirz and Smolderen	• Numerical Methods in Fluid Dynamics
	PROCEEDINGS
Hoogendoorn and Afgan	• Energy Conservation in Heating, Cooling, and Ventilating Buildings: Heat and Mass Transfer Techniques and Alternatives
Keairns	• Fluidization Technology
Spalding and Afgan	• Heat Transfer and Turbulent Buoyant Convection: Studies and Applications for Natural Environment, Buildings, Engineering Systems
Zarić	• Thermal Effluent Disposal from Power Generation

A publication of the International Centre for Heat and Mass Transfer
Belgrade

ENERGY CONSERVATION IN HEATING, COOLING, AND VENTILATING BUILDINGS

Heat and Mass Transfer Techniques and Alternatives

Volume 2

Edited by

C. J. Hoogendoorn

Applied Physics Department
Delft University of Technology
Delft, The Netherlands

and

N. H. Afgan

University of Belgrade
Belgrade, Yugoslavia

HEMISPHERE PUBLISHING CORPORATION

Washington   London

ENERGY CONSERVATION IN HEATING, COOLING,
AND VENTILATING BUILDINGS

1 2 3 4 5 6 7 8 9 0   D O D O   7 8 3 2 1 0 9 8

Library of Congress Cataloging in Publication Data

Main entry under title:

Energy conservation in heating, cooling, and ventilating
   buildings.

   Lectures and papers presented at a seminar sponsored
by the International Centre for Heat and Mass Transfer,
held at Dubrovnik, Yugoslavia, Aug. 29–Sept. 2, 1977.
   Includes indexes.
   1. Buildings—Energy conservation—Congresses.
2. Heat—Transmission—Congresses. 3. Mass transfer—
Congresses. I. Hoogendoorn, C. J. II. Afgan, Naim.
III. International Center for Heat and Mass Transfer.
TJ163.5.B84E528   697   78-1108
ISBN 0-89116-095-7 (vol. 2)

CONTENTS

VENTILATION AND AIR MOVEMENT INSIDE BUILDINGS

MATHEMATICAL MODELLING AND EVALUATION OF ENERGY REQUIREMENTS IN HEATING AND COOLING OF BUILDINGS

CONTENTS

PREFACE

The International Centre for Heat and Mass Transfer (ICHMT) organizes an annual seminar on a subject related to transport phenomena. The one on "Heat and Mass Transfer in Buildings," held at Dubrovnik, Yugoslavia, August 29–September 2, 1977, was the forum for presentation of the lectures and papers contained in these volumes. Dealing with the energy requirements for heating, cooling, and ventilating buildings, these papers are from heat and mass transfer scientists, from engineers working in the field of heating and ventilation, and from building engineers.

The recent awareness of the need to conserve our primary energy resources has emphasized the necessity of energy conservation strategies in industrialized societies. Many studies on the energy problem have been made. The 1975 ICHMT seminar, "Future Energy Production—Heat and Mass Transfer Problems,"* followed the energy crisis of 1973 and stressed the need to find new sources of energy. A recent report from the Workshop on Alternative Energy Strategies (WAES) stresses that in the next 25 years decline in the supply of primary energy will cause serious problems. Various energy scenarios have been made for the period 1985–2000. All show a decline in oil and natural gas production as well as the need to change over to drastically different energy policies. As these changes often require lead times of five to ten years, timely decisions are needed to avoid catastrophic consequences of present policies. This is, in fact, the greatest challenge our technological societies have to face in the next 25 years.

As use of primary energy for heating and cooling buildings is about one-third of total consumption for many countries, we have here an important field for changes in energy policies. Building practice was until recently based on a large and relatively cheap energy supply. In the future we will still have sufficient energy, certainly if we include that derived from solar and fusion sources, but it will not be cheap or easily available. If we are to introduce new energy-conserving techniques on a large scale in building practice, we must make timely decisions as necessary. As lifetimes of buildings are from 25 to over 100 years, decisions taken now will have to be based on future energy requirements. This is certainly very important, because our well being and the full development of our communities strongly depend on the environmental control of our living and working areas. Not only technical, but also economic and human factors are involved.

This volume contains papers on the many factors that the energy requirements of buildings involve. These range from human comfort conditions and economic aspects to heat and moisture transfer in building materials, ventilation of buildings, mathematical modelling, and new techniques like heat pumps and solar energy.

We would like to acknowledge the seminar committee members who cooperated in

---

*Published as J. C. Denton and N. H. Afgan (eds.), *Future Energy Production Systems*, vols. 1 and 2, Hemisphere, Washington, 1976.

setting up the programme, selecting the papers, and organizing the various sessions of the meeting: V. P. Motulevich, F. Kreith, A. F. E. Wise, K. Gertis, and B. Givoni. They all contributed review lectures in their special fields, which are included in this volume.

C. J. Hoogendoorn
Chairman, Seminar Committee

N. H. Afgan
Scientific Secretary, ICHMT

NEW TECHNIQUES IN HEATING AND COOLING OF BUILDINGS

TECHNICAL AND ECONOMIC BASES OF DISTRICT HEATING DEVELOPMENT

L. S. KHRILEV

Siberian Power Institute
Siberian Branch of the USSR Academy of Sciences
Irkutsk, USSR

District heating is treated as centralized heat supply from the power stations on base of the combined electric power and heat generation. Such power plants are usually called heating and power plants (HPP). Systems of the heat supply, including sources of heat (heating and power plants, heating and industrial boiler plants), main heat networks linking the sources of heat energy with large consumers and different installations by the consumers are called district heating systems.

In the USSR district heating is widely developed. It can be explained by many reasons: concentration of heat consumption by means of planned allocation of industry, mainly manystoreyed buildings of towns, relatively rigorous climate in the main regions of the USSR, the tendency of using comparatively low-grade fuel for heat supply (if the sanitary conditions are acceptable enough), comparatively high cost of fuel in many regions of the country. Nowadays about 1/3 of demand of the national economy of the country for heat energy is satisfied at the expense of district heating. It gives the fuel economy about 35–40 mln tons of coal equivalent per year, that can be compared with the fuel economy obtained from hydroelectric power plants (about 40–45 mln tons of coal equivalent per year).

The annual input of the electric capacity at the heating and power plants in our country reaches 2,5–3 mln KW, and the capacity of some of them - 1000–1500 MW. In future these indices will be much higher. The large equipment was developed and is widely used for heating and power plants: back-pressure turbines, extraction turbines (with one or with two extractions), having unit capacity 100–250 MW, the heat supply from them being equal to 160–330 Gcal-h, power and large water boilers. This circumstance allows to accomplish the systematic introduction of new heating and power plants in different regions of the USSR.

The fundamental scheme of an industrial heating plant on organic fuel is given in Fig.1. Both the composition of main equipment and complex relations between it and the consumers of hot water and steam are shown here.

---

* According to the International System of Measures 1 Gcal-h corresponds to 4, 187 GJ or 1,163 MW.

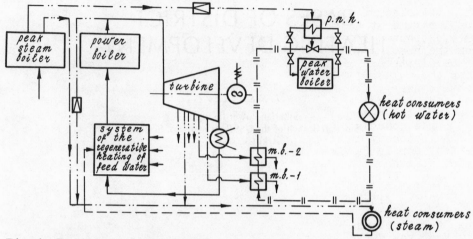

Fig.1. Fundamental scheme of industrial heating and power plants
on organic fuel.
        Legends:
————-steam;— — — –condensate from the industrial
enterprises;———„——-hot water;————- -feed water;
p.n.h. - peak network heater; m.b. - 1 - main boiler N 1,
m.b.- 2 - main boiler N 2.

The principles and special features of development of the
district heating in the USSR. The main significance of the district heating for the national economy consists in the fuel economy achieved by generation of: a) the electric energy in accordance with district heating regime (on the heat consumption -W_h),
under which the latent heat of steam generation is usefully utilized; it is a chief source of heat losses in electric energy
generation according to the condensational regime (W_c); b) the
heat for the exterior heat supply by means of increasing the
annual (net) efficiency of the boiler plants of the HPP (l_β =
0,80 + 82) in comparison to large boiler plants (l'_β = 0,76 + 0,78).
Heating and power plants equipped with steam extraction turbines
overexpend the fuel in comparison with the condensational electric station (CES) at the electric energy generation according to
the condensational regime (mainly because of the lesser unit capacity of the generators).
    Generally, the fuel economy at the heating and power plants
(tons of the coal equivalent), referred to 1 Gcal., which is supplied by the HPP for the whole year, is defined as follows [3]:

$$\Delta b = \frac{0{,}143}{l_\beta}\left[y_e\left(q_c - q_h\right)\delta_{year} - \frac{W_c}{Q_{year}}\left(q_c - q_{CES}\right)\right] + 0{,}143\left(\frac{1}{l'_\beta} - \frac{1}{l_\beta}\right) \quad (1)$$

where: Q_{year} - the yearly supply of heat from the HPP; y_e -
the specific generation of electric energy on the exterior heat
consumption of the HPP, i.e. taking into account the regenerative cycle q_h, q_c , q_{CES} - the average yearly specific expenditure of heat on the supplied MW-h according to the district heating and condensational regimes at the HPP and on the substituting
condensational electric station (CES) respectively ;

- 3 -

W_c - the electric energy generation according to the condensational regime at the HPP, MW-h; α year - a fraction of the steam heat, extracted from the turbines in the yearly supply of heat from the heating and power plant.

The first term of the given expression in square brackets defines the value of the fuel economy, got from the generation of electric energy at the HPP according to the district heating regime; the second term in the brackets means the fuel overexpenditure from the generation of the electric power at the HPP according to the condensational regime; the third term defines the additional economy of fuel at the HPP from the centralization of the heat supply (at the expense of the higher efficiency of the boiler plants of the HPP in comparison to the district boiler plants).

The value α year depends on the value of the rated coefficient of the district heating ($\delta_{z.h.}$), which is one of the most important indices and specifies the fraction of the steam heat, taken from turbines ($Q_{z.h.}$) in total calculated hourly supply of heat from the HPP ($Q_{z.e.}$). Accordingly the value I $-\delta_{z.h.}$ defines the fraction of the heat supplied directly from the boiler plant to cover the peak part of the calculated heat load of the HPP. The character of yearly duration curve of the heating load and the corresponding dependance α year $= f(\delta_{z.h.})$ are represented in Fig.2. A hatched part of this figure signifies the heat supply from the peak boiler plant of the HPP.

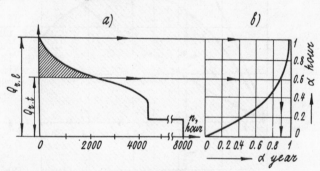

Fig.2. Yearly duration curve (a) and integral curve (b) of heat load.

The influence of the calculated coefficient of the district heating on the fuel economy obtained at the heating and power plant (kg/Gcal of the yearly heat supply from the HPP), characterizes the composition of the Fig.3 for the invariable heat load of the HPP. Changes of this economy (Δb_e) at the generation of electric power are shown on this figure above the axis of absciss, below this axis - at the generation of heat (Δb_h). Line 1 defines the theoretical economy of the fuel obtained from the generation of the electric energy by district heating regime depending on the value α year; line 2 shows the overexpenditure of fuel from the generation of the electric energy at the condensational regime in comparison to the generation of the electric power at the compared condensational electric station; line 3 specifies the difference between the ordinates of the lines 1 and

2, that is the real fuel economy from the generation of electric power at the HPP (Δb_e). The ordinate Δb_h defines the additional economy of fuel obtained at the HPP from the centralization of the heat supply. The mutual arrangement of the lines 1, 2, 3 may be essentially changed in dependence on the type and regime of using the compared turbines of the heat and power plant and condensational electric station.

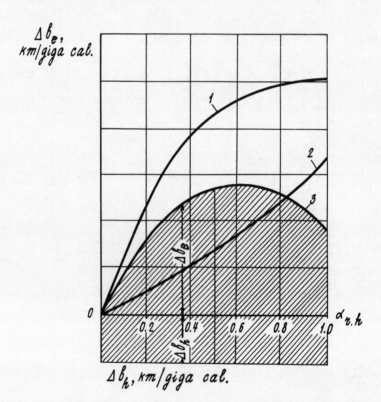

Fig.3. Character of change in the heat economy of district heating versus the value $\delta_{z.h}$.

One can draw two important conclusions out of the Fig.3:
   1. The assumed value of the calculated coefficient $\delta_{z.h}$ essentially influences on the heat economy of the district heating; the more is the difference between the capacities and the initial parameters of steam of the turbines at the heating and power plant and substituted condensational electric stations, the less must be the value of the calculated coefficient of the district heating under the conditions of obtaining the largest fuel economy, and vice versa. As a rule it changes from 0,5 to 0,7 for the steam-turbine heating and power plants on organic fuel.

2. The economy of district heating essentially changes (using the district heating turbines of the same types in dependence on the type and parameters of the substituted turbines of the condensational electric stations. Therefore, the growth of the capacities of the HPP is an important direction of increasing district heating economy. Its facilities depend directly on the concentration of the heating loads and on the heat network extension. Nowadays the distance of steam transport is about 2-3 km. The decrease of the pressure in the network usually limits this radius, because it is necessary to increase the pressure of steam taken away from the turbines of the HPP. As a result the electric energy generation on the heat consumption reduces, that brings to the decrease of the economy of the HPP. The distance of the hot water transmission in existing two-pipe scheme of the network is 10-15 km (one-pipe - for feed water, the other one - for the return cooled water to the HPP); the increase of this distance in two-pipe scheme of the network comes usually to a considerable rise in price of the heat network. Such considerable distance of the heat transmission at the existing loads of industrial complexes and town building may provide the heat load of the HPP to 1500-2000 t/h in steam or 3000-4000 Gcal-h in hot water. This corresponds to the electric capacities of the HPP on organic fuel in 600+800 and 1000+1200 MW respectively. District heating electric stations of such capacity are used in the USSR, and are also projected and built.

The main features of the district heating development in the USSR are:

1) growth of unit capacities of the main equipment and the heating and power plants themselves;

2) use of high and overcritical parameters of steam at the HPP;

3) installation of turbines of different types (with back pressure, turbines with one or two steam extractions), that allows to satisfy the needs in heat energy of different kinds of consumers;

4) construction, as a rule, of large out-of-town HPP (to provide the so called "pure use of energy" in towns) with a wide branched scheme of the main and distribution heat networks;

5) use of hot water as a heat-carrier for heating and so on.

Principles and efficiency of the use of nuclear district heating. The computations show that in case of building only nuclear condensational electric stations (NCES) in the USSR one can substitute with nuclear fuel less than 10% of total expenses of organic fuel (even in the long term perspective) [4]. To increase the fraction of the nuclear fuel in the fuel and energy complex of the country it is necessary to use widely this fuel for the purposes of the heat supply, keeping in mind the expediency of production of hot water and technological steam, first of all, at the nuclear heating and power plants (NHPP). It is important to consider the fundamental differences in the development of the district heating on nuclear and organic fuel, which are following. As is known the specific cost of generation of any kind of energy (z) is defined by the expression:

$$\mathit{z} = \frac{\rho \cdot c}{h} + \alpha \beta, \qquad\qquad (2)$$

where: C - the cost of the building of a generator per unit of
its capacity (productivity); p - the fraction of the so-called
constant part of the yearly costs from the installation cost including the normative coefficient of the efficiency of the investment; h - the yearly number of the hours of use of the installation capacity; a - the cost of a unit of the used fuel; b - the
specific expenditure of fuel for the generation of a unit of
energy.

The technical progress in power engineering leads to two basic
regularities, applied to different types of installations: to
constant decrease of b and to increase of c , which is the
less, the more is the unit capacity of the main equipment of the
installation. In consequence the cost of the generated energy z
decreases comparatively slowly, and to the mind of some specia-
lists, it can even increase. For example, one can give comparative data about condensational electric stations on organic and
nuclear fuel. So, for the NCES the value of c is 25-30% higher,
and the cost of the fuel component in the regions of the European
part of the USSR is 3 times lower, than for the condensational
electric station. The second typical feature of the NCES is the
essential decrease of the specific cost value by the increase of
the unit capacity of the nuclear reactors. According to foreign
data transition from the reactors of 500 MW (el.) to the reactors with unit capacity of 1000 MW (el.), leads to the decrease
of their specific cost (average) to 25-30%, and from the last
reactors to the reactors of 2000 MW (el.) to 15-20%.

Proceeding from the special features of the expenditures on
the nuclear power plants (NPP) one may come to two important
conclusions:

1) in view of the lower fuel component at the district generators of NHPP (in contrast to the HPP on organic fuel) for the
sake of the economy of the organic fuel it is necessary to generate the maximum possible amount of electric power at them according to the condensational regime;

2) in view of the essential dependence of the specific cost of
the nuclear reactor on its productivity one must install reactors
of the critical capacity at the NCES and at the NHPP. Consequence
of these conclusions is the decision about necessity of the building of the atomic electric stations of the combined condensa-
tional-district-heating type, equipped with reactors of the maximum capacity (in this period of time), condensational turbines
(type K) and district heating turbines with great attached condensational capacity * and with a constant steam expenditure
(type TK). Such decision is of most economy for the generation of
the heat energy, because it will be fulfilled on the reactors of
the highest capacity, and the proportional division of the expenses for NPP for the generation of electric and heat energy gives
the least value of the last one. At the same time the given solution allows to cover the heat loads by the NPP on a large scale,

---

* The attached condensational capacity of a turbine (N_{ca}) is
the capacity, generated at the maximum calculated steam extraction (for exterior heat supply) at the expense of transmission of steam into the capacitor. At this electric capacity of a turbine the more is N_{ca} the less is the possible
steam extraction from a turbine for heat supply.

as a minimum, the load of the settlement, where operating staff
of the nuclear electric stations lives and as a maximum, the load
of large energy and industrial complexes, built on the base of
NPP and obtaining from it electric and heating energy. That is
why NPP may essentially influence the rational location of industry and related building of living houses.

In some cases the electric capacity of the NPP, located for
the heat supply near big cities must be limited (by sanitary
norms, conditions of water supply and so on). It can result in
decrease of the number of the installed reactors or their unit
capacity, that in turn would bring to the decrease of the district heating economy. As the estimates show, in such cases the
nuclear district heating will be effective at the heat load of
500-1000 Gcal-h and higher (it depends on the cost of the substituted organic fuel and on the regimes of using the NHPP).

While considering the problems of development of the nuclear
district heating one usually takes into account the supply of the
hot water only from the nuclear power plants. Such point of view
is unjustified, as about one half of the heat is used by the national economy of the USSR in the form of the technological
steam. Therefore, the NHPP should supply steam for technological
purposes and hot water.

The practical realization of the idea of the nuclear district
heating requires profound studying of questions about the profile
and composition of the main equipment of the nuclear heating and
power plant, of the heat supply schemes, security and safety of
the steam supply of consumers and others. While solving these
problems one must distinguish two periods in the creation of the
equipment for the nuclear district heating: the first period is
the near term when the nuclear district heating and power plants
will use mastered water - graphite and water cooled reactors; the
second period is the perspective one, when there will be developed a special type of the reactor for the construction of nuclear
district heating stations, allowing to obtain steam with high parameters. In this case the calculated coefficient ($\sigma_{z.h.}$) district heating and the calculated heat supply from the extraction
($Q_{z.h.}$) are assumed to be the most important factors, influencing the choice of the profile of the district heating turbines
for the nuclear district heating electric stations. As a composition of the main equipment for these nuclear electric stations
one understands the number of nuclear blocks (n_ℓ) and the number
of district heating (n_h) and condensational (n_c) turbines,
which are installed together with the reactors. As a nuclear district heating electric station is an element of the district heating system, so the search of the optimized parameters must be
organized taking into consideration special features of projecting and development not only of the nuclear district heating
electric stations, but also of the other elements of this system,
including district and peak boiler plants and main heat networks.
Besides one must consider the maximum electric capacity of the
NHPP as a predeterminated value (as a version - in dependence on
the number of the installed blocks), and the heat load of a town
($Q_{z.\ell.}^{NHPP}$) which is expedient to join to the NHPP is a variable
(by every value n_ℓ). It is naturally that with the increase of
$Q_{z.\ell.}^{NHPP}$ the productivity of the district boiler plants will
diminish and the radius of the heat supply from the NHPP will

increase and consequently the expenditures of the town main and distribution heat networks ($\mathcal{E}_{h.n.}$) will increase too. So if the value $q_{z.e.}^{NHPP}$ changes, the total expenses ($\mathcal{E}_i$) in the district heating system being equal tp

$$\mathcal{E}_i = \mathcal{E}_{NHPP(i)} + \sum_{\gamma=1}^{z} \mathcal{E}_{d.6.(i\gamma)} + \mathcal{E}_{p.6(i)} + \mathcal{E}_{h.n.(i)}^t + \mathcal{E}_{h.n.(i)}^m + \mathcal{E}_{e(i)}, \text{rbl} \quad (3)$$

will also vary,
where $\mathcal{E}_{NHPP(i)}$ — discounted expenditures on the NHPP at the i-th value of $q_{z.e.}^{NHPP}$, rbl; $\mathcal{E}_{d.6.(i\gamma)}$, $\mathcal{E}_{p.6(i)}$ — the same, on the γ district and peak boiler plant, rbl; $\mathcal{E}_{h.n.(i)}^t$ — the same on the transit and main heat networks respectively, rbl; $\mathcal{E}_{e(i)}$ — discounted expenditures on the substituted electric energy, rbl; z — the number of district boiler plants.
  While choosing the profile of the main equipment of the NHPP the task is to find such values of $\delta_{z.h.}(0 < \delta_{z.h.} \leq 1), q_{z.h.}$, by which the expenses, computed by expression (3) will be minimal. In addition to that one must take into account the work of the peak and district boiler plants in covering the given heat load: at $\delta_{z.h.} = 1$ the peak boiler plant is considered to be only as a reserve source of heat supply (if the nuclear block goes out of action); at $\delta_{z.h.} < 1$ – as a main source, and at $\delta_{z.h.} = 0$, i.e. at a separate scheme of energy supply district boiler plants are taken into account. The boiler plants on organic fuel are assumed as peak and district boiler plants, in so far as now the question of construction of the atomic boiler plants is still being developed (one must take into account the variable regimes of nuclear reactors operation).
  While solving the task of choice of the profile, composition and fields of applying the nuclear stations with turbines of type TK, it is the most difficult to find out dependences of electric capacity of the HPP on the variable value of heat supply from the extractions of the turbines. To define such dependence a special method helping to analyse all the possible regimes of the turbines is developed. Herewith, the computation of the boiler unit is performed; parameters by the sections of turbines are defined; the capacity of separate sections and the whole turbine are computed; the heat supply from the regulated extractions is found[6].
  The results of computations for the NHPP are illustrated in Fig.4. Nuclear heating and power plants with the blocks BBЭP-1000 and turbines of type TK-400/500-60 and the initial conditions that are typical for the main regions of the European part of the USSR were studied. Fig.4 shows the changes in the discounted costs on the district heating system with NHPP depending on the index $\delta_{z.h.}$ at different values of the heat load. The upper envelope line of hatched parts corresponds to the maximum expenses and the lower envelope line – to the minimum ones on the organic fuel. From Fig.4 it follows that the optimum value of the district heating coefficient changes somewhat in dependence on the value of the heat load, but in general it is in the range of 0,6÷0,7. So it is worthwhile to cover only the basic heat load from the NHPP. The peak of such a curve must be supplied with heat from the peak

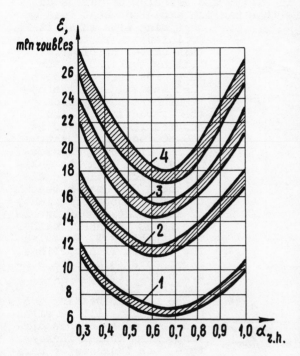

Fig.4. Variation of discounted costs on the heating district system with the NHPP depending on the calculated coefficient of district heating at different values of heating household load ($Q_{z.\ell}^{NHPP}$); line 1 - $Q_{z.\ell}^{NHPP}$ =1000 Gcal-h, line 2 - 1500, line 3 - 2000, line 4 - 2500.

boiler plant on gas and situated within the town or near it. It permits to fulfill the most secure and economic development of the district heating system of the town in general, herewith, first of all one must build large boiler plants. But at the increase of heat loads the turbines of type TK are introduced at the nuclear district heating stations; the basic part of the heat load curve is covered from them and the boiler plants cover the peak heat loads.

Two possible initial conditions were taken into account in the calculations while studying the nuclear heating and power plants of the combined type (with turbines of type TK and K) and at its comparison with purely condensational nuclear electric station: a) apart from the type of the nuclear electric station one and the same site is fixed for its construction; b) nuclear heating and power plants are located nearer to the centers of heat and electricity consumption than the nuclear condensational power plants, having - as a rule - limited possibilities for building special reservoirs - coolants. Under such conditions one must adopt the return system of technical water supply with water cooling towers for the NHPP. At the same time for the nuclear condensational power plants which have relatively greater freedom for the choice of sites one can consider the system of the technical water supply with water reservoirs - coolants. It brings to the necessity of taking into account additional expenses on the system of technical water supply for the NHPP in comparison to the atomic power plants.

The first initial premise shows the efficiency of the construction of the combined type of atomic electric stations at the maximum possible electric capacity (by the local conditions and the safety conditions). The computations performed at the second premise show that in this case it is efficient to build the nuclear district heating electric plants with turbines of type TK. This thesis is illustrated in Fig.5. The crossing of curves on this figure allows to define this minimum expedient heat load beginning with which it is efficient to install a certain number of district heating turbines at the NHPP. The analysis shows that the value of the optimum heat supply from the extractions of the turbine TK-400/500-60 is 450-600 Gcal-h, id est, it appears to be considerably lower than its maximum possible value equal to about 900 Gcal-h that testifies about efficiency of installations of turbines TK at the NHPP, rather than turbines T. Herewith the version with turbines TK appears to be more economic than the one with turbines TK+K.

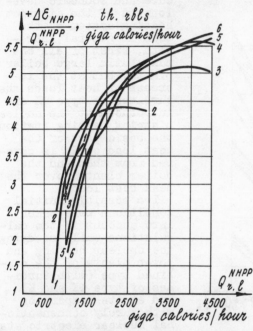

Fig.5. Changes of the economy of specific discounted costs on the combined scheme of energy supply with the NHPP in comparison to the separate scheme depending on the heat load value and the composition of nuclear blocks BBƏP-1000 (n_ℓ); district heating turbines of type TK-400/500-60 (n_h) and condensational turbines of type K-500-60 (n_c): line 1 – n_b=1, n_h=1, n_c=1; line 2 – n_b=1, n_h=2; line 3 – n_b=2, n_h=3, n_c=1; line 4 – n_b=2, n_h=4; line 5 – n_b=3, n_h=5, n_c=1; line 6 – n_b=3, n_h=6.

The nuclear heating and power plants will be situated at a considerable distance from towns. Therefore, the increase of parameters of feed water may be more effective for them than for the HPP on organic fuel ($t_{1(z)}$). The computations performed confirm this, their results for the two values of heat load of the nuclear heating and power plants are given in Fig.6. From this figure it follows that with the inrease of the difference between the NHPP and heat consumers the transition to the increased temperature curves will be efficient ($t_{1(z)} > 150°C$). In this case it will be advisable to use the one-pipe scheme of the heat network from the nuclear heating and power plants working mainly under basic regime with hot water supply which is approximately equal to its expenses on the household water supply. The other sources of heat being situated in a town and working on organic fuel mainly cover the half-peak and peak heat load. The reality of such a scheme depends first of all on the possibility of obtaining required water expenditure on the

NHPP site and on the distribution and capacity of the heat sources on organic fuel.

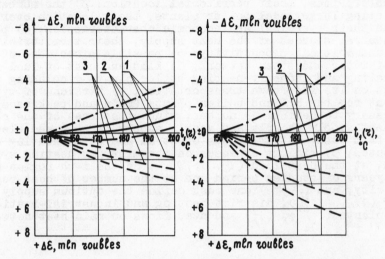

Fig.6. Economy (+) or overexpenditure (-) of discounted costs
on the NHPP, heat networks and in general on district
heating system by increasing the calculated temperature
of the feed network water ($t_{1(2)}$); a) $q_{z.\ell}^{NHPP}$ =2000 Gcal-h;
b) $q_{z.\ell}^{NHPP}$ =3500 Gcal-h. Line 1 - $\mathcal{L}_{h.n.}$ =10 km; line 2- $\mathcal{L}_{h.n.}$
=15 km; line 3 - $\mathcal{L}_{h.n.}$ = 20 km; $\underline{\quad\quad}$ - nuclear HPP, $-\,-\,-$
- transit heating networks, $\underline{\quad\quad}$ - heating on the
whole.

In case of heat supply both for heating and technological purposes it seems to be advantageous to install at the NHPP the direct heating turbines ПТК, in particular, the turbine TK-400/500-
60 that can supply about 450 t/h of heat in steam (with pressure
1.2-1.5MPa) and 130 Gcal-h in hot water. Still it is necessary
to study specially the efficiency and security of steam supply
from the NHPP.

Methods of complex optimization of the district heating system
development. Due to complexity of these systems and duration of
time needed to substantiate and realize the principal solutions
they are studied for the perspective of 10-15 years. Such is the
projecting practice in the USSR. The so-called schemes of heat
supply of the towns and industrial centres are developed for such
a long-term perspective. While preparing these schemes one per-
forms the complex analysis of different sources of the heat energy and versions of heat transportation from them to the consumers
that depend on productivity and location of these sources. Such
an approach results in a choice of the optimal solution on the
number and capacity of sources of the centralized heat supply and
on the direction of the routes of main heat network.

In general the problem of complex optimization of the district
heating system development (at the stage of working out the heat
supply scheme) can be formulated in the following way (on the
example of the HPP on organic fuel). There are given: 1) heat

regions of a town, hourly and yearly loads for each one; 2) hourly and yearly heat loads (for steam and hot water) of large industrial enterprises, their territorial location; 3) the number of
the existing large local boiler plants, their location over the
city and their technical and economic characteristics; 4) possible central sources of the heat supply, their territorial location and critical productivity; 5) possible directions of the
route of water and steam network; 6) kind and cost of fuel for
HPP, heating and industry-heating boiler plants (including the
expenses on its intertown transportation; 7) versions of type of
equipment for the HPP and boiler plants (main and peak ones). It
is necessary to define: 1) number, kind of sources of the centralized heat supply, hourly and yearly supply of heat and electric
energy from them; 2) directions of steam and water heating network routes; 3) minimum expenses on the district heating system.

Thus, at the heat load ($q_i^1, \ldots, q_i^\tau$) of the i-th region given
by the years of studied period and at the number of heat regions
of the city, one ($i = \overline{1, n}$) one must define the optimum supply from
the HPP ($x_{NPP}^1, \ldots, x_{NPP}^\tau$), district heating and industrial-heating
boiler plants ($x_\delta^1, \ldots x_\delta^\tau$) and heat flows on main heat networks
($x_{h.n.}^1, \ldots, x_{h.n.}^\tau$).

Find

$$\mathcal{E} = min \left\{ \mathcal{E} \left[x_{NPP}, x_\delta, x_{h.n.} (x_{NPP}) \right] \right\} , \qquad \text{rbl} \qquad (4)$$

where

$$x_{NPP} = x_{NPP}^1, \ldots, x_{NPP}^\tau ;$$

$$x_\delta = x_\delta^1, \ldots, x_\delta^\tau ;$$

$$x_{h.n.} = x_{h.n.}^1, \ldots, x_{h.n.}^\tau ;$$

$$\left\{ \begin{array}{l} x_\delta^1 = \sum_j x_{\delta(j)} + \sum_\gamma x_\delta^1 (\gamma) ; \\ \cdots \cdots \cdots \cdots \cdots \\ x_\delta^\tau = \sum_j x_{\delta(j)} + \sum_\gamma x_\delta^\tau (\gamma) ; \end{array} \right.$$

$$x_{h.n.} = \sum_\ell x_{h.n.(\ell)}^1 + \cdots + \sum_\ell x_{h.n.(\ell)}^\tau ;$$

$$j = \overline{1, p} ; \quad \gamma = \overline{1, c} ; \quad \ell = \overline{1, \mathcal{L}} ; \quad t = \overline{1, \tau}$$

$$\mathcal{E} \left[x_{NPP}, x_\delta, x_{h.n.} (x_{NPP}) \right] = \sum_{t=1}^\tau \xi^t \left\{ \mathcal{E}_{NPP} (x_{NPP}^t) + \right.$$

$$+ \sum_\ell \mathcal{E}_{\delta(j)}^t (x_{\delta(j)}^t) + \sum_\gamma \mathcal{E}_{\delta(\gamma)}^t (x_{\delta(\gamma)}^t) +$$

$$\left. + \sum_\ell \mathcal{E}_{h.n.(\ell)}^t (x_{h.n.(\ell)}^t) + \mathcal{E}_e^t \left[W_3^t (x_{NPP}^t), \mathcal{Z}_e (h_e^t) \right] \right\}, rbl$$

$$(5)$$

While solving the problem the following conditions and constraints must be taken into consideration:

a)
$$
\begin{cases}
x'_{NPP} + x_{\mathcal{B}}' = \sum_{i=1}^{n} q_i^1 \; ; & 0 \le x^t_{NPP} \le q_{NPP} \\
\cdots\cdots\cdots\cdots & 0 \le x^t_{\mathcal{B}(j)} \le q_{\mathcal{B}(j)} \qquad (6) \\
x^\tau_{NPP} + x^\tau_{\mathcal{B}} = \sum_{i=1}^{n} q_i^\tau \; ; & 0 \le x^t_{\mathcal{B}(\gamma)} \le q_{\mathcal{B}(\gamma)}
\end{cases}
$$

where $\mathcal{E}^t_{NPP}$ - the discounted expenditures on the HPP in the year t , rbl; $\mathcal{E}^t_{\mathcal{B}(j)} \mathcal{E}^t_{\mathcal{B}(\gamma)}$ - the same on the j-th district heating and j-th industrial heating boiler plants, rbl; $\mathcal{E}^t_{h.n(i)}$ - the same on the i-th section of the heat network, rbl; $\mathcal{E}^t_e$ - the additional expenses on the substituted electric energy (at the decrease of the electric capacity of the HPP) in the year t, rbl; $x^t_{\mathcal{B}(j)}, x^t_{\mathcal{B}(\gamma)}$ - the supply of heat from the j-th district heating and γ-th industrial heating boiler plant in the year t , respectively, Gcal-h; $q_{NPP}, q_{\mathcal{B}(j)}, q_{\mathcal{B}(\gamma)}$ - the maximum possible calorific power of the HPP, j-th heating and γ -th industrial heating boiler plants, respectively Gcal-h; ρ, c - the number of these boiler plants respectively; $\mathcal{Y}$ - the number of sections of heat networks; W^t_s - the substituted electric energy; h^t_j - the yearly number of the hours of using the substituted electric capacity, hours; $\mathcal{z}^t_e$ - the specific shadow prices of the electric energy in the year t , rbl/MWh.

There are no methods of solving the examined problem, that permit to find an absolute optimum. The difficulty is in the necessity of taking into account: a) the integer number of the unit capacities of the turbines, boilers, diameters of the pipelines; b) nonlinearity caused by the nonlinear dependences of the investments on the capacity of the heat supply sources and by other dependences of nonlinear character; c) dynamics, that is to say development of the district heating system at the gradual growth of heat loads in the years of the studied period.

The Siberian Power Institute has developed approximate methods based on using mathematical modelling and electronic computers, that do not guarantee obtaining a really optimum solution in all cases, but the solution, found with their help, may be near to optimum or may coincide with it. By means of a variety of main initial indices one can define the stability of obtained solution.

Example of computation. There is given a conditional town (Fig.7), consisting of 8 heat districts with total heat consumption 1600 Gcal-h (the share of hot water supply load-15%). As possible heat sources are considered: a new heating and power plant with different versions of its location (HPP-1, HPP-2, HPP-3 in Fig.7(a)), the out-of-town condensational electric station, which can cupply heat from bleeds of the condensational turbines, four district boiler plants with limited productivity (by the local conditions), two electric boiler plants and six large local boiler plants.

Fig.7(a) shows the position of listed heat sources and directions and the length of the routes of the main heat networks. On the whole Fig.7(a) represents the initial ("redundant")scheme

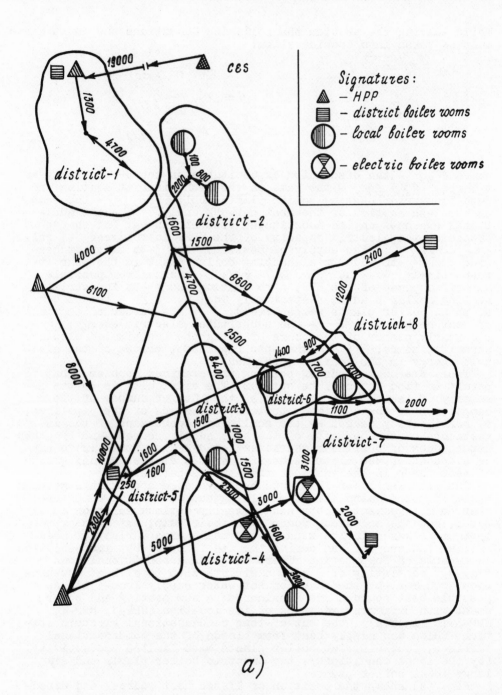

Fig.7(a)

554

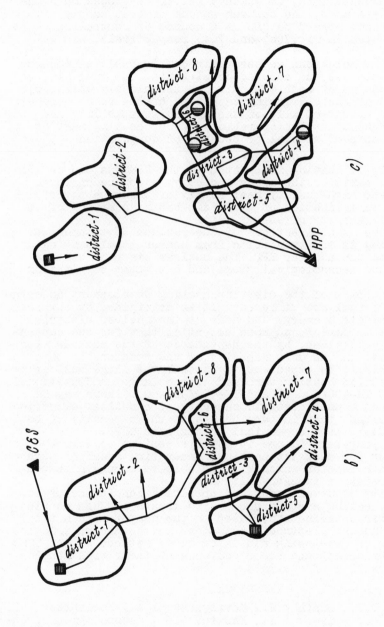

Fig.7. Initial (a) and optimal schemes of heat supply of the town with (b) and without (c) consideration of the CES as a source of centralized heat supply.

Fig.7(b,c)

of the heat supply of a town. As a result of the carried out optimization computations, in which a set of developed mathematical models were used, the optimum scheme of heat supply with regard and without regard to CES as a source of centralized heat supply represented in Fig.7(b) and 7(c) respectively, was obtained.

Comparing the solutions we see their considerable dependence on initial conditions and adopted constraints. One must also say that for the choice of both economic and technically admissible solution one must check "working capacity" of the found scheme of heat supply (first of all, of the hydraulic stability and reliability of heat network operation). Such checking must be done with the help of special procedures of computations [5].

Conclusions

1. The district heating is treated in the USSR as one of the most important parts of the energy economy, a source of large fuel economy and high-quality heat supply of towns, a powerful means for the normalization of their air basin, a method of improving the reliability of energy supply.

2. One of the main directions of the further development of district heating is the transition from conventional heating and power plants to the nuclear HPP. The nuclear district heating must provide the technological steam and hot water supply from NHPP.

3. The principles of the district heating development on organic and nuclear fuel are different. It is determined by low fuel component of electric energy and heat at the nuclear electric stations, that is direct and more essential than for the conventional electric stations, by the dependence of the nuclear electric station cost on the unit capacity of the reactor. In this case at the NHPP it is expedient to install the large nuclear reactors and district heating turbines with a large condensational capacity (type TK-400/500-60 and ПTK-400/500-60). There may be such cases, when a unit electric capacity of a nuclear electric station is limited, that in turn brings to the necessity of building comparatively less economical specialized NHPP.

4. For the long-term future it seems expedient to develop a special reactor for district heating, generating steam of high parameters and a reactor which will be able to supply high temperature processes in industry with heat.

5. Complexity of development and operating conditions of modern district heating systems require to create and use complex methods of their optimization, based on the use of mathematical modelling and modern electronic computers in the design practice. The experience of developing and using these methods in the USSR shows their efficiency and the expediency of their wide application in the practice of designing.

REFERENCES

1. Budnyatsky D.M., Bunin V.S., Kovilyansky Y.A., Koritnikov V.P., Levental G.B., Smirnov I.A., Khrilev L.S., Technological problems of the district heating on the base of nuclear fuel. "Teploenergetika", (in Russian), 1974, N 11, pp.10-16.
2. Khrilev L.S., Zinger N.M., Smirnov I.A., Optimization of district heating systems. Report N A/5, at the Third International Conference for district heating (in Russian). Warsaw,1976,28 p.

3. Melentyev L.A., Optimization of development and control of large energy systems. Publishing House "High School" (in Russian), Moscow, 1976, 336 p.
4. Melentyev L.A., Principles of nuclear district heating., "Teploenergetika" (in Russian), 1976, N 11, pp.6-9.
5. Methods of mathematical modelling in energetics. Under edition of Melentyev L.A., Belyaev L.S. (in Russian). East Siberian Publishing House, Irkutsk, 1966, 432 p.
6. Smirnov I.A., Fedyaev A.V., Khrilev L.S., Choice of profile for main equipment of nuclear heating power plants with turbines of type TK. In the collection "Methods of mathematical modeling and optimization of parameters, type of technological scheme and profile of the equipment for nuclear condensational and heating power plants" (in Russian). Irkutsk, 1976, p.90-104.
7. Sokolov E.Y., District heating and heat networks, Publishing House "Energia" (in Russian), Moscow, 1975, 376 p.

HEAT STORAGE IN BUILDINGS:
AN OVERVIEW

BARUCH GIVONI

Institute for Desert Research
Ben-Gurion University
Beer Sheva, Israel

ABSTRACT

The paper discusses the methods of heat storage in buildings and its effect on thermal performance, human comfort and energy consumption of air-conditioned and non-conditioned buildings. The role of heat storage in buildings heated and cooled by solar and other renewable 'natural' energies is analysed in relation to different types of buildings.

INTRODUCTION

The capacity of the building structure, or some specialized part within the buildings, to store thermal energy has a pronounced impact on the sensitivity of the building to changes in the environmental conditions, and on the indoor temperatures, human comfort and the energy demand of the building.

In general, a higher heat storage capacity (i.e. a heavier structure) provides the building with a higher thermal inertia. The response of the building's temperatures to sharp changes in ambient thermal conditions or to internal heat generation becomes smoother and delayed in time.

The quantitative impact of the building's heat storage capacity depends, however, on many factors, notably:
- the ambient conditions (type of climate, season, etc.).
- whether or not the building is air conditioned.
- the kind of the energy (conventional, solar, etc.).
- the diurnal pattern of occupancy of the building, depending on its type (residential, office, school, etc.).

Heat storage in the building can be provided either in the mass of the structure or in a specialized space under, above or within the building. The performance characteristics of thermal storage within the structure depends on the location of the main mass of the building, for example:
- external envelope, exposed to the environment.
- external envelope, insulated from the environment.
- internal floors, partitions, etc.

Heat storage in a specialized space can be provided by a water tank, a space filled with gravel, eutectic salts, etc.

The following discussion of the role of heat storage capacity will proceed according to the above aspects.

While the significance of storage in buildings utilizing conventional energies is discussed in this article, the main emphasis is on buildings heated and cooled by solar and other renewable (natural) energies. This is because of the crucial role of thermal storage in such buildings.

IMPACT OF HEAT STORAGE UNDER DIFFERENT CLIMATIC CONDITIONS

The impacts of the building's heat storage capacity on indoor temperatures, human comfort and energy requirements depend greatly on the outdoor climatic conditions: air temperature, humidity, winds, etc., which characterize different climatic types and seasons. [6]

In regions and seasons with large diurnal (as well as short-term day-to-day) changes in outdoor temperature, and with intense solar radiation, a high heat storage capacity means a substantial reduction in the ranges of the indoor air and radiant temperatures. Indoor temperatures tend to stabilize around the diurnal average of the sol-air temperature (a combination of outdoor air temperature and solar energy absorbed in the building).

The significance of this effect is maximized in summer in arid regions, where outdoor climatic conditions often fluctuate around the human comfort zone. During the nights outdoor temperatures are often in the "cool" zone (e.g. 15-20°C) while during the day the air temperature (e.g. 30-35°C) together with the impact of solar radiation, form a combination resulting in a heat stress.

Under such conditions a building with a high heat capacity, closed during the day and ventilated during the nights, can stabilize the indoor temperature around the comfort zone (e.g. 22-26°C).

On the other hand, in warm-humid regions or seasons, which are character-ized by small diurnal temperature range, the main discomfort may be experienced in late afternoons and the evenings, when the winds decline. In such climatic conditions the possibility of utilizing heat capacity for lowering the daytime temperatures is greatly reduced because of the physiological need for ample ventilation for comfort in the humid environment with the result that the indoor air temperature follows closely the outdoor pattern. During the evening and nights, however, due to the decline in winds, the indoor temperature can be appreciably above the outdoor level. Under such conditions a building with a high heat capacity, which cools off more slowly, has higher indoor air, and mainly radiant, temperatures leading to a higher thermal stress than that which is experienced in a light-weight, but insulated, buildings.

In winter, in most regions, outdoor temperatures are usually below the comfort zone, ranging from cool to very cold in different climates. Diurnal, as well as day-to-day changes in the outdoor temperature result in variations in the energy needs for heating. Equipment has to accomodate the most extreme conditions, leading to large sizes of the heat generation and distribution systems. A large heat capacity of the building smooths these variations and enables a reduction in the sizing of the equipment.

The importance of heat storage capacity is evident in most climatic regions in the transition seasons: spring and fall. In many places these seasons are characterized by cool nights and warm days. Under this pattern of temperature the thermal inertia of a building tends to stabilize its indoor temperatures just at the comfort level, thus minimizing the need for heating during the nights and for cooling during the day.

UN-CONDITIONED VERSUS AIR-CONDITIONED BUILDINGS

Thermal storage plays different roles in buildings which are un-conditioned and in thode which are either heated or cooled.

Un-conditioned buildings

When buildings are not heated (in winter) or not cooled (in summer) the indoor temperatures, in relation to given outdoor climatic conditions, are determined by the thermo-physical properties of the building (its thermal resistance, heat storage capacity and solar absorption) and the ventilation conditions.

The effect of heat storage capacity on indoor temperatures and comfort depends on the conditions with respect to the other factors, and mainly on the ventilation pattern. [6].

When the building is not ventilated the indoor air temperature follows closely the average surface temperatures of the walls and the roof. A high heat capacity leads to low rates of heating during the day and of cooling during the night. The indoor average temperatures tend to stabilized around the average level of the sol-air temperature.

When a building with a high heat capacity is effectively ventilated, the indoor air temperature may follow the outdoor air pattern while the temperatures of the internal layers of the external walls and roof may be quite different.

With light coloured external surfaces the temperature of the internal surfaces will be below the outdoor air level during the day and be above it during the night. With dark coloured external surfaces the internal temperatures will be elevated. In both cases of external colour the indoor air temperature, in ventilated buildings of high heat capacity, may differ significantly from the average indoor surface (radiant) temperature.

In lightweight buildings with low heat storage capacity, expecially those with a relatively high thermal resistance of the envelope, the internal surface temperatures follow closely the indoor air temperature. In ventilated buildings the patterns of the internal surfaces and the indoor air temperature both follow closely the pattern of the outdoor air temperature.

It is possible to utilize these characteristics in adapting buildings, from the heat storage acapcity viewpoint, to different climatic conditions and patterns of occupation so as to maximize the comfort of the inhabitants in unconditioned buildings.

Air conditioned buildings

In buildings which are air-conditioned the heat storage capacity has little effect on the comfort of the occupants, as the mechanical systems take care of it. Nevertheless, storage capacity may have a significant impact in reducing the energy demand of the building. This impact can take place in two situations:

a) Energy conservation due to the possibility to provide indoor comfort without the operation of the conditioning equipment. This is possible when the outdoor climatic conditions are such athat the indoor temperatures are stabilized, by the heat storage capacity of the building, at a level close to the comfort zone.

b) Energy conservation while the conditioning equipment is in operation, due to the smoothing of energy expenditure for heating and cooling during different hours of the day. This mode of impact is effective mainly in reducing peak demands for energy under extreme heating or cooling loads, such as those occuring under a sharp drop in temperature in winter or a sharp heat wave in summer.

In summer a high heat storage capacity enables also to store "coolness" within the structure by mechanical ventilation during the night hours, thus reducing the load on the cooling system during the daytime hours.

HEAT STORAGE LOCATION WITHIN THE STRUCTURE

As mentioned in the introduction, storage of thermal energy within the structure can be provided in the external envelope (walls, roof and floor), either exposed to or insulated from the environment, as well as in the internal partitions and floors. These different locations affect greatly the impact of the thermal storage on comfort and energy demand.

Storage in the external envelope, exposed to the environment

This form of heat storage, mainly in load bearing walls and heavy roofs of reinforced concrete or stone, bricks, etc. is the traditional way of minimizing the effects of diurnal variations in the outdoor climatic conditions. When the storage capacity is located in the external envelope in such a way, the heat flow from the environment into the walls during the day and out of them during the night is maximized. This is expecially the case with respect to heat gain from solar radiation into dark coloured external walls and the roof.

It is difficult to control this mode of heat transfer and to adjust it to human comfort and energy conservation requirements, except by orientation of the heat storing walls. For example, a southern heavy and dark coloured wall in the northern hemisphere absorbs most of the solar heat in winter and much less of it in summer. However, the efficiency of this mode of heat gain in winter in southern walls is small, as most of the absorbed heat is lost back to the environment by radiation and forced convection.

Considering the roof and the western and eastern walls, storage of solar energy in them occurs mainly in summer, increasing discomfort and energy demand for cooling, mainly during the evenings. Consequently, they should be treated so as to reject solar energy (e.g. by light colours), although their mass can still be useful in reducing the impact of large fluctuations in the outdoor air temperature.

Storage in the external envelope, insulated from the environment

When heavy (load bearing) external walls and the roof are covered by an insulating layer on their external side, the heat exchange between them and the outdoors is reduced markedly while they can exchange heat more easily with the interior (indoor air and surfaces). The rate of heat exchange between the thermal store and the outdoor environment is governed by the thermal resistance of the "mantle" of insulation, while the heat exchange with the interior is governed by the areas and the respective surface coefficients of the internal surfaces of the various elements of the building envelope. It is possible to control the rate of heat transfer between the heat storing building elements and the interior air by such devices as heavy curtains,although the effectiveness of this way of control is quite limited.

In summer such storage capacity can be utilized to reduce the rate of indoor temperature elevation during the daytime hours by limiting the heat inflow by preventing penetration of solar radiation and minimizing ventilation during the hot hours. The heat which enters the interior (mainly through the windows) can readily be absorbed within the high-mass walls and floors. During the evening the rate of heat flow out of the building, and consequently of its cooling, can be increased by cross-ventilation.

In winter solar energy penetrating through windows can be absorbed in the mass of the building, with little heat loss to the environment due to the insulation surrounding that mass. The absorbed heat is released back to the interior space at night. This point will be discussed in more detail in the section on the role of structural storage in buildings heated by solar energy.

The effect of mass location in external walls (either inside or outside an insulation layer, in a cold climate, has been studied by an analysis of the energy demand of the Energy Conservation Demonstration Building of the G.S.A. in Manchester, N.H., U.S.A., which has been done through computer simulation by T. Kusuda et al. [9] before the detailed design of the building.

The walls were of concrete, 15 cm. thick, with insulation yielding a total U value of 1.7 Watts/m^2·K. Comparison of the energy demand with the two location in shown in the following table:

Insulation location	Max. heating load 10^6 Btu/hr	Max. cooling load 10^6 Btu/hr	Total annual heating 10^9 Btu	Total annual cooling 10^9 Btu	Total annual energy 10^9 Btu
inside	4.340	3.063	4.21	1.70	13.19
outside	3.760	2.69	3.86	1.58	12.55

It should be realized that in regions with a less cold winter, and especially in hot arid regions, the effect of placing the insulation outside the mass would yield a much larger effect in reducing the energy demand in airconditioned buildings.

The effect of heat storage located in specialized spaces will be discussed later on, in the section on buildings heated and cooled by "natural renewable energies.

COMBINED EFFECT OF HEAT STORAGE AND THERMAL RESISTANCE

Both the thermal resistance of the envelope and its heat storage reduce the rate of heat flow between the external and internal surfaces when they are subjected to fluctuating thermal conditions and a temperature difference exists between them, although the mechanisms of this effect are different.

Increasing the thermal resistance of a wall, while keeping its heat storage capacity constant, reduces the heat flow from the warmer to the colder surface without changing the heat content of the wall itself. On the other hand, increasing the heat storage capacity of a wall while keeping its thermal resistance constant, reduces also the rate of heat flow from the warmer to the colder surface, but throught another mechanism. This reduction results from the "trapping" of part of the heat, which had penetrated the warmer surface, within the wall during the heating stage of the thermal cycle and thus preventing it from reaching the interior surface before it flows back to the outdoors during the cooling stage.

The combined effect of the heat storage capacity and the thermal resistance of walls composed of several layers of different materials depends on the relative locations of the layers, those providing the storage and those providing the resistance.

Under cycling external conditions, when a resistance layer is on the exterior side and a heat storage layer is on interior side, the heat flow into the wall during the heating stage of the cycle is reduced by the resistance layer. The heat which nevertheless has passed through this layer is absorbed in the masses of the layers with a high storage capacity, causing a low rate of temperature elevation. During the cooling stage of the cycle, the flow of the heat which has been stored in the internal mass to the outdoor is retarded by the thermal resistance of the exterior layer. The result is an effective damping of the external temperature wave.

Quantitatively, the combined effect of thermal resistance and heat storage capacity of a composite wall, taking into account the relative locations of layers with different properties, is expressed by the "Thermal Time Constant" (TTC) of the wall.

The concept of the TTC was first suggested by Bruckmayer [1], further developed by Raychandury and Chandhury [12] and expanded by Givoni and Hoffman [5].

The TTC of a homogeneous wall is defined by the multiplication of the heat capacity of the wall by the thermal resistance from the external air to the center of the wall, i.e.:

$$TTC_{(w)} = (Q_w) (R_w)$$

$$Q_W = d\, \rho\, c$$

$$R_W = \left(\frac{1}{h_o} + \frac{d/2}{\lambda}\right)$$

when:
- d = thickness of the wall (meters)
- ρ = valumetric weight (tons/m^3)
- c = specific heat of the material (Kcal/Kg·K or KJ/Kg·K)
- h_o = external surface coefficient, e.g. h_o = 20.
- λ = thermal conductivity of the material (Kcal/hr·m·K or watts/m·K)

When a wall is composed of several layers the TTC_W of the wall is the sum of the TTC values of the different layers, $TTC_W = \Sigma\, TTC_i$, when the TTC of each layer is the multiplication of the heat capacity of that layer i times the thermal resistance from the external air until the center of the layer i.

$$TTC = (d_i \rho_i c_i)\left(\frac{1}{h_o} + \frac{d}{\lambda_1} + \frac{d}{\lambda_2} \ldots + \frac{d_i/2}{\lambda_2}\right)$$

The TTC of the whole wall is then given by:

$$TTC_W = \Sigma\, TTC_i$$

The TTC_W of a wall determines the damping of the temperature wave of the external surface ($\Delta t_{(i)}$), i.e. the ratio: $\Delta t_{(i)}/\Delta t_{(o)}$

Consequently, the temperature amplitude at the internal surface of the wall, under given amplitude of the external surface temperature, can be estimated by the formula [5]:

$$\Delta t_{(i)w} = \Delta t_{(o)w}(1 - e^{-\frac{1}{TTC_W}})$$

HEAT STORAGE IN BUILDINGS HEATED OR COOLED BY SOLAR (AND OTHER RENEWABLE) ENERGIES [3, 4, 7, 8, 10, 11, 13, 14]

Systems relying on natural energies have to incorporate as an integral part of the system a considerable capacity for energy storage in order to allow for a continuous operation.

Solar heat has to be accumulated and stored while it is available on clear days, to provide energy for nights and cloudy days. In the same way, nocturnal cooling energy sources, such as outgoing longwave radiation, evaporative and convective cooling at night, are intermittent and require the provision of a certain storage capacity for daytime cooling.

Basically, thermal energy can be stored in two forms:
a)   as sensible heat, by changing the temperature of inert materials such as water, gravel, etc.
b)   by latent heat storage of change of state in reversible chemical and physico-chemical reactions in phase-change materials (PCM).

In general, a storage for solar energy system should fulfill several requirements. Absorption and storage of the surge of energy has to be made rapidly while available, and on the other hand this energy should be released promptly when required. Also, undesirable heat losses have to be minimized.

The main problem associated with the storage of sensible heat is that of the space requirements. A cubic meter of water stores 1000 kcal for 1°C change in temperature, while a similar volume of rocks contains about 400 kcal per 1°C. The structure of the building stores about 200 kcal per degree per ton of its mass. In high rise buildings the weight itself may constitute a problem.

One of the main advantages of sensible heat storage is that the same material and accessories can be used both for the storage of heat in winter and the storage of cold in summer. It is usually also less expensive than latent heat storage. However, storage of latent heat requires much less space than that of sensible heat. Thus, for $NaSO_410H_2O$ (Glauber salt hydrate), which has a melting point at about 32°C, one cubic meter can store about 4500 kcal while changing its phase from a solid to liquid.

In the following some of the design aspects of the use of the various storage materials will be discussed.

Thermal storage in water

Water has the highest heat capacity per unit weight of the sensible heat storage materials, although the difference on the basis of volume is less marked because of the greater volumetric weight of the other materials.

The main problem concrning the use of water for storage of thermal energy is the cost of the container. For heating and cooling of a dwelling unit several cubic meters of water are required and the storage tank should be rust-resistant over a long period. The cost of such tanks is the main expenditure in the use of water as a storage material. On the other thand the comparatively small volume of the space needed for storage in water (compared with rocks) reduces the cost of the insulation required to conserve the energy within the storage space.

Water solar collectors are usually more complex and expensive than air collectors and, as the cost of the collectors constitute the main expenditure of solar energy systems, the use of a whole water system may result in a higher total cost. On the other hand such systems can more easily be combined with a conventional energy supply and distribution system, such as fan-coils for heating and cooling.

From the design details viewpoint a large water storage tank is the simplest arrangement. It calls, however, for very careful treatment against leakage. It is easy to check the tank under pressure during the initial installation. However, corrosion may cause leakage at a later time and because of the huge quantity of water stored in the tank, the results may be quite severe. Therefore, the need for very high quality of materials and waterproofing of the tank.

Thermal storage in gravel

Gravel and pebble beds have many advantages for storing thermal energy when ample space is available (e.g. in a basement or underneath the building). The transfer of the energy to and from the storage is done usually by air flow. A blower is necessary to induce the air flow to the storage and across it. The required power depends mainly on the resistance of the gravel pile to the air flow. As the size of the gravel is reduced, the specific surface area of the storage mass is increased and the heat transfer in and out of the material is facilitated but, on the other hand, the resistance to the air flow within the gravel pile increases.

Using uniform size of the gravel reduces the resistance. A size of 3 to 5 cm is a convenient one for short-time storage (1-2 days). For longer periods of storage, e.g. 5-7 days larger stones can be used as there is more time for the heat flow in and out of the stones.

The specific heat of gravel is about 0.22 kcal/kg/°C. The gross density of the storage mass is about 1.5-2.0 kg/litre, depending on the type of the rocks used. Thus, the volumetric heat capacity of the gravel pile is about 0.4 kcal/litre/°C (400 kcal/m^3/°C).

When the sotrage space is of a sufficient size and is located underneath the building, it is possible to utilize also the heat capacity of the earth under the gravel as part of the storage system. In such a case the bottom of the storage area should not be insulated but insulation is needed around the periphery of the ground area extending beneath the storage level.

Water-gravel combination

When the solar collection system is based on water but the energy distribution within the building is based on air flow, it is possible to use a combination of an un-insulated water tnak and a gravel bed surrounding it for the storage of the thermal energy.

With such a combination the rate of heat flow from the collectors to the water storage tank can be much faster than the rate of heat flow from the water tank to the surrounding gravel bed. This later rate is limited by the low coefficent of surface heat transfer (mainly by natural convection) and the relatively small surface area of a large water tank.

However, the duration of energy flow from the collectors to the storage is limited to only several hours per day, or even with interruptions due to cloudiness. On the other hand, the energy flow from the water tank to the surrounding gravel bed is a continuous process. Therefore, the total energy flows, over the whole day, may balance, and such a combination can perform satisfactorily. From the maintenance point of view care should be taken in the design of such a combined system to provide an access to the water tank for repairs, cleaning, etc. The heat is distributed within the gravel bed by natural convection currents between the tank and the gravel. At the upper layer of the storage space, hot air accumulates and can be withdrawn by a fan and distributed within the occupied space.

During the cooling season, cold water is stored in the tank and the colder air accumulates at the bottom of the storage space. Provision should be made for its withdrawal by perforated ducts located underneath the gravel. From these aspects the design details of the energy withdrawal from a water tank-gravel combination are similar to those of a complete gravel storage, although the details of the energy input into the storage are different.

Latent energy storage in phase change materials

Physical changes of state and physico-chemical reactions in PCM produced by heat absorption and release involve greater amounts of energy,per volume as well as per weight, than the sensibly temperature changes of inert materials. The changes must be reversible for a very large number of cycles. In addition, the kinetics of the reactions should enable rapid heat flow in and out of the storage system.

Change of state involves the formation and dissolution of crystals. It some cases the crystalization may not proceed before the liquid is supercooled, unless suitable nucleating agents are added.

The main problems associated with latent heat storage are: segregation of the two phases and the geometry of the heat-exchange container which must promote the heat and mass transfer during the dissolution-solidification cycles. Different salt hydrates or their eutectics are required for the storage of heat and cold, according to their melting points. Thus, 30-40°C will be suitable for heating in winter and about 5-15°C for cooling in summer. The temperature required for efficient operation of absorption-cooling system might be about 90°C. It seems to the author that, at the present, latent energy storage has not reached a stage of development sufficient for application, except for experimental buildings.

Location and design of the storage space

One of the main design problems associated with the storage of thermal energy is its location, either within the space of the building or adjacent to it, e.g. in the basement. In one or two storeyed buildings the storage space can be provided within the space of the basement or on the roof. In the former case the transport of the energy from the collection point to the storage requires longer routes, but the storage space is better protected from heat loss by suitable insulation. Storage over the roof area is the least efficient from the energy conservation viewpoint, as the temperature gradient between the storage medium and the ambient environment is maximized.

From the energy conservation point of view, the best location of the storage is within the inhabited space, as thus all the heat flow in and out, across the container envelope, is utilized for heating or cooling. However, such a location may be the most difficult from the viewpoint of area utilization and might conflict with the fuctional planning considerations. In the case of multistoreyed buildings the storage space may have to be provided within the inhabited space. In such a case salt hydrates may be more practical because of the saving in valuable space. The mass of the structure may provide partial solution for the storage problem in high rise buildings. A residential unit of concrete structure may contain a mass of over 100 tons. Such a mass releases 20,000 kcal through the change of 1°C in its average temperature. However, to be effective as a heat storage the mass has to be concentrated mainly within the interior of the building, i.e. in the floors and internal partitions, while the external envelope should be of high insulating capacity. Therefore, heat storage within the structure calls for a specific approach to the design and structure of the building.

Another design detail of the storage space affecting its performance is its geometrical shape. For a given volume of storage space the heat loss will be lower as the surface area is smaller. The cost of insulating the storage also is proportional to the surface area of the space.

An exception to these considerations is the storage space is located within the occupied area of the building. In this case all the "losses" from the storage are "gains" for the occupied space. Therefore, the geometrical configuration is of minor importance. A good solution in such a case is to provide the storage at the center of the occupied space, preferably at the junction of several rooms, and thus facilitating the transport of the heat from the storage to the different rooms.

A recent interesting development is the invention by D. Chahroudi, [2] of a concrete in which phase change materials are imbeded. Walls and floors built of such concrete (when the material is commercially available) would be able to store, and then to release large quantities of heat while keeping a near constant temperature.

INTERACTIONS OF THERMAL STORAGE WITH THE TOTAL SYSTEM

In the choice of thermal storage materials and the design of the storage subsystem the following factors should be considered:
- Compatibility with solar collector type (air or liquid) and distribution system.
- Compatibility with cooling method.
- The type and size of the building, which affect the space availability for the storate.
- Initial, operating and maintenance costs.

Compatibility of storage with collector type and distribution system

The type of the collector, whether of the liquid or air heating type, affects the options with regard to the storage medium and the coupling of the collectors with the storage.

Liquid type collectors can be coupled, of course, with a liquid storage. The coupling can be direct, when the fluids of the storage circulate through the collectors, or indirect when the fluid from the collector transfers heat to the storage by means of a heat exchanger. In this case the transfer of energy from the storage and its distribution within the conditioned space is usually performed by water flow.

A liquid type collector can be coupled also with a combination of an uninsulated water tank surrounded by gravel (such as in the Thomason's system).

In such a combination the fluid from the water tank, or several small water tanks coupled in parallel or in series, circulate through the collector during sunshine hours. The unisulated water tanks transfer heat by convection and radiation to the surrounding gravel mass, which in turn is insulated on its periphery. This heat transfer takes place continually, as long as the water in the tanks are warmer (or colder in the case of cooling) than the surrounding gravel.

With this combination of a water tank and gravel storage, heat is removed for utilization in the form of hot air, with forced or natural convective flow, or by direct conduction through the floor above the storage.

Liquid type collectors can be coupled also with thermal storage within the mass of the building, e.g. in the walls and/or the floor and ceiling. In this case water pipes are imbedded within the sturctural elements and the liquid from the collectors is circulated in the pipes, trasnferring heat to the mass of the building elements.

The storage capacity in this method of storage is limited by the heat capacity of the storing mass and by the upper limit of comfort, which is determined by the combined physiological effects of the temperatures of the indoor air and the surfaces of the heated elements.

When external walls or the roof are utilized for thermal storage, they should be well insulated against heat loos to the environment. To this effect the insulation should be located on the outside of the heat storing layer.

An air-type collector can be coupled with a gravel storage and with containers filled with water or phase change materials. It is also possible to couple air collectors with heat storage by building elements of high heat capacity such as walls and floors.

When the storage is in gravel, the hot (or cold) air is blown from the collectors through the gravel. Heat,or cold, is withdrawn by blowing air across the gravel and into the conditioned space. When the gravel space is under the floor (in a ground floor unit) the whole floor can be turned into a heating panel with heat flow across it by conduction.

When the storage is within the mass of building elements, open channels or cavities should be provided within the building element, where warm air from the collector circulates.

When the storage is performed by containers filled with heat storing material (e.g. water or eutectic salts) provisions should be made to direct the air flow in such a way that it will "flood" the containers. With such a system the limiting factor is the surface resistance of the containers to heat flow in and out of the storage material.

The heat flow from the storage and its distribution in the conditioned space is performed in this system by air flow.

Comapribility of storage with the cooling system

Three cooling systems can be coupled with a thermal storage, namely:
- absorption cooling
- night radiation, convection and evaporation
- air-conditioning (or heat pump) operating at night and using off-peak electricity.

Absorption cooling

Present absorption cooling machines can be coupled only with water type collectors. The storage type, however, will depend on whether the heat required for generation, or the cold produced, are stord.

When solar heat is stored, e.g. when it is not needed immediately to operate the absorption cooler, the storage can only be in the form of a hot storage.

When the cold is stored, fir instance, in cases where the main cooling needs are in the evening, the absorption cooler can cool a gravel storage during the mid-day hours and the cold produced by the evaporator can cool a gravel storage for subsequent use later on.

As cooling needs occur most frequently during the daytime, the former stor¬
age system (in water) would be the more common one.

Night cooling
          Night cold can be stored either in the form of cold water or as air-cooled
gravel. In both cases the storage is not coupled with solar collectors.
          When a water storage is contemplated to store night cold the cooling can
be obtained by several means:
- by water flow through unglazed plates facing the sky
- by water flow over the roof (either a flat roof or an inclined one)
- by water flow through a cooling tower.
          When a gravel storage is used the cooling of the gravel can be obtained
at night by blowing through the storage space the night air divectly, or air
which has been colled before by contact with plates radiating to the night sky
and/or sprinkling water on the gravel and evaporating it at night by air flow
through the gravel.

Type and size of the building

          The type and size of the building affect the space required for the sto-
rage and its availability, either under the building or within it.
          For low-rise buildings, e.g. of one or two storeys, which cover a large
area relative to their volume, and in which the distance between the roof and
the ground is small, gravel storage under the floor might be the most suitable.
In such buildings and storage system, large quantities of thermal energy, heat
or cold can be stored inexpensively, as the marginal cost of increasing the
storage capacity is very small. It involves mainly the cost of the additional
gravel itself (for a higher height of the storage space) as well as the
additional size of the end walls and their insulation. The larger the building
ground area, the lower will be the relative cost of increasing storing capacity.

          As air heating collectors can generally be less expensive than water type
collectors, heating and night cooling of low-rise, large area buildings by air
systems might be more economical than with water types of natural energies.

          High rise buildings are characterized by small ratio of the ground area
which they cover to their volume. In addition, the distance from the roof to
the ground is large, a factor which makes the transfer and distribution of
energy by air flow more expensive and space consuming.

          The 'ideal' system for such buildings would be a combination of phase
change materials for thermal storage and south-wall air collectors. Unfortunately, the unsolved technical problems involved with the use of PCM for
storage do not enable, at present, their reliable and economical use.

          Where the structural scheme of the building is such that its mass can
provide adequate heat capacity, the structure, in the form of load baaring
walls and massive floors, can be utilized for thermal storage. To perform
this function, provision should be made for heat flow from the collectors to
the mass of the building. This heat flow can be either by direct conduction
or by water (in pipes) or air (in channels) flow within the building elements.
Energy transfer by water flow calls, of course, for water type collectors, and
air flow calls for air-type collectors.

          When the structural mass is small or is not utilized for thermal storage,
then a centralized large water tank under the building might be a suitable
storage solution. In this case water-collectors should be the appropriate
ones.

If it is desirable to decentralize the energy system, then an individual storage can be provided for every floor, or for every unit in the building which has a southern exposure, unshaded in winter by other buildings. In such cases, water tanks would be more suitable than gravel bins, because the weight of the former, for a given thermal capacity, would be only about 20% of the latter.

Sizing the thermal storage

Sizing the thermal storage is related to both the thermal characteristics of the building and the characteristics of the storage system.

The most important building characteristic which determines the storage needs is the diurnal coefficient of daily heat loss, which expresses the thermal quality of the building. It is defined in terms of Kcal of heat which are required per day per degree difference between 18°C (the threshold temperature) and the average outdoor diurnal temperature.

The diurnal coefficient of daily heat loss depends on the thermal resistance of the external envelope of the building and on the ratio of the envelope surface to the volume of the building.

Several factors influence the decisions regarding the required thermal capacity and physical dimensions of the storage space, among them the following should be mentioned:

-       The desired length of the storage period.
-       Upper temperature level of the storage (depends on collector's type).
-       Lower temperature level required for utilization (depends on the energy distribution system).
-       Marginal cost of the storage space (depends on type and design details).
-       Heat capacity of the storing material.
-       Heat loss from the storage (depends on the insulation of the storage space and its location in relation to the building).
-       The economic aspects of energy storage.

The desired length of thermal storage period

The decisions with regard to the desired length of the storage period should take into account several factors. The most important among them are:
-       The pattern of usage of the building, whether it is continuous (e.g. housing or a hospital) or only during few daytime hours (e.g. school, or kindergarten).
-       The pattern and sequence of cloudiness during the heating (or cooling) season.

Determination of the length of storage period

Determining the length of the thermal storage period is not a simple matter. Usually it is measured in terms of the number of days of 'carrythrough' of the thermal storage system (i.e. the number of cloudy days for which the storage would provide sufficient heat. This number, however, is somewhat ambiguous. It depends not only on the heat loss characteristics of the building and on the thermal capacity of the storage (which can be computed relatively accurately) but also on the ambient meteorological conditions. The ambient conditions, for a given locality, are not constant during the heating (or cooling) season, but change from month to month. Even if the coldest month (e.g. January) is assumed in the calculations, there might be great differences between the statistical average conditions and the actual conditions which could be encountered during an extreme cold period. As a result, the actual number of 'carrythrough' days of a given system might be very different from the assumed number of days.

Another way to define the length of the period in which thermal energy can be extracted from the storage, is to express it in terms of the calculated degree-days for which the stored energy would suffice. This measure takes into account both the heat loss characteristics of the building and the thermal capacity of the storage over a specified temperature range. For example, a thermal storage with a total capacity of 500,000 kcal, (e.g. 20,000 Kcal/°C over 25°C of temperature range) serving a building with a daily heat loss coefficient of 10,000 kcal/degree day would provide heat for 50 degree-days. In a region and a period with an average outdoor temperature of 3°C, and threshold temperature of 18°C (for computing the degree-days), it would mean a storage length of 50/15, namely a little over 3 days. The same storage capacity would suffice for only two days when the average outdoor temperature would be -7°C, or if the building would have a heat loss coefficient of 15,000 kcal/degree-day and the average outdoor temperature would be 3°C daily.

The pattern of usage of the building.

The pattern of usage of the building, whether continuous or intermittent, has great impact on the relative importance of provision of storage capacity and its economic implications.

Any storage system involves heat losses from the storage, whether or not it is in actual use. Therefore, storing energy in buildings which are in use for short periods, and in particular when there is substantial gap between the time of energy collection and the time of its utilization, inherently involves thermal inefficiency and unrecovered expenses.

In buildings which are occupied only during a few hours in the day, such as as kindergartens or some offices, storage can be eliminated altogether. Solar energy can be utilized whenever available, with an auxiliary energy source providing heating whenever solar energy is unavailable or insufficient. This approach calls for an air heating system, with short time-lag of response, and in some cases this may be the most economical approach.

On the other hand, in buildings which are occupied continuously during the day and night, such as residences, hospitals, etc., energy storage is essential, for at least one day, in any solar heating system. Likewise, in regions and in seasons where cooling is required in summer and where night radiation, convection and evaporation are the sources of the cold collected, storage of cold is essential in buildings which are used continuously, or mainly during the daytime, such as schools.

Pattern and sequence of cloudiness

The considerations for storage capacity beyond one day depend on the pattern and sequence of cloudiness.

In regions with clear sky during the winter (e.g. the plateau of South Africa), storage is needed only for the few hours of the night, as solar heating can usually be counted on for the next day.

In climates where the duration of most cloudy periods is one day, and they are usually 'sandwiched' between clear days, thermal storage for one and a half days would be the most logical.

On the other hand, in climates where long spells of cloudiness are frequent with a duration of about a week or more, it would be too costly to provide storage large enough to supply the heating during such long periods. The total cost would actually be much higher than the cost of the storage itself. In fact, the collectors would also have to be much larger, to be able to 'charge' the storage to full capacity during the sunny periods. In such climates it would be more reasonable to store heat for two or three days, and to rely on the auxiliary energy source to provide the heating during the rest of the cloudy period.

References

1.   Bruckmayer, F. The Equivalent Brick Wall. Gesundheit Ingenier, 63 (6) pp 61-65.

2.   Chahroundi, D. and Wellesley-Miller, S., 1976. Thermocrete Heat Storage Materials: Applications and Performance Specifications. In: Proceedings, Sharing the Sun - Solar Technology in the Seventies. Joint Conference, August 15-20, Volume 8 pp. 245-259. Winnipeg, Canada.

3.   Close, P.J. 1965. Rock Pile Thermal Storage for Comfort Air Conditioning, Mechanical and Chemical Engineering. Transactions of the Institution of Engineers, Australia, MC1, pp 11-12.

4.   Duffie, T.A. and Beckman, W.A. 1974. Solar Energy and Thermal Processes John Wiley and Sons, N.Y.

5.   Givoni, B. and Hoffman, M.E. 1972. Prediction of the Thermal Behaviour of Full Scale Buildings. Research Report, Building Research Station, Technion, Haifa, Israel.

6.   Givoni, B. 1976. Man, Climate and Architecture. 2nd edition. Applied Science Publication, London.

7.   Givoni, B. Paciuk, M. and Weiser, 1976. Natural Energies for Heating and Cooling of Buildings - Analytical Survey. Research Report. Building Research Station, Technion, Haifa, Israel.

8.   Kreider, J.F. and Kreith, F. 1975. Solar Heating and Cooling. McGraw-Hill Book Co., N.Y.

9.   Kusuda, T., Hill, J.E., Lin, S.T. Barnett, J.P. and Bean, J.W. 1975. Pre-Design Analysis of Energy Conservation Options for a Multi-Storey Demonstration Office Building. National Bureau of Standards Building Science Series 78, Washington, November.

10.   Lof, G.O.G. and Hawley, R.W. 1948. Unsteady-State Heat Transfer between Air and Loose Solids. Industrial and Engineering Chemistry. Vol. 40, No. 6 pp 1061-1070.

12.   Raychaudhury, B.C. and Chaudhury, N.K.D. 1961. Thermal Performance of Dwellings in the Tropics. Indian Construction News, Dec. pp 38-42.

13.   Telkes, M. 1949. Storing Solar Heat in Chemicals. Heat and Vent. Nov. pp. 80-86.

14.   U.S. National Science Foundation, 1973. Proceedings of the Solar Heating and Cooling for Buildings Workshop. NSF/RA/N-73-004. Washington.

DOMESTIC HEAT PUMPS FOR HEATING ONLY

C. J. BLUNDELL

Electricity Council Research Centre
Capenhurst, Chester CH1 6ES, United Kingdom

ABSTRACT

In North European countries there are few houses that require summer cooling, and so a domestic, air-to-air heat pump does not have to fill two roles. Choice of a capillary for refrigerant control, electric resistance defrost, suitably controlled, and heat exchangers sized to give maximum efficiency in the heating cycle, give a considerable improvement in heating performance, and a reduction in capital costs, compared with units designed for heating and cooling.

Evaporator and condenser sizes must be decided from an economic optimisation. Experimentally verified calculations show that an improvement in seasonal coefficient of performance from about 2 (a value typical of present units) to about 3, may be obtained.

The heat pump test facility used in this work consisted of two separate spaces, each with its airflow temperature controlled and monitored, so that the performance of a heat pump, operating between the two spaces, could be deduced

INTRODUCTION

Domestic sales of packaged heat pumps in the USA are climbing steadily. Well over a million units are now installed. However, the main reason for their popularity in the USA does not apply to most North European countries. A heat pump can, by use of a reversing valve, provide either domestic heating or cooling as desired, and in many parts of the USA the cost of the system can be justified, since air conditioning is necessary for comfort in the summer. The efficient heating provided in the winter is thus virtually a free bonus. Few house owners in North European countries have air cooling systems, and so the purchase of a heat pump must be justified only on reduced heating costs. In fact, in many climates, a cooling facility would not be used even if installed. In British field trials of American reversible equipment only 15 hours of cooling were recorded in three years of monitoring (1).

Fortunately, some advantages may be obtained by designing a heat pump for heating only, both in terms of improved efficiency, and reduced capital cost. This paper discussed the design of air-to-air heat pumps for domestic heating

HEAT EXCHANGER SIZING

Figure 1 is a simplified diagram of an air-to-air heat pump

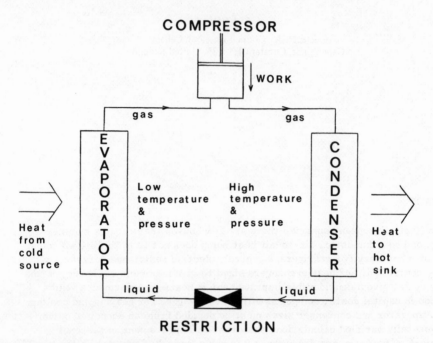

Figure 1. A simple heat pump.

It is most important when designing an efficient heat pump, to correctly size the evaporator and condenser, since the sizes of the heat exchangers affect evaporating and condensing temperatures, and thus overall efficiency, directly. Other aspects of design may make a small but significant contribution to improving the system.

As has been previously pointed out (2), sizing the evaporator and condenser for optimum efficiency in heating brings a substantial improvement in performance over units designed for a compromise between heating and cooling. A computer model of a heat pump was developed and used to show how much improvement could be expected from increasing the area of the heat exchangers. Figure 2 is a typical example of the results from this analysis.

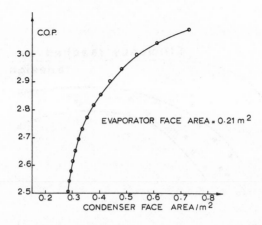

C.O.P. FOR 5.0 kW. OUTPUT

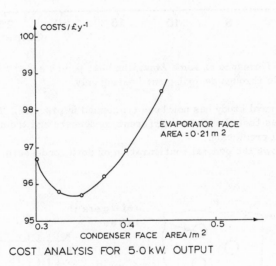

COST ANALYSIS FOR 5.0 kW. OUTPUT

Figure 2. Heat exchanger sizing for the condenser of a 5kW heat pump.

Since the performance continues to improve as larger heat exchangers are used, the final optimum areas must be found by an economic analysis. This work has already been published (2). Figure 3 shows the expected improvement obtainable in the performance of several reversible American heat pumps. The expected coefficient of performance (c.o.p.) falls at small outputs due to the low efficiency of small compressor motors. C.O.P. is defined as heat output divided by total electrical input.

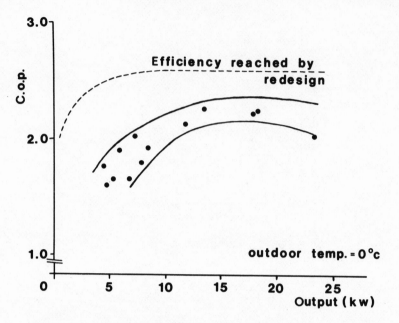

Figure 3. The performance of some American heat pumps and the improvement possible through designing for heating only

This theoretical study has now been confirmed in practice. The performance of a heat pump was tested with two different condensers and the difference compared with a theoretical prediction.

Figure 4 shows the general configuration of both condensers.

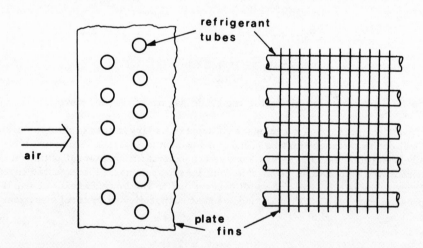

Figure 4. Continuous fin heat exchanger

Details of the two heat exchangers used, and of the performance figures obtained were as follows:

Condenser 1:

Face area	=	0.171 m^2
Depth	=	0.095 m
Fin pitch	=	0.003 m

Total heat transfer surface area = 11.4 m^2

Performance measured:

Output =	4.81 kW)	Source - sink temperature
C.O.P. =	2.55)	difference of 21°C

Condenser 2:

Face area	=	0.323 m^2
Depth	=	0.06 m
Fin pitch	=	0.0022 m

Total heat transfer surface area = 16.0 m^2

Performance predicted using computer model:

Output =	4.93 kW)	Source - sink temperature
C.O.P. =	2.74)	difference of 21°C

Performance measured:

Output =	4.97 kW)	Source - sink temperature
C.O.P. =	2.72)	difference of 21°C

The condenser sizes were chosen simply to test the computer model, not as economic optima of any kind.

As can be seen the improvement in c.o.p. obtained using the larger condenser was less than predicted by 12%, but since this is only .02 in 2.7 this error is well within the accuracy of the performance measurements.

These measurements were made in a calorimeter room split into two by an insulating wall. Air was circulated through the two halves separately, and each air stream drawn from the room was cooled to a set temperature (see figure 5). Before being returned to the calorimeter, the air was reheated by a bank of heaters controlled by the temperature of the air leaving the room. Thus any heat created within the calorimeter reduced the power used by this bank of heaters, and so the performance of a heat pump operating between the two halves could be measured.

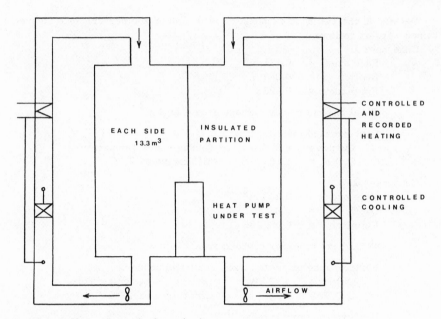

Figure 5. The airflows through the calorimeter.

OTHER DESIGN CONSIDERATIONS

Because of the relatively high capital cost related to its output, a heat pump can only sensibly be installed in a house that has been insulated to a fairly high standard. This means that the largest European market could well be for units of 5 kW output or less. For instance, a typical British house of 80m² floor area, with loft and cavity wall insulation would have a design heat loss of 4 - 5 kW. For these small units a capillary expansion device may be used, saving the cost of a thermostatic expansion valve (t.e.v.) Although a capillary cannot control the refrigerant flow rate over such a wide range of conditions as can a t.e.v., it should be satisfactory for a unit designed for heating (3).

The moisture condensed on the evaporator of a heat pump freezes when the evaporator is below 0°C. This frost formation will be greatest when the amount of moisture contained by the air is high. That is at the relatively mild temperatures of 0-5°C. British field trials of American heat pumps have shown the importance of an efficient defrosting method and control system (1). The fact that winter temperatures are often around 0-5°C with high humidities in European climates means that many more defrost operations will be needed for European installations than is typical for American experience. Figure 6 shows the number of defrost operations monitored in the field trials.

The most commonly used method of defrosting is to reverse the refrigerant cycle for a short period, cooling the indoor heat exchanger and warming the evaporator. The indoor air is kept at a reasonable temperature by switching on direct electric supplementary heat. This is a fast and efficient method of removing frost, but, unfortunately, the sudden reversal of pressures on the compressor when the defrost

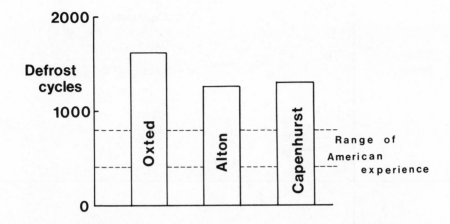

Figure 6. Comparison between British and American experience of annual defrost
         operations.

is initiated and terminated cause very serious reliability problems. For a heating only
heat pump the cost of a reversing valve may be saved, and this reliability problem
avoided by choosing a different defrosting method, such as using direct electric
heaters. In this method electric heating elements are switched on in the evaporator
and the compressor is switched off, during the defrosting period. This method has
the same efficiency as the reverse cycle defrosting described above, but causes no
compressor reliability problems.

   Figure 7 illustrates another defrosting problem highlighted by the field
trials. Many defrost operations occurred at high temperatures ($\sim 10°C$). This
was due to an unsuitable control system sensing the need for a defrost. It is
essential to have a positive sensing method. Probably the simplest is to sense the
temperature difference between the heat exchanger coil and the air, defrosting
when this exceeds a certain value indicating poor heat transfer caused by frost
formation.

   It is also important that, in a heating only design, all sources of heat
(compressor and fan motors) should be in the indoor air stream, so making full
use of all possible heat.

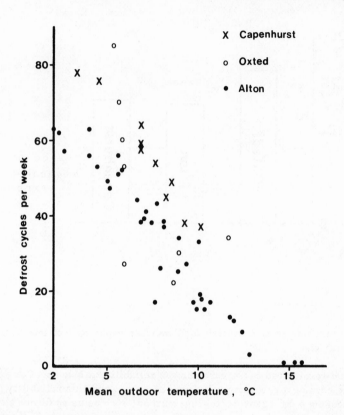

Figure 7. Number of defrost operations as a function of outdoor temperature

CONCLUSIONS

1. Houses in many European climates do not require air cooling and therefore a
    heat pump installation must be economically justified by reduced heating costs.

2. A heat pump designed for heating only will cost less than a reversible unit due
    to omission of reversing valve, simplified pipework, and possibility of using
    capillary control.

3. A heat pump designed for heating only will be more efficient due to optimum
    sizing of heat exchangers, and placing of compressor and fan motors in the
    indoor air stream.

4. The reliability of a heat pump may be improved by avoiding reverse cycle
    defrosting, particularly in a relatively wet, mild climate.

5. A computer model of a heat pump, described elsewhere, has been developed
    which predicts an improvement in c.o.p. from about 2.0 to about 2.7 for a $21^{\circ}C$
    source-sink temperature difference, by designing for heating only.

6. This model has been experimentally confirmed to an accuracy of about 12%.

REFERENCES

1. Heap, R. D. 1977. American Heat Pumps in British Houses. Elektrowarme International Edition A 35, A2: A77-81.

2. Blundell, C. J. 1977. Optimising Heat Exchangers for Air-to-Air Space-heating Heat Pumps in the United Kingdom. Int. J. Energy Res. $\underline{1}$, 69-94

3. Kowalczewski, J. J. 1961. Performance of Refrigeration Systems with fixed Restriction Operating under Variable Evaporator and Condenser Conditions. The Journal of Refrigeration, Nov./Dec. 122-128.

A LABORATORY INVESTIGATION OF A STIRLING ENGINE-DRIVEN HEAT PUMP

DAVID DIDION

National Bureau of Standards
Washington, D.C., USA 20234

BARRY MAXWELL

Bucknell University
Lewisburg, Pennsylvania, USA

DAVID WARD

National Bureau of Standards
Washington, D.C., USA 20234

ABSTRACT

An experimental investigation was conducted on an air-to-air heat pump powered by a single-cylinder, seven-horsepower, water-cooled Stirling engine. The steady-state part-load performance of the engine-driven heat pump system was determined in both the heating and cooling modes of operation. The unit was operated over a broad range of outdoor temperatures and corresponding co-efficients of performance (COP), and seasonal performance factors (SPF) were determined. The energy rejected to the engine's cooling water was measured and included in the heating mode calculations.

NOMENCLATURE

A/F	air-fuel mass ratio
bsfc	brake specific fuel consumption (Kg/kWh)
COP	coefficient of performance of heat pump unit
COP_{eff}	effective coefficient of performance of engine-driven heat pump system
DD	degree (fahrenheit) day
HP_c	compressor horsepower
N_e	engine speed (rpm)
N_c	compressor speed (rpm)
ODT	outdoor temperature (°C)
N_i	number of hours in ith bin
$\bar{P}$	mean operating pressure (atm)
SPF	seasonal performance factor
T_{HTR}	heater temperature (°C)
T_i	median temperature of the ith bin
β	fraction of fuel energy recovered
η_t	thermal efficiency (%)

INTRODUCTION

The theoretical concept of using engine driven heat pump systems for space conditioning dates back to Kelvin, when he rightly contended that this system has the potential to utilize a primary fuel more efficiently than any other building heating system. However, the trade-off between the energy savings and the pragmatic problems associated with the relatively complex machinery has seldom favored on-site prime movers. In recent years, ever increasing fuel costs have motivated researchers to reexamine several systems heretofore considered feasible but uneconomical or impractical. This paper discusses one such study.

The application of an engine to a heat pump system has the inherent advantages of engine waste heat recovery and increased compressor efficiency due to speed modulation. The effect of the waste heat recovery may be accounted for in the heating mode by expanding the usual definition of heat pump efficiency (COP) to an effective coefficient of performance:

$$COP_{eff} = \frac{\text{Heat Pump Capacity + Recovered Waste Heat}}{\text{Fuel Energy}} \tag{1}$$

The compressor modulation generally increases compressor efficiency as speed decreases. Thus the advantage of this procedure is best realized through matching the system output to building load as closely as possible. These two advantages are quantitatively evaluated for a Stirling Cycle Engine Rankine Cycle Heat Pump.

In principle, many prime movers could be used to drive a heat pump; the selection of the Stirling engine was based on two general considerations. First, it is one of the most advanced prototype heat engines available today; and second, its inherent attributes seem to favor a buildings application. Specifically, the Stirling Engine's basic advantages appear to be:

(1) Potential for maximum thermodynamic efficiency
(2) Low pollutant emissions
(3) Low maintenance requirements
(4) Quite operation
(5) Multi-fuel capability
(6) Most of the waste energy is in the cooling water.

It is the authors' opinion that the inherent disadvantage is the cost of the engine. It would appear that even under mass production conditions the high temperature materials and the complex geometry required to transfer the external combustion heat into the working medium chamber will always dictate higher manufacturing costs than a comparable internal combustion engine. However, for a buildings application even this disadvantage is less visible since mechanical equipment expenditures normally represent a small fraction of the overall building cost.

A Philips 1-98 Stirling engine was selected for this study because to the authors' knowledge, it is the only tried and proven Stirling engine of the 10-HP power range available in the world today. And since this study was intended to focus on the system or application rather than on engine development, it was essential to use a reliable prototype. Figure 1 is a cross-section of the 1-98 engine. The displacer piston reciprocating motion is a constant phase angle ahead of the power piston, so that the working medium is in the upper high-temperature area during expansion and in the lower low-temperature area during compression. The working medium regenerator and the combustion air preheater are the heat exchange components that are key to the high thermal efficiency of the engine. The principles of operation are explained in detail in references (1) and (2).

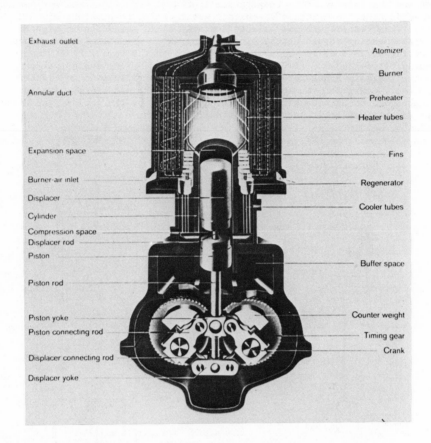

Figure 1. Philips 1-98 Stirling Engine

EXPERIMENTAL PROCEDURE

A schematic diagram of the Stirling engine-driven heat pump system illustrating the major components is shown in Figure 2. For measurements convenience, propane (C_3H_8) was used as fuel in this study. Its flow rate was measured with a rotameter and was thermostatically controlled so that a constant heater temperature of 608°C (1126F) was maintained. The air flow was measured with a rotameter and was manually adjusted for an optimum air-fuel mass ratio of 27 to 1 (72% excess air). Helium, used as the working fluid, was stored in external high pressure tanks and delivered to the engine through a regulating valve at pressures ranging from 30-100 atmospheres while operating in a speed range of 1000 to 3000 rpm. Power control was accomplished by manually admitting more helium to the cylinder or by venting excess helium

to the atmosphere. Under field conditions the helium circuit would be a
closed loop with a pump and auto-controller. The energy rejected to the
cooling water was determined by measuring the water flow rate through the
engine and its corresponding temperature increase. The flow rate was manually
adjusted in order to maintain a constant water outlet temperature of 60°C
(140F). This temperature was selected because it is sufficiently high for
many space heating requirements. The cooling water was circulated through
a conventional automobile radiator and returned to the engine at a temperature
typically between 43.3°C (110F) and 48.9°C (120F). During engine performance
tests the engine's brake torque and speed were determined with an electric
dynamometer, timer and revolution counter. Since the water pump, starter,
alternator and combustion air pump were not powered by the engine, the measured
torque did not reflect these parasitic energy losses.

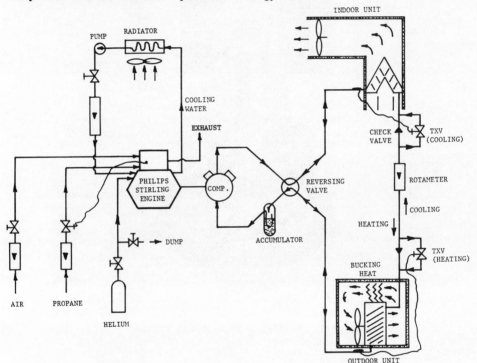

Figure 2   Schematic Diagram of Stirling Engine-Driven
           Heat Pump System

     The heat pump was a uniquely constructed apparatus whose major components
(i.e. compressor, indoor and outdoor coils) were commercially obtained. The
outdoor coil and fan, and three stages of electrical resistance bucking heat were
placed inside a specially constructed insulated box, and the contained air was
circulated around a loop and over these components. During the heating mode,
low temperatures were established in the outdoor unit by the evaporator removing energy from the air. Once the desired outdoor air temperature was
achieved, the bucking heaters were used to automatically control it by trans-
ferring energy to the air at a rate sufficient to replace the energy extracted
by the evaporator. During the cooling mode, high outdoor temperatures were

achieved by partially opening the box and drawing in ambient air and diverting
it through the resistance heaters to raise its temperature to the desired level.
During all phases of testing, the engine and heat pump were located in a large
chamber which simulated indoor conditions by maintaining a 21.1°C (70F) and 50%
relative humidity environment. In either mode of operation, outdoor temperatures
could be controlled to within 0.28°C (0.5F). The indoor unit consisted of a
specially constructed L-shaped duct which contained the indoor coil and fan.
Since the compressor had to be externally powered by either the Stirling engine
or the dynamometer, a reciprocating, 4-cylinder, belt-driven, open compressor
designed for Refrigerant-12 was used. Based upon the range of engine and
compressor speeds anticipated during the study, the pulley ratio between the
dynamometer or engine and the compressor was set at 1.41. Control of the
heat pump unit in both modes of operation was accomplished with conventional
thermal expansion valves (TXV) and a reversing valve.

Heat pump instrumentation included thermocouples distributed over the
tubewalls of the four circuits of the outdoor coil, and the six circuits of
the indoor coil. In addition, thermocouples were placed so that the temperatures of both the refrigerant and the air entering and existing both coils
could be measured. Additional instrumentation included provision for measuring
compressor suction and discharge pressure and temperature, indoor coil pressure,
electrical input to the bucking heaters and to the indoor and outdoor fans,
and refrigerant mass flow. Thermocouple readings for the entire engine-driven
heat pump system were either continuously recorded with strip-chart recorders,
or with an automatic data scanning and acquisition system.

The experimental study was divided into three major phases. The first
phase was devoted to determining the performance of this specific Philips
engine over its entire speed and power range. The engine drove the dynamometer
during this phase. In the second phase, the performance capability of this
unique heat pump unit was determined for both operating modes as a function
of outdoor temperature and compressor speed by having the dynamometer drive
the heat pump. Based upon these separate engine and heat pump performance tests,
the combined performance of the total system was determined by matching the
engine's performance characteristics with the power requirements of the
heat pump unit. In the third phase (that of the engine driving the compressor)
the predicted system performance was experimentally verified in the heating
mode by operating the system over operating schedules defined for two
hypothetical building loads.

EXPERIMENTAL RESULTS

The experimental results are presented in Figures 3 through 6. Figure 3
presents a map of Stirling engine performance which relates brake horsepower,
brake specific fuel consumption, engine speed, and mean operating pressure.
The horsepower requirement of the heat pump during the heating mode is
superimposed on the map and is presented as a function of outdoor temperature.
As noted earlier, the measured engine power does not reflect parasitic losses
from auxiliaries which the manufacturer's data indicated could absorb as much
as 9% of the engine's power at any given speed. A more detailed examination of
the data indicates that the variation of horsepower is approximately a linear
function of mean operating pressure, which was limited to 100 atm during these
tests. An increase in that limit would have directly increased the maximum
horsepower. A more detailed performance map than Figure 3 would indicate that,
at any operating pressure, the most fuel-efficient range of engine operation
is from 1500 to 1700 rpm. At 1700 rpm, the brake specific fuel consumption
varies from approximately 0.49 kg/kWh to 0.28 kg/kWh. The corresponding

variation of thermal efficiency is approximately 16% to 28%, when based upon the lower heating value of the fuel. The results at a given power indicate that fuel consumption remains relatively constant over a broad range of engine speeds. This fuel use efficiency at part loads is typical of Stirling engine performance, and contrasts sharply with part-load performance of many conventional reciprocating engines where the efficiency decreases rapidly when operating outside of a relatively narrow speed range.

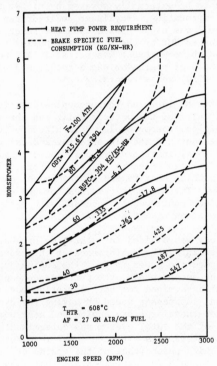

Figure 3. Stirling Engine Performance

The power requirement of the heat pump does not include the power consumed by the indoor and outdoor fans. In general, for the type of compressor used, lowering its speed increases its efficiency, with 900 rpm being the limit due to oil pressure. A pulley ratio of 1.41 between engine speed and compressor speed was selected so each would tend to operate in its most efficient speed range a large portion of the time. If a different speed ratio had been chosen, the heat pump power curves in Figure 3 would have been shifted to higher or lower engine speeds and different fuel efficiencies. Since the engine operated most efficiently at 1500 to 1700 rpm, a variable speed ratio pulley could have been used so that the engine would always operate in its most efficient speed range. However, it was felt that the additional mechanical complexity was not merited for this concept study.

Figure 4 illustrates the variation of energy rejected to the cooling water expressed as a percentage of fuel energy, and is plotted as a function of engine power and speed. Since the combustion process is external to the working fluid and is relatively efficient, most of the energy rejected from the engine is transferred to the cooling water. The results show that generally 50 - 60%

of the Stirling engine's fuel energy is rejected to the cooling water.
These high heat rejection rates are in direct contrast to those in conventional
internal combustion engines, where 15 to 30% of the fuel energy is typically
rejected to the coolant. The rejection of large amounts of heat to an engine's
coolant system might ordinarily be a disadvantage, particularly in closed
systems where the high heat load must be rejected to the environment. However,
in applications where the heat can be usefully employed, such as in a heat pump
system, it is an advantage. This is one of the attractive features of the
Stirling engine as a power source for building applications.

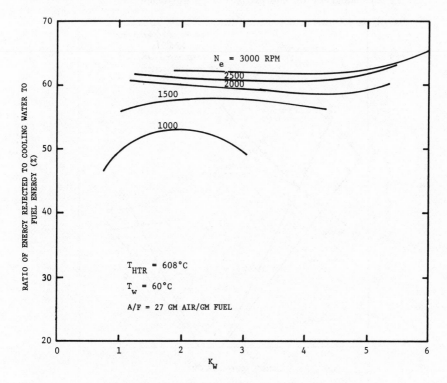

Figure 4.   Percentage of Fuel Energy Rejected to
           Cooling Waste

    Figure 5 illustrates the measured heating and cooling capacity of the
heat pump unit itself. It also illustrates the projected and measured heating
capacity of the engine-driven heat pump system. The results are based upon a
constant indoor temperature and relatively humidity of 21.1°C (70F) and 50%,
respectively, and a maximum cooling water temperature of 60°C (140F). Coil
frosting during the heating mode did not occur because of the small amount of
moisture in the air which was recirculated in the outdoor unit. The system
heating capacity is equal to the heat pump capacity plus the energy rejected
by the engine to the cooling water. The impact of rejected engine heat in the
heating mode is observed to be significant. Capacity increases range from 47%
at 15.6°C (60F) and 900 rpm, to approximately 135% at -17.8°C (0F) at 1800 rpm.
For purposes of clarity, the results have been plotted only for speeds of 900
and 1800 rpm.

Since the speed ratio between the engine and the compressor was fixed, the performance of the engine-driven system in the heating mode can be projected by matching the engine's power capability with the heat pump's power requirements. This process was carried out for all outdoor temperatures and compressor speeds and is the basis for the system capacity curves in Figure 5. In order to confirm the validity of the engine-heat pump matching, a series of engine-driven heat pump tests were performed at selected outdoor temperatures and compressor speeds. The results of these tests are indicated on the figure, and the agreement between the projected and actual performance is sufficiently good to justify the accuracy of the projected performance.

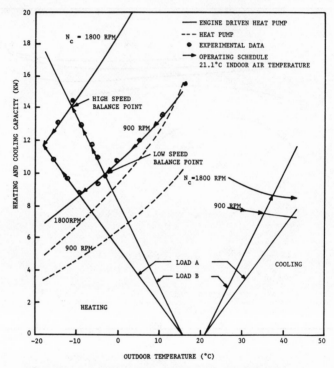

Figure 5. Heat Pump Heating and Cooling Capacity

In order to assess the effects of capacity modulation on system performance, two typical building energy loads were assumed. They are illustrated in Figure 5 as straight lines A and B, and are characterized by thermal conductances of 351.2 watt/°C (666 Btu/hr-F) and 527.4 watt/°C (1000 Btu/hr-F), respectively. The superposition of the system's capacity on these load curves results in fixed high speed and low speed balance points at -17.8°C (0F) and -9.4°C (15F), respectively, when operating against load A, and at -11.1°C (12F) and -3.3°C (26F) when operating against load B. The corresponding fixed balance points, without recovered engine heat, are -6.1°C (21F) and -1.7°C (29F) for load A, and -1.4°C (29.5F) and 2.5°C (36.5F) for load B. Two variable speed operating schedules, matched to loads A and B respectively, are shown on Figure 5 for the heating mode.

The projected variation of the effective COP in both modes is presented in Figure 6 as a function of outdoor temperature and compressor speed. Again for reasons of clarity, the results are only plotted for speeds of 900 and 1800 rpm. Experimentally obtained data is superimposed on the projected data, and as before, the agreement between the projected performance and the actual performance is very good. The variable-speed operating schedule described above is illustrated on Figure 6, and indicates that when operating in the heating mode at temperatures above the low speed balance point, the maximum efficiency occurs at the lowest compressor speed. At temperatures below the balance point, capacity modulation results in a penalty in effective COP. Despite the penalty, the effective COP remains greater than 1.0 and increases as either the compressor speed decreases or the outdoor temperature increases. This indicates that capacity modulation should only be employed when operating at temperatures below the low speed balance point, and that the speed should only be increased sufficiently to meet the building load. A variable-speed operating schedule is also shown for the cooling mode, and indicates that a speed increase is required only when temperatures exceed the low speed balance point. The optimum efficiency operational paths are indicated by the arrows in Figure 5 and 6.

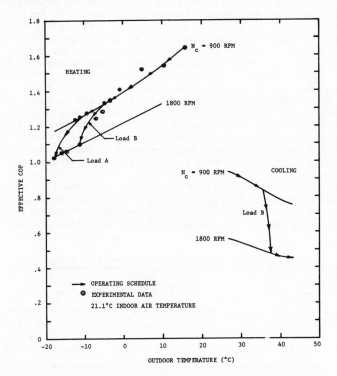

Figure 6   Engine-Driven Heat Pump Effective COP

As previously noted, the power consumed by the indoor and outdoor fans was not included in the system performance calculations. Fan power consumption data was taken during the testing program and was approximately constant at 1.1 horsepower. The impact of fan power on the effective COP in the heating mode can be estimated by considering the approximate increases in fuel energy and system capacity resulting from inclusion of an additional 1.1 HP load on the engine. Assuming that the engine has an average thermal efficiency of 25% and that it rejects 60% of its fuel energy to the coolant, then 1969 watts

(6719 Btu/hr) of additional energy is contained in the cooland and is delivered
to the heated space. Consideration of the fan horsepower results in a
15 - 24% decrease in the effective COP at 900 rpm and an 8 - 12% decrease at
1800 rpm.

Comparative Performance

Insight into the comparative performance of differently-powered heat pumps
may be obtained by expressing equation (1) in terms of component efficiencies
and the energy recovery fraction. Noting that (a) heat pump capacity =
HP_c · COP, (b) fuel energy = HP_c/η_t, and (c) recovered waste energy =
β · (HP_c/η_t), equation (1) becomes:

$$COP_{eff} = \eta_t \, COP + \beta. \tag{2}$$

For an electrically-powered heat pump there is no energy recovery (β=0) and
η_t denotes the efficiency of electrical power generation and transmission.
Performance comparisons of differently-powered heat pumps were estimated using
equation (2) by combining assumed values of β and η_t with typical heat pump
COP data. Table I shows the comparisons made for (a) an electrically-driven
heat pump, (b) a "typical" diesel engine-driven heat pump, and (c) the
capacity modulated Stirling-driven heat pump system examined in this study.
A value of η_t = 30% was used for the electrically-driven unit. The thermal
efficiency chosen for the comparable size Diesel engine was 25%. A typical
Diesel rejects approximately 30% of its fuel energy to the coolant, and
approximately 30% of the fuel energy is also contained in the exhaust gases,
and should not be neglected. Therefore, the recovery fraction for a typical
Diesel engine was chosen as 0.45, and assumes that all of the energy rejected
to the coolant, and one half of the energy rejected to the exhaust is
recovered. It must be emphasized that the values chosen for thermal
efficiency and recovery fraction are simply representative, and may vary
significantly with speed, load, outdoor temperature, and specific heat pump
application. The summary of the results clearly indicates the heating
mode superiority of engine-driven heat pumps relative to electrically-driven
units, and the superiority of the capacity-modulated Stirling system relative
to the Diesel.

Table I - COP_{eff} Comparisons for Differently-Powered Heat Pumps

Power Source	Outdoor Temperature (°C)						
	-17.8	-12.2	-6.7	-1.1	4.4	10	15.6
Constant Speed Electric Motor (η_t = 30%)	.54	.60	.68	.77	.79	.89	.93
Typical Diesel Engine (η_t = 30% β=.45)	.90	.95	1.01	1.09	1.11	1.19	1.22
Capacity Modulated Stirling Driven (Fig.6,Load A)	1.04	1.23	1.32	1.38	1.46	1.55	1.65

Defrost Mode Energy Requirements

During a "normal defrost" cycle the heat pump is temporarily reversed and placed in the cooling mode. The total energy required by the heat pump system during a defrost period is an important consideration when evaluating the energy use of engine-driven heat pump systems. In addition to the fuel energy supplied to the engine during this period, supplemental resistance or waste heat must also be supplied to temper the energy extracted from the building by the indoor coil. The total energy requirements during this brief, but perhaps frequent, defrost period can therefore become appreciable. Typical residential installations often require from 5 to 10% of the heating season's energy input for defrost.

In order to compare the energy requirement of several modes of defrost operation, the total energy required during an engine-driven "normal defrost" period with heat recovery was measured for several frosted coil conditions. A comparison was made with the energy required during a normal defrost when the heat pump is electrically-driven.

In addition, when the heat pump is driven by a heat engine, it is also feasible that the defrost operation be accomplished with recovered engine heat. The heat pump is temporarily stopped, the engine is operated under little or no load, and the waste engine heat may be diverted across the outdoor coil to melt the frost. Under these conditions, the only additional energy needed is that required by the unloaded engine. The total energy requirements during this engine-heat defrost mode might therefore be less than during the normal defrost mode. To checkout this possibility, defrost was accomplished by simulating with heating coils the introduction of recovered engine heat into the outdoor coil airstream. The total energy consumption was measured during this defrost period and was compared with the normal defrost mode energy requirements.

The results are summarized in Table II and are based upon source energy requirements. As before, the efficiency of electrical power generation and transmission was assumed to be 30%. At all three operating points there is less total energy consumed during the engine-driven "normal defrost" mode than during either of the other two modes. The results indicate that the engine-driven heat pump requires less energy than the electric driven unit because the availability of waste heat lessened the resistance energy required to temper the indoor air. Also the engine heat defrost technique proved expensive due to the ineffectiveness of melting coil ice from the outside.

Table II - Defrost Mode Source Energy Requirements

Operating Condition (°C/RPM)	Normal Defrost Engine-Driven (kWh)	Normal Defrost Electrically-Driven (kWh)	Engine Heat Defrost (kWh)
-1.1/900	5.1	6.2	7.2
-1.1/1500	3.1	4.8	7.3
-9.4/900	4.1	5.4	6.3

Seasonal Performance Factors

The effective COP of the engine-driven heat pump system is a valuable indicator of a system's efficiency at a particular indoor temperature, outdoor temperature and compressor speed. However, for determining the amount of energy required by the system to meet a given load over a given heating

season it is necessary to weight the different system efficiencies in proportion to the amount of time they are operated at that condition. This weighting
procedure allows for the account of hourly variation in building loads,
relative sizing of system capacity and any supplemental heating the system may
require. For this study, the seasonal performance factor (SPF) is defined
as follows:

$$\frac{1}{SPF} = \frac{\sum\limits_{i=1}^{I} N_i \text{ Load } (T_i) \frac{1}{COP}_{eff}}{\sum\limits_{i=1}^{I} N_i \text{ Load } (T_i)} \tag{3}$$

where the right hand side's numerator is the seasonal energy input and the
denominator is the system's required seasonal output.

    In order to assess the seasonal efficiency of the engine-driven heat pump
system under different conditions, the heating SPF was calculated for several
operating schedules at each of three geographical areas. The areas chosen,
and their degree-days, were Washington, D.C. (4224 DD), Chicago, Ill., (5882
DD), and Minneapolis, Minn., (8382 DD). The building heat loss, the system
capacity, the fuel energy input to the system, and any required supplemental
resistance heat were calculated for building loads A and B for each 2.78°C
(5F) increment, or bin (Reference 3), of the outdoor temperature range. This
data was combined with U.S. weather information data on the number of hours
per year that fall within each temperature increment for each city to determine
the SPF. The operating schedules chosen were: (a) constant 1800 rpm compressor
speed (corresponds to maximum system capacity and minimum COP_{eff}), (b) constant
900 rpm compressor speed (corresponds to minimum system capacity and maximum
COP_{eff}), and (c) the capacity modulated load-matched schedules noted on
Figures 5 and 6. The SPF of the heat pump when electrically-driven was also
determined to show the impact of engine heat recovery and system modulation.
All electrical energy inputs were converted to the equivalent source energy
requirements. The efficiency of electrical power generation and transmission
was assumed to be 30%. The results are summarized in Table III. The total
seasonal heating requirement for the Washington, D.C. area was 17,700 kWh
and 26,600 kWh for loads A and B, respectively. The corresponding heating
requirement for the Chicago, Ill., area was 24,900 kWh and 37,300 kWh,
and was 28,500 kWh and 42,800 kWh for the Minneapolis, Minn., area.

    The engine-driven heat pump system was sized to meet the heating design
loads at -18°C (0F) and -11°C (12F). Physically this means it has been
assumed that the same system in the same building is being moved from city to
city. Since each of the example cities has approximately the same cooling
design temperature, the net effect is that the ratio of the cooling capacity
to the load at the design cooling condition is held constant, which is the
usual practice today. In heat-recovery heat-pump systems of the future,
this may not be the best sizing criterion.

Table III - Seasonal Performance Factor Based Upon Source and Energy Requirements

Power Source and Operating Schedule	LOAD A			LOAD B		
	Wash., D.C.	Chic., ILL.	Minn., MINN.	Wash., D.C.	Chic., ILL.	Minn., MINN.
Electrically Powered 1800 RPM	.76	.70	.63	.70	.62	.54
Engine Driven 1800 RPM	1.24	1.21	1.18	1.24	1.16	1.03
Engine Driven 900 RPM	1.38	1.26	1.11	1.23	1.03	.83
Engine Driven Capacity Modulated	1.37	1.37	1.27	1.29	1.28	1.16

The results indicate that: (a) capacity-modulated engine-driven systems require less total energy input than comparable unmodulated engine-driven systems. (The data of Table III shows a 9-12% savings relative to the 1800 rpm engine-driven case), (b) the energy consumption of the engine-driven capacity modulated system is 46-53% less than the electrically powered heat pump operating at 1800 rpm. (A theoretical off design point study (reference 4) of a Stirling engine-driven residential heat pump with heat recovery in the Chicago, Ill., area indicated a 43.6% energy savings relative to a 3500 rpm constant speed electric heat pump. Their heating season results showed SPF's of .84 and 1.5 for electric and Stirling engine-driven heat pumps, respectively), (c) capacity modulation is more effective in Northern climates where the system encounters more low temperature operation. It results in a reduction in the balance point and a corresponding decreased requirement for supplemental resistance heat (d) regardless of the power source or the mode of operation, the heat pump system becomes less efficient as the building load increases because of an increased requirement for supplemental heat, (e) as the climate becomes more severe, an unmodulated engine-driven system must be operated at a higher constant speed. This results in a penalty in COP_{eff}.

Summary of Results and Conclusions

The Stirling engine performance tests indicated that part-load fuel consumption remained relatively low over a wide speed range. The engine rejected generally 50-60% of its fuel energy to the cooling water.

A comparison of projected and actual engine-driven heat pump performance indicated that the performance of the system can be satisfactorily estimated based upon the engine's power capability and the heat pump's power requirements. Recovery of the energy rejected to the cooling water resulted in a 47-135% increase in system heating mode capacity over the range of outdoor temperatures and compressor speeds examined. Capacity modulation in the heating mode results in a reduction in the high speed balance point. Modulation should only be employed when operating at temperatures below the low speed balance point, and the speed should then be increased sufficiently to meet the building load. Performance comparisons of differently-powered heat pumps in the heating mode showed that capacity-modulated Stirling-driven systems are more fuel efficient than electrically-driven and diesel-driven units. Defrost mode tests also indicated that less energy is consumed during a normal defrost period by an engine-driven heat pump than by an electrically-driven unit.

SPF calculations indicated that engine-driven heat pump systems have significantly lower fuel consumptions than electrically-driven systems, and that capacity modulated systems require less fuel energy than comparable unmodulated systems and are more effective in northern climates. However, two-speed systems would probably be almost as energy effective as a fully modulating one and may be preferable if mechanical or control simplicity is gained significantly.

REFERENCES

1. Walker, S., "Stirling Cycle Machines", Clarendon Press, Oxford, England 1973.
2. Meijer, R.J., "The Philips Stirling Engine", DeIngenieur, 621.41, pp.1-23.
3. ASHRAE Handbook, 1976 Systems Volume, Chapter 43.
4. Wurm, J. and Panikker, G., "Evaluation of Engine-Driven Heat Pump Systems of Small Capacities", Environment and Energy Conservation Symposium of EPA, Denver, Colorado, November 1975.

MOISTURE CONTROL IN BUILDINGS: OPPORTUNITIES FOR A HEAT PUMP DEHUMIDIFIER

G. W. BRUNDRETT

Electricity Council Research Centre
Capenhurst, Chester CH1 6ES, United Kingdom

ABSTRACT

Criteria for moisture control are reviewed. These vary from personal factors such as comfort and absence of electrostatic shock to mould growth which can insidiously attack the building fabric. Field studies on housing are examined to quantify the magnitude of the problem. These show that for a mild, damp climate such as Britain's the risk of condensation problems is large. These are avoided in practice by judicious manipulation of background heating and ventilation control. However the energy cost can be high.

An alternative approach is proposed which first seeks to reject as much moisture from clothes drying and cooking at source. The bulk of the remaining moisture is then extracted by a heat pump dehumidifier. Performance curves are given for a typical unit.

INTRODUCTION

The two reasons for heating a building are firstly to protect the building fabric and secondly to provide thermal comfort in cold weather. Building protection, even for unoccupied buildings, is really to prevent condensation and hence avoid the corrosion or mould growth which is associated with it. Thermal comfort is much more dependent on the air temperature matching the activity and clothing insulation of the individual. Humidity is a minor but not insignificant factor particularly at extremes of moisture levels when very dry or very moist. Recent studies (1) show that the ventilation habits of British housewives follow a strong seasonal pattern which is linked to the moisture levels of outdoor air, figure 1. This may not be the complete explanation for the behaviour.

This paper sets out to identify criteria for moisture control to quantify sources of moisture generation and then to compare present methods of control with that of a heat pump dehumidifier.

CRITERIA FOR MOISTURE CONTROL

Comfort

While it is accepted that high humidities exaggerate the effects of overheating, there is still no clear agreement on a permissible comfort zone in the normal sedentary non-sweating situation. Early studies by the League of Nations in 1937 (2) showed a wide difference in recommendations between countries and this is still true today. Conditions of 20% and 75% relative

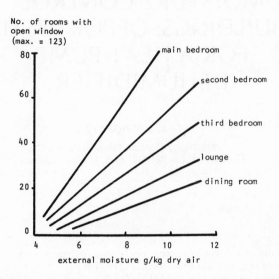

No. of rooms with
open window
(max. = 123)

Figure 1.   Relationship between windows open and external moisture

humidity at ordinary room temperature are less comfortable than at 50% relative
humidity (3).   However while British recommendations favour a band 40-70%
relative humidity (4) the Americans propose that it is not relative humidity
but vapour pressure which is the key factor (5).   They suggest an upper comfort limit for people to be 14mm Hg (i.e. 80% relative humidity at 20°C) and a
lower limit of 4mm Hg for other reasons (23% relative humidity at 20°C).

    The second factor is electrostatic shocks (6).   In general these are a
function of the room humidity and the distance walked.   The electrical
resistance of a carpet determines the leakage rate for an electrostatic charge
on a person to leak away.   If the electrical resistance to earth is high
($>10^{12}$ ohm cm^2) then an electrostatic shock is possible.   The electrical
resistance of a carpet is strongly influenced by the relative humidity in the
room, increasing as the relative humidity reduces.   The influence of relative
humidity on personal voltage level has been measured by several authors and
their findings are summarised in figure 2.   The normal permissible voltage
limit for people is 2 kV which on average can occur in rooms below 40% relative
humidity.   The chances of a shock increase with distance walked.   Shocks are
therefore more probable in large buildings than in small rooms in houses.

Building Protection

    Warm, damp conditions favour mould growth (7).   Organic materials such as
paper, cotton and wool, take up moisture progressively with increasing relative
humidity of the ambient air.   If the relative humidity exceeds 80-90% for long
periods mould growth is probable.   It can occur if the relative humidity
exceeds 70%.

    This limit for mould determines the permissible temperature limits for
houses which contain rooms which are occupied and rooms which are not.   The
water vapour pressure will tend to be constant within the house and hence the
relative humidity will increase in those areas which are cooler.   This is
illustrated in figure 3.   The shaded zone is assumed to be the comfort band in

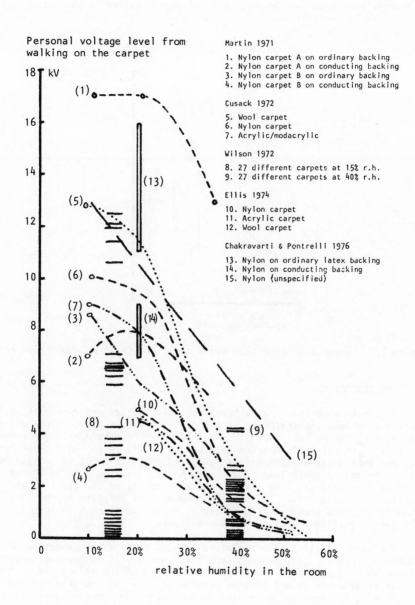

Figure 2.   Recently published data on electrostatic potential created by
            walking on a carpet

the occupied room.   If comfort is at the lowest humidity of 40%, then other
rooms will need to be warmer than 12°C if high relative humidity and mould
risk are to be avoided (<70%).

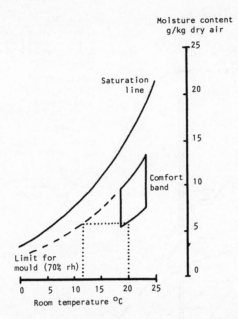

Figure 3.   Temperature limits to avoid mould growth in an unoccupied room

MOISTURE GENERATION

Personal evaporation rates have been well studied and vary with activity
level and ambient temperature.   For normal activities, temperature values lie
between 40-80 g/hour per person.   Basic data on household moisture generation
rates is rare.   Smith's American study in 1948 (8) is the most comprehensive
one, and subsequent assessments are compared in Table 1 (9, 10).   Estimates
for an average moisture load in a family house vary from 7 to 11 kg per day,
of which approximately 1 kg is from the by-products of the gas from the gas
cooker.   Portable paraffin fires add an amount of moisture approximately
equal to the volume of paraffin burned.   Houses using such devices can therefore have significantly higher moisture loads.

The major single item of moisture release comes from the drying of
clothes.   This not only depends upon the frequency of laundering, and type and
size of material but also upon the effectiveness of the spin dryer.   The
British tendency is to dry clothes outside for most of the year but the winter
period is more likely to be unsuitable and make indoor drying essential.
Under these circumstances 5-12 kg of water vapour are released into the house
on top of the normal quantity due to people and cooking.

MOISTURE CONTROL

Most of the moisture released in a house is diffused through the house.
Very little moisture is rejected at source.   Dishwashing machines blow vapour
directly into the house.   Tumbler clothes dryers, now becoming established in

Table 1.   Moisture generation rates in houses

Description	Author			
	Smith 1948[7] USA	Fournol 1957[8] France	Conklin 1958[9] USA	Loudon 1971[6] England
Family size	4	unspec.	4	5
Personal evaporation kg per day	5 kg	1.2-1.9 kg	2.5 kg	1.7 kg
Floor mopping	1 kg	-	1.1 kg	-
Clothes washing	2 kg/day	-	2 kg/day	0.5 kg/day
Clothes drying	12 kg/week	-	12 kg	5 kg/day
Dish washing	-	-	0.5 kg/day	
Cooking by gas	15 kg/week*			3 kg/day
breakfast	0.4 kg		0.4 kg	
lunch	0.5 kg		0.5 kg	
dinner	1.2 kg		1.2 kg	
Shower	0.2 kg		0.2 kg	) 1 kg/day
				) including
Bath	0.05 kg		0.1 kg	) dishes
Daily quantity				
washday	25 kg	10-42 kg/day	21.9 kg	14.4 kg
average	11.4 kg		7.9 kg	7.2 kg

*42% from food, 58% from gas cooker

British homes (11), have provision to reject the moist air to the outside but this opportunity is seldom taken. Moisture from cooking is more likely to be restricted to the kitchen. About 12% of British houses have an extractor fan to help control this moisture, the remainder of houses rely on opening the window. The use of an open window to ventilate the kitchen may simply blow the moisture into the rest of the house under some conditions of wind speed and direction.

British houses have a basic infiltration rate which depends upon the sealing of the building and the local wind and temperature. Recent research (12) found ventilation rate to vary directly with wind speed and under average wind varied from house to house from 0.3-1.3 air changes per hour. The actual leakage rates were not always the obvious ones of openable cracks in doors and windows. Air infiltration actually occurred through numerous fine cracks, particularly where timber was adjacent to masonry (13). Any additional

ventilation above this uncertain basic rate has to be provided by opening
windows.   A typical increase in ventilation rate in rooms when a window was
opened to its first fixed position was four times that when the window was
closed.

The actual amount of ventilation needed to remove the moisture depends
upon the outdoor humidity.   In cold weather the outdoor air will contain
little water vapour and so relatively small quantities of fresh air will be
needed.   As the weather warms up, the outdoor water vapour increases and so
more fresh air will be needed in the house to remove the same amount of water
vapour.   In Britain's temperate island climate, the outdoor humidity is
closely linked with outdoor temperature and the relative humidity stays
constant at about 90% figure 4 (14).   This means that the ventilation need

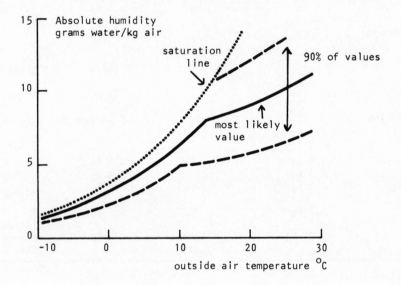

Figure 4.   Variation of humidity with outdoor air temperature in Britain

for moisture will increase rapidly with milder weather, figure 5.   This shows
that in a well heated house (19°C) the ventilation rate needed for moisture
control exceeds the normal design rate in mild weather.   If the house, or
parts of it, were normally much cooler than 19°C, then correspondingly more
fresh air is needed.   Measurements from field trials (15) show a seasonal
pattern of this type, which is in agreement with the more recent window opening
behaviour study (1).

Estimates of the energy cost of this extra seasonal ventilation are difficult but could account for an additional 4000 kWh/year out of a seasonal total
of 12,000 kWh.   This is in addition to the design estimate of one air change
an hour (16).

An alternative method of moisture control is dehumidification and the
favoured technique is a heat pump dehumidifier.   Such equipment is already
available for drying new houses, or for eliminating damp from cellars.   The
room air is drawn over the evaporator coil of the heat pump and cools down.
In cooling down, some of the moisture is condensed and drips from the coil.
The drier, cooler air then passes over the condenser coil and is reheated.

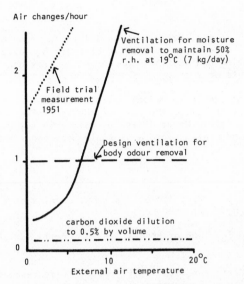

Figure 5.   Ventilation requirements for a British family house

It emerges into the room at a higher temperature than when it entered the unit. This is partly due to the compressor energy and partly to the transfer of the latent heat of the water vapour into sensible heat to the air.

The performance of such heat pumps varies widely with the ambient conditions extracting more water at higher relative humidities and to a lesser extent at higher temperatures. A typical moisture extraction rate is given in figure 6. A small dehumidifier operating in the coolest zone of the house has the unique property of not only removing the potential problem of water vapour but also providing free background heating with its latent heat.

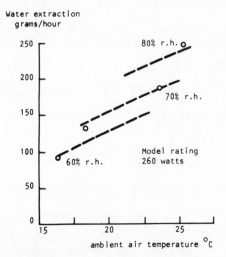

Figure 6.   Water extraction rate of a domestic dehumidifier

CONCLUSIONS

1.   Buildings in mild, damp climates such as Britain's, run the risk of
condensation problems.   These are avoided in practice by judicious manipulation of background heating and ventilation control.   This practice can use
deceptively large amounts of energy.

2.   An alternative approach is to reject whatever moisture can be rejected
directly at source and then remove the bulk of the remaining moisture with a
heat pump dehumidifier.   This dehumidifier is unique in translating the
potential problem of condensation into a sensible heat bonus towards background heating.

REFERENCES

1.   Brundrett, G.W. 1976.   Ventilation: a behavioural approach.   CIB
     Conference on Energy, Watford.

2.   Housing Commission 'Hygiene of houses' 1937.   Bull. Health Org. 6 (4):
     505-550.

3.   McIntyre, D.A. & Griffiths, I.D. 1975.   Subjective response to
     atmospheric humidity.   Environmental Research, 9: 66-75.

4.   IHVE Guide Book A, 1970.

5.   Nevins, R.G. & McNall, P.E. 1972.   ASHRAE Thermal Comfort Standards.
     CIB 45 Symposium on Thermal Comfort, Watford.

6.   Brundrett, G.W. 1976.   A review of the factors influencing electrostatic
     shocks in offices.   ECRC/M1046.

7.   Loudon, A.G. 1971.   The effects of ventilation and building design
     factors on the risk of condensation and mould growth in buildings.
     Architects Journal, 153 (20): 1149-1159.

8.   Smith, J.M., Blome, C.E., Hauser, H., Eades, A. & Hite, S.C. 1948.
     Research in home humidity control.   Purdue Univ. Proj. DR8C Res. Ser. 106.

9.   Fournol, A. 1957.   Ventilation et condensations.   CSTB Report 28.

10. Conklin, G. 1958.   The weather conditioned house.   Reinhold.

11. Electricity Council 1976.   Handbook of Electricity Supply Statistics.

12. Warren, P.R. 1974.   Preliminary studies of domestic ventilation.   Proc.
     Conf. on Controlled Ventilation: Aston University.

13. Skinner, N.P. 1974.   Natural infiltration routes and their magnitude in
     houses.   Proc. Conf. on Controlled Ventilation: Aston University.

14. Heap, R.D. 1973.   Heating cooling and weather in Britain.   ECRC/M631.

15. Dick, J.B. & Thomas, D.A. 1951.   Ventilation research in occupied houses.
     JIHVE, 19: 306-326.

16. Brundrett, G.W. 1975.   Some effects of thermal insulation on design.
     Applied Energy, 1: 7-30.

HEAT AND COOLING SUPPLY
OF BUILDINGS
IN COASTAL ZONES

O. SH. VEZIRISHVILI AND R. A. KHACHATURJAN

(Authors' address can be obtained from V. P. Motulevich, National Committee
for Heat and Mass Transfer, Academy of Sciences of the USSR, Moskva V-71
Lenjinsky prospekt 14, USSR)

ABSTRACT

This report deals with the novel system of heat and cold
supply of the buildings in the coastal zone, with the employment
of the heat pump units utilizing the heat of sea water.
The report presents the results of experimental and power-
-economical researches which have proved a high efficiency of
the system developed by the authors.

NOMENCLATURE

Q_o refrigerating capacity   (kW)

Q_h heating capacity   (kW)

Q_e effective power   (kW)

φ conversion coefficient

λ actual volumetric efficiency

η_i indicated compressor efficiency

η_e net efficiency

ζ reduced costs   (roubles)

$И$ annual production expenses

K capital investments

E_H authorized coefficient of capital investment efficiency

Subscripts

o evaporation

h heating

e effective

i indicated

1 commencement of compression

2 termination of compression

The introduction of summer air conditioning in the Southern regions, when the rated consumption of cold in summer is equal to or greater than the rated consumption of heat in winter, justifies the employment of air conditioners at the heat pump duty. This is a way for obtaining new heating sources without additional capital expenditure for plant construction, and a further contribution to the cleanness of the environment.

By way of example, in the Black Sea resort zone of the Caucasus the maximum rate of cold consumption in summer is 10 to 25% higher than the maximum rate of heat consumption in winter. This has been one of the reasons for the construction in Sukhumi of the experimental-industrial heat pump unit operating to the "sea water – air" cycle. The purpose of construction of the world's first installation of this type is to assess the feasibility and technical-economical effectiveness of employment of the series-manufactured refrigeration compressors at the heat pump duty.

The arrangement diagram of the heat pump unit developed in the Soviet Union is presented in Fig.1.

In winter sea water is fed by the shore pumping station (4) to the evaporators (3) of the heat pump unit, where low-temperature heat is removed from the sea water which then is discharged back to the sea. City water is heated to the desired temperature in the condensers (2) of the heat pump unit, enters the air conditioners (6) and heats the air being supplied to the premises.

In summer city water is cooled in the evaporator of the heat pump unit and, flowing through the same pipelines and air conditioners, cools the premises. Sea water is fed to the condenser and cools the latter.

The heat pump unit is changed over to the desired working duty by manipulating the respective gate valves (7,8).

The function of the heat pump is performed in this case by the Soviet series-manufactured refrigeration machines XM- YY80 operated on Freon-12. To cool the air in summer and to heat it in winter, vertical conditioners KH-10 and KH-20 are used.

In the course of researches the following parameters of the heat pump unit have been studied:

– heating capacity, refrigerating capacity, effective power and conversion coefficient, at Freon-12 evaporation temperatures from 0 to 10°C and condensing temperatures from 35 to 60°C;

– heating capacity ranges, at hot water temperature 55°C at the condenser outlet and sea water temperature 8, 10 and 15°C at the evaporator inlet, as well as at various rates of water flow through the condenser and evaporator;

– volumetric and power characteristics of the refrigeration compressor operated at the heat pump duty.

Besides, the wearing parts of the compressors have been subjected to micrometric measurement before and after the tests.

During the period of experimental operation, no breakdowns of the suction or discharge valves have occurred. The degree of wear of compressor main units and parts has been within the allowed limits.

The main initial data and results of the researches are given in table 1.

Table 1. Initial Data and Main Results of Experimental Researches on YY-80 Refrigeration Compressor Operated at Heat Pump Duty, Using Freon-12, in Accordance with "Sea Water - Air" Cycle

No.	Description	Unit of measurement	Summer duty		Winter duty	
			max. rated	mean	max. rated	mean
1	Sea water temperature	deg.C	+ 30	+26	+ 6	+12
2	Ambient ait temperature	deg.C	+ 32	+25	- 2	+6.8
3	Heating agent temperature	deg.C	+ 6	+10	+50	+45
4	Freon-12 evaporation temperature	deg.C	+ 1	+5	+ 1	+7
5	Freon-12 condensing temperature	deg.C	+35	+30	+55	+50
6	Refrigerating capacity (Q_o)	kW	160	185	115	150
7	Heating capacity (Q_h)	kW	193	220	140	180
8	Effective power (N_e)	kW	46	53	43	45
9	Conversion coefficient (φ)	-	4.2	4.4	3.2	4.0
10	Actual volumetric efficiency (λ)	-	0.76	0.8	0.67	0.75
11	Indicated compressor efficiency (η_i)	-	0.75	0.79	0.68	0.72
12	Net efficiency (η_e)	-	0.66	0.69	0.60	0.63

The results of the tests have proved that the heating and refrigerating capacity of the machine at t_o from 0° to 10°C and t_k from 35° to 55°C has been in compliance with the rated values.

Fig.2 illustrates the interdependency of the heating capacity of the heat pump unit and variations of the condensing temperature, at Freon-12 evaporation temperature 0, 5 and 10°C.

On the basis of the researches, the actual conversion coefficient has been determined for various operating conditions of the heat and cold supply system (Fig.3). The mean conversion coefficient for the heating season proved to be four.

The tests have made it possible to find out the interdependency of the actual volumetric efficiency of the compressors and the pressure ratios, when the compressors are operated at the heat pump duty and at the refrigeration duty (Fig.4). The interdependency of the actual volumetric efficiency and pressure ratios of the compressor operated at the heat pump duty is practically similar to that of the compressor operated at the refrigeration duty. This proves that the series-manufactured refrigeration machines can be employed at the heat pump duty.

Fig.5 illustrates the dependency of indicated compressor efficiency η_i and net compressor efficiency η_e upon the ratio between condensing pressure P_k and evaporating pressure P_0. As can be seen from Fig.5, indicated compressor efficiency reaches its maximum in the zone $\sigma = \dfrac{P_2}{P_1} = 4-5$. Efficiency decrease at lower values of σ is caused by the losses at the valves when drawing high-density vapour, while at $\sigma > 5$ the losses caused by heating, reverse expansion and bypassing increase.

The nature of the curves η_i to a great degree predetermines the nature of the interdependencies of net efficiency (η_e) of the compressors.

The results of the experiments and researches have proved that the volumetric and power characteristics of the refrigeration compressors operated at the heat pump duty have not undergone any considerable changes. Moreover, long-term operation of refrigeration machines at the heat pump duty has shown that their employment on a wide scale is definitely practical.

As regards the economy of energy, the effectiveness of employment of heat pump units on the Black Sea coast is proved by the data on consumption of electric power and fuel required to generate 1 Gcal of heat.

In accordance with the existing system of energy supply (from boiler houses), to generate 1 Gcal of heat it is necessary to use 190 - 250 kg of reference fuel. In accordance with the system developed by us and based on the employment of heat pump units, the consumption of electric power for generation of 1 Gcal of heat is 290 - 300 kW·h or 115 - 120 kg of reference fuel (assuming that to produce 1 kW·h it is necessary to use 0.4 kg of reference fuel).

Consequently, the introduction of heat pump units ensures a major economy of fuel.

On the grounds of wide-scale researches, in 1975 at the Pizunda Resort the then-existing system of air conditioning was converted into the heat pump system in accordance with the diagram shown in Fig.1. The refrigeration compressors ensure the necessity effective heating of the cinema and concert hall. The thermotechnical and power-economical characteristics obtained in the process of long-term service operation of the heat pump units have proved that the employment of heat pumps on the Black Sea coast is fully justified.

In accordance with the data furnished by the Operating Technical Board of the Pizunda Resort, heating and hot-water supply of the resort complex proper and the dwelling houses of the attendants, using the existing direct systems of electric power supply, involves a mean specific consumption of electric power per one bed in the order of 10 000 kW·h/year. The introduction of the heat pump systems of heat and cold supply, with the utilization of heat of sea water, has brought about the reduction of electric power consumption to 2500 kW·h/year per one bed, which gives an economy of approx. 75000 kW·h/year per one bed.

The development and prolonged operation of the experimental--industrial heat pump unit, the assessment of actually obtained technical and economical results and the study of power service characteristics provide for making a comparison of power and economical properties of the heat pump system "sea water - air"

and other existing or planned systems of heat and cold supply.
In making the above-mentioned comparison, it is assumed that the
boiler-furnace fuel used by the alternative plants is natural gas.

The comparison is based on the study of five variants of
heat and cold supply (table 2) to three groups of consumers:

- a supermarket with an annual consumption of heat (for
heating of 1050 Gcal and cold of 300 Gcal;

- a resort building with an annual consumption of heat (for
heating and hot water supply) of 3700 Gcal and cold of 340 Gcal;

- a resort complex with an annual consumption of heat of
190 000 Gcal and cold of 26 000 Gcal.

Table 2   Comparison of Variants of Heat and Cold
                 Supply Systems

Variant	Heat supply	Cold supply
1	Gas-fired boiler house	Compression refrigerating plant
2	Gas-fired boiler house	Absorption plant
3	Electric boiler room	Compression refrigerating plant
4	Accumulation electric boiler room	Compression refrigerating plant
5	Heat pump unit	

The power-economical comparison of the variants tabulated
above has been made judging upon the reduced costs calculated
from the expression:

$$z = И + E_н K \quad (\text{roubles}),$$

where: z - reduced costs,

$И$ - annual production expenses,

K - capital investments,

$E_н$ - authorized coefficient of capital investment
efficiency, equal to 0.12.

The obtained results of the calculations are given in relative values in table 3. The correlations of the reduced costs
give a sufficiently clear picture of the economical advantages
to be gained through the employment of heat pumps.

Table 3.   Correlation of Reduced Costs for
            Variants of Heat and Cold Supply of Consumers

Consumers	Variants				
	1	2	3	4	5
Supermarket	1.25	1.30	1.35	1.60	1.0
Resort building	1.25	1.30	1.55	1.80	1.0
Resort complex	1.15	1.20	1.60	1.90	1.0

Among the variants which have been compared with the heat
pump system, the most efficient one is the system of heat and
cold supply using the gas-fired boiler house and refrigeration
plant. However, even this seemingly efficient system, in case of
the scattered consumers (the supermarket, resort building) is
25% more expensive, in terms of reduced costs, than the heat
pump system.

In case of heat and cold supply to major closely-situated
consumers the extra cost of 25% is somewhat reduced, but nevertheless it remains as high as 15%.

Side by side with the saving in reduced costs, we estimate that wide-scale employment of heat pump plants in the littoral
zone of the Black Sea coast of Georgia alone will provide an
annual economy of reference fuel of approx. 450 - 500 thousand
tons.

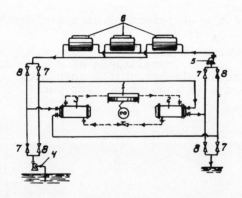

Figure 1.

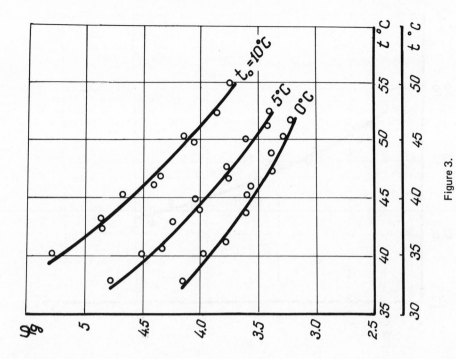

Figure 3.

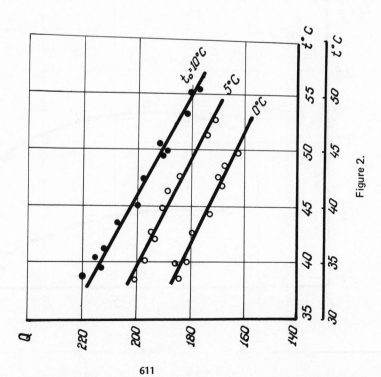

Figure 2.

611

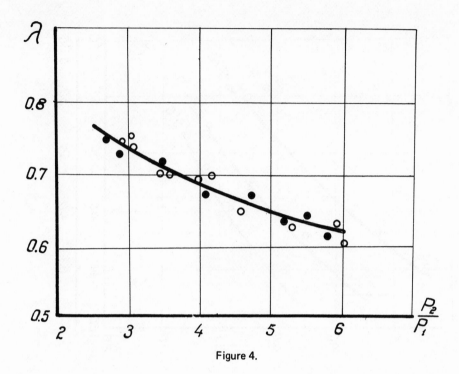

Figure 4.

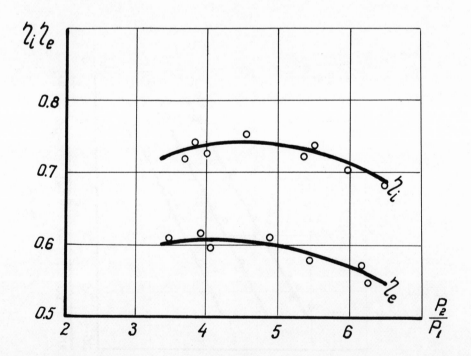

Figure 5.

DESIGN OF LOW COST VENTILATION AIR HEAT EXCHANGERS

R. W. BESANT, E. E. BROOKS, G. J. SCHOENAU, AND R. S. DUMONT

University of Saskatchewan
Saskatoon, Canada S7N 0W0

ABSTRACT

The design of a low cost ventilation air heat exchanger proposed which utilizes plastic sheets as the heat transfer surface is presented. Laboratory tests on such a counter flow heat exchanger have demonstrated very high values of overall heat transfer coefficients, heat exchanger effectiveness, and temperature recovery when the flow is laminar and buoyancy effects assist the heat transfer. The costs of such heat exchangers would make them attractive for many building applications.

NOMENCLATURE

A   heat exchanger area

C_p specific heat at constant pressure

C   cost

D   distance between heat exchanger plates

g   gravitational acceleration

Gr_D Grashof number

   $= g\rho^2\beta\Delta T\ D^3/\mu^2$

h_{ig} heat of sublimation or vaporization for water

K   present value factor

   $= \dfrac{(\frac{1+x}{1+r})^N - 1}{(\frac{x-r}{1+r})}$

m   mass flow rate

N   number of years

$NTU = \dfrac{UA}{m_c C_{Pc}}$ number of transfer units

q   heat rate

Q   total heat required for ventilation air per year

r   discount rate for money

T   temperature

U   overall heat transfer coefficient

V   average velocity

x   annual inflation increase in fuel prices

β   expansion coefficient

ΔT   change in temperature from mean surface to mean bulk

ε   heat exchanger effectiveness

η_f fuel burner efficiency

μ   viscosity

ρ   density

ω   humidity ratio

Subscripts

c   cold air

f   fuel

h   hot air

HE   heat exchanger

TE   thermal energy

VA   ventilation air

1   warm air inlet

2   warm air exhaust

3   cold air exhaust

4   cold air inlet

INTRODUCTION

In the past, heat recovery from ventilation air exhaust has not usually been considered for small buildings due to its large capital costs relative to the small savings incurred from reduced heating loads. Now with more expensive fuel costs, heat recovery from ventilation air may be one of the most cost effective ways to cut fuel consumption. This paper describes the design, cost analysis and testing of a ventilation air, counter flow, heat exchanger which would be suitable for use in houses and other small buildings.

The efficacious application of a heat recovery system for ventilation air demands that a building be "tight" so that uncontrolled air flow leakage is reduced to a small fraction of the ventilation air flow rate. This may be done by incorporating an effective vapor and air flow plastic seal or barrier on all the inside wall, ceiling, and floor surfaces and by eliminating such things as open chimneys and cracks in doors and windows. While new buildings can and should be built to have such features at a modest extra cost, older buildings might require extensive retrofitting to be suitable for ventilation air heat recovery. In addition, heating systems that are directly compatible with ventilation air heat recovery are restricted to electrical heating, central steam or hot water heating, and solar heating. By isolating the combustion supply and chimney for fuel burning in buildings, one is able to make efficient use of a ventilation air heat exchanger.

HEAT EXCHANGER DESIGN

For the purpose of laboratory testing, a ventilation air heat exchanger has been designed and constructed as shown in Figure 1. The heat exchanger's $11.36 \ m^2$ of heat transfer surface was constructed of eleven 0.15 mm thick poly-ethylene plastic sheets such that each 50x203 cm sheet was placed 1.27 cm apart. The heat exchanger was designed to operate in the vertical position such that cold air entered at the bottom and flowed 2 m up through alternate sheets and the warm air which entered at the top and flowed down. The exterior frame of this counter flow heat exchanger consisted of 1.27x5 cm plywood stripes covered with 5 cm of polystyrene insulating board which gave an effective thermal re-sistance in excess of $1°C \ m^2 \ W^{-1}$ on all sides of the heat exchanger.

HEAT EXCHANGER TESTS AND RESULTS

Tests were done on the heat exchanger over a range of flow rates from 0 to $3 \ m^3$ per minute and the temperature recovery fraction, heat exchanger effective-ness and the overall heat transfer coefficient were calculated using the follow-ing equations for the heat transfer rate:

$$q = m_c C_{Pc}(T_3-T_4) \tag{1}$$

$$= m_h [C_{Ph}(T_1-T_2) + (\omega_1-\omega_2)h_{ig}] \tag{2}$$

$$= \varepsilon \ m_c C_{Pc}(T_1-T_4) \tag{3}$$

$$= UA \ \overline{\Delta T} \tag{4}$$

where

$$\overline{T} = LMTD = \frac{(T_1-T_3) - (T_2-T_4)}{\ln(\frac{T_1-T_3}{T_2-T_4})}$$

For thermal energy recovery in ventilation air the most important thermal para-meter is the temperature recovery factor, $(T_3-T_4)/(T_1-T_4)$, or heat exchanger effectiveness, ε. The results from tests presented in Figure 2 suggest that very high values of heat exchanger effectiveness exist at low flow rates. In addition, the heat exchanger effectiveness as a function of number of transfer units, NTU, may be compared to the theoretical results for the case $m_c C_{Pc}/m_h C_{Ph}$ = 1 as shown in Figure 3. That is,

$$\varepsilon = \frac{NTU}{1 + NTU} \tag{5}$$

when $m_c C_{Pc}/m_h C_{Ph}$ = 1 for a counterflow heat exchanger where NTU is defined by

$$NTU = \frac{UA}{m_c C_{Pc}} \tag{6}$$

The overall heat transfer coefficient, U, is presented as a dimensionless Nusselt number $Nu = 2 UD/k$ as a function of Gr/Re^2 in Figure 4 where Gr/Re^2 has been shown to be the appropriate dimensionless parameter in forced/natural convection problems. It can be seen from these results that the Nusselt number increases substantially beyond that for a fully developed laminar flow in long ducts with constant wall heat flux. An approximate correlation of the experimental results is given by the equation

$$Nu = 3.6 + 3.4x10^4 (Gr/Re^2)^2 \tag{7}$$

It would appear that buoyancy effects, as given by the Grashof number, Gr, increase as forced convection effects, as given by the Reynolds number, Re, decrease. Attempts to operate the heat exchanger against the gravitation forces which indicated heat exchanger effectiveness of much less than .5 would add further evidence to this theory. Flows against gravity not only produced low heat exchanger effectiveness values, but they are inherently unstable and the results are not necessarily reproduceable with slightly different experiments. Hence, no further investigations were made of the inverted mode of operation.

In cold climates heat exchangers may condense water and freeze up if warm moist inside air is cooled down below the local dew point or freezing point of water. When freezing occurs, the layer of frost forms on the inside air flow channels which restricts the flow passages and reduces the heat transfer rate. Long term tests showing the reduction in temperature recovery and flow rate as a function of time are presented in Figures 5 and 6. Again, if the flow rate is small, the efficiency is seen to maintain a high value prior to gradually dropping to very low values over a period of days. Freeze-up of the heat exchanger causes the deterioration in performance, but periodic defrosting of the heat exchanger rapidly restores the heat exchanger ot its original performance.

COST CONSIDERATIONS

The total present value cost of introducing a ventilation air heat exchanger which is financed over a number of years of operation, N, consists of the initial cost of installing the heat exchanger, C_I, plus the cost of thermal energy over the same number of years required to preheat the required ventilation air up to room temperature, C_A. That is

$$C_T = C_I + C_A \tag{8}$$

where the heat exchanger cost may be assumed to vary directly with its heat transfer area A such that

$$C_I = C_{I1} + C_{I2} A$$

and   (9)

$$C_A = K \frac{C_f}{n_f} (1-\varepsilon) \int_{year} m_c C_{Pc}(T_1-T_4) \, dt$$

or

$$C_{TE} = \frac{K C_f Q}{n_f} (1-\varepsilon) \tag{10}$$

It can be seen from (5) that

$$1-\varepsilon = \frac{1}{1 + NTU} \tag{11}$$

For a constant rate of air flow required for ventilation NTU will vary directly with the UA product in (6) where

$$U = \frac{Nu\ K}{2D}$$

The Nusselt number was shown to vary in Figure 4 with the parameter Gr/Re^2 in a quadratic manner as given by (7). The parameter Gr/Re^2 is given by $Dg\beta\Delta T/V^2$ where both the temperature difference ΔT and the flow velocity tend to vary inversely with the heat exchanger area, A, over finite changes in 1/A and for a specified flow rate. Combining these equations for NTU in (11) finally gives approximately

$$1-\varepsilon = \frac{1}{1 + .20\ A + 1.0\ A^3} \tag{12}$$

where A is in m^2. The condition for minimum cost is given by

$$\frac{d\ C_T}{dA} = 0 \tag{13}$$

or

$$\frac{C_{I2}}{K\ C_f\ Q}\ [1 + .20\ A + 1.0\ A^3] = .20 + 3.0\ A^2 \tag{14}$$

which gives an optimum heat exchanger area of approximately 50 m^2 for a typical house. The corresponding effectiveness as given by equation (12) is nearly 100% under normal flow conditions. Dust, condensation, frosting, and variable flow would likely given lower values. Furthermore, natural convection will cause a small flow rate under most conditions so that Nu and ε will be limited. Different load cost and climatic conditions would yield other sizes.

DISCUSSION AND CONCLUSIONS

The design of a low cost ventilation air heat exchanger has been presented. Such a heat exchanger could be used to recover most of the thermal energy from ventilation air which must be brought into buildings by fans. The experimental results indicate that such a heat exchanger should be operated so that gravitational forces assist the flow and hence the heat transfer when the flow rates are low and the flow is laminar. The method of optimizing costs over a specified period of time presented indicates that the use of low cost materials such as plastics permits the designer to use very high heat exchanger effectiveness values. In very cold climatic conditions where frosting occurs periodic defrosting must be provided for in the design.

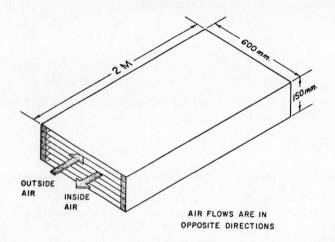

AIR-TO-AIR HEAT EXCHANGER SCHEMATIC
Figure 1

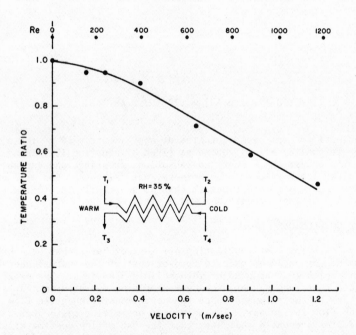

AIR TO AIR HEAT EXCHANGER PERFORMANCE

TEMPERATURE RATIO $\dfrac{T_3 - T_4}{T_1 - T_4}$ VS AIR VELOCITY

Figure 2

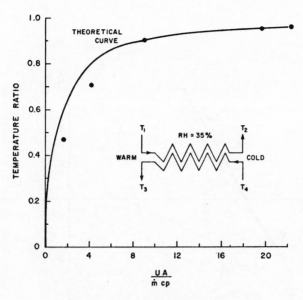

AIR TO AIR HEAT EXCHANGER PERFORMANCE

TEMPERATURE RATIO $\dfrac{T_3 - T_4}{T_1 - T_4}$ VS $\dfrac{U A}{\dot{m}\, c_p}$

Figure 3

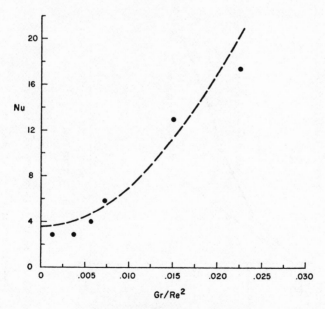

PLOT OF NUSSELT NUMBER VERSUS
GRASHOF/(REYNOLDS)^2

Figure 4

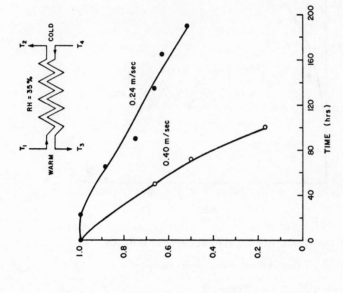

AIR TO AIR HEAT EXCHANGER PERFORMANCE

FLOW EFFICIENCY Q_1/Q_2 $Q_{hot\ side}/Q_{cold\ side}$

Figure 6

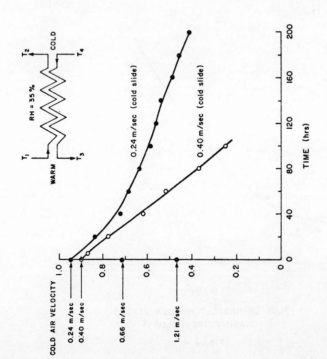

AIR TO AIR HEAT EXCHANGER PERFORMANCE

TEMPERATURE RATIO $\dfrac{T_3 - T_4}{T_1 - T_4}$ VS TIME (hrs)

Figure 5

RADIANT COOLING AND
COMFORT WITH HIGH
AMBIENT AIR TEMPERATURES

S. R. MONTGOMERY AND S. HART

University College London
London WC1E 7JE, England

ABSTRACT

The place of radiation in providing comfort conditions without the additional
need to reduce the ambient air temperature in an occupied space is discussed
and details are given of radiant cooling panels designed to operate with
surface temperatures lower than the ambient dew point temperature.
Experimental data on full scale panels provides information on the relative
merits of various designs.

INTRODUCTION

Studies on thermal comfort have been carried out for many years but it is
only comparatively recently that any real attempt has been made to correlate
all the interacting effects of the various controlling parameters. In one
of the most comprehensive studies Fanger (1) lists six basic parameters,
four being properties of the thermal environment and two properties of the
human subject, and he also points out that the heat loss from the individual
should not be too asymmetrical. Several of the papers presented at a
symposium on thermal comfort and moderate heat stress (2) compare the various
criteria used for assessing human comfort including the ASHRAE standard
55-66 (revised 1972)(3) which is widely used for the design of thermal
control systems in buildings. It was noted at the symposium that one of the
original objectives of this standards was to "provide systems which satisfy
the occupants all the time, without regard to cost"(4).

Standards such as this may no longer be universally valid when factors associated with energy conservation are included. To cite a specific example,
the normal design conditions adopted for heating office accommodation are
those appropriate to occupants wearing relatively light clothing. If lower
temperatures were to be provided in buildings the occupants could still be
comfortable if they wore heavier clothing. Twenty five years ago it was a
common practice in the United Kingdom to wear a woolen sweater or cardigan
indoors in winter and to keep on one's jacket : perhaps energy costs may
force us to readopt such practices in the future.

At the other end of the temperature range one has to consider the problem
of comfort under conditions of high ambient temperature. Here there is no
easy way to provide comfort by simply modifying one's clothing but there
may well be good reasons for relaxing the conditions currently accepted

as necessary for design purposes especially if this will reduce the demand
for energy. There are a number of ways in which this could be achieved:

1. By choosing design conditions which achieve a more restricted degree of
    comfort e.g. on the ASHRAE rating scale (3) a "slightly warm" condition
    rather than a "neutral" one or "comfortably warm" on the Bedford rating
    scale (5) rather than "neither warm nor cool".

2. By providing local regions where comfortable conditions are provided
    rather than a global provision throughout a room or a building. This
    concept has already been used in a number of extreme instances (e.g.
    space suits, cool zones near blast furnaces,) but the concept is adap-
    table to more conventional situations.

3. Temporary provision of comfort conditions only in occupied areas. One
    example in current practice is the provision of package room air condi-
    tioners which are only operated while the room is occupied.

4. Design of new buildings to make best use of natural aids e.g. shading of
    windows, use of natural ventilation. Attempts to import "standard
    design" rather than adapting local techniques may unreasonably increase
    the energy required to create comfort conditions.

The work described in this paper envisages a cooling system which incorporates
all these characteristics and which enhances the effect of one particular
parameter in the physical environment, namely the mean radiant temperature.

PREVIOUS WORK ON RADIANT COOLING PANELS

While the use of radiant panels as a means of heating a room is well established in the United Kingdom and elsewhere, relatively little use has been
made of similar panels for cooling. If the panel surface temperature is
less than the local dew point temperature water will condense on the surface
and ultimately form unacceptable pools of water below the panel. One must
either prevent the humid air from reaching the cold surface or control the
condensation by some other means.

Leopold (6) has described an early example of a system consisting of combined
convective and radiant cooling panels set in the ceiling and on the wall.
The panels were maintained at a temperature above the dewpoint and never
less than 18°C for a room air temperature of 25°C. The system does not
appear to have been widely adopted, probably because radiation contributes
little to the overall cooling effect.

One of the few detailed studies of the relationship between comfort and heat
losses by radiation to a cold panel was that reported by Morse and Kaletzky
(7,8). They took great care to prevent condensation on the panel surface by
covering it with one or more layers of polythene which is relatively transparent to radiation of the wavelengths characteristic of room temperatures.
The temperature of the outer layer of polythene was maintained above the dew
point of the room air so that no water was condensed on to the surface.

Tests were carried out on 16 subjects in a test chamber maintained at a dry
bulb temperature of 33.4°C and wet bulb temperature of 29°C and equipped with

large cooling panels at temperatures of either 7°C or -8°C.

The results showed that the energy transfer between a subject and the panels amounted to 70W and 105W for the higher and lower panel temperatures respectively. This compares with an average metabolic rate for a man at rest of 120W. Subjective assessments of comfort indicated that the subjects still generally felt "warm" rather than "neither warm nor cool" but this is not surprising in view of the low air velocity and the severe thermal condition of the environment.

A very similar idea was proposed by Kjerulf-Jensen(9) using polyethylene rather than polythene films over the panel. The main disadvantage of using such films is that the construction technique is relatively complicated and the films, which must be thin if they are to remain transparent,are consequently rather fragile.

The alternative scheme proposed here is believed to be a novel one. Experiments have shown that certain fabrics attached to a cold surface appear to limit the amount of condensation so that although the surface becomes saturated with water no drops are formed and some form of dynamic equilibrium is established. This process allows a simple cooling panel to be operated at temperatures well below the ambient dew point temperature.

COOLING PANELS

The basic concept behind the construction of the cooling panels is that of a flat surface behind which a cooling fluid can be passed and over which an absorbent layer of cloth material can be securely attached. The rear surface of the cooling panel is well insulated to prevent unwanted heat gain.

Preliminary studies have been carried out using a conventional steel radiator. For convenience in attaching the cloth the particular model chosen has a completely flat front surface with headers on the rear face for the attachment of cooling fluid supplies. The individual flow channels have a rectangular cross section and are welded together to form a panel with a groove between each channel. The details of the panel are given in Table 1. The rear face is insulated with 50 mm of expanded polystyrene held within a wooden frame around which the cloth covering the front face is passed and fixed.

The cloth is attached to the front face by means of .05mm thick double sided adhesive tape (Scotch plate mounting tape # 463DAZ). Great care is taken to ensure that the cloth is adequately stretched over a subsidiary frame prior to its application to the adhesive tape.

This conventional metal radiator is relatively heavy and expensive and preliminary experiments have also been carried out using a light weight polypropylene panel constructed from extruded sheets. These sheets Corruflex, manufactured by Corrux Ltd. have relatively small passages (see Table 1) and require a header into which the end of the sheet is attached by welding. (A similar design has recently become available commercially for use as a flatplate solar collector). Compared with the metal panels, the plastic panels are much less rigid and have a semi-cylindrical section at each end which must either be covered in cloth or be adequately insulated. Because of difficulties encountered in attaching stiffening ribs to the rear face

of the polypropylene sheet the first panels have made use of thin bands of
material to clamp the sheet against a wooden frame; these bands are sufficiently thin and narrow to allow cloth to be fixed to the front face without
creating a focal point from which condensed water can drip. The rear face
is again insulated with expanded polystyrene.

Material		Steel	Polypropylene
Panel length	m	2.0	2.0
Panel width	m	1.25	1.25
Channel width	mm	70	5.5
Channel depth	mm	11	4
Header dimensions	mm	33 x 33	25 (Dia)
Channels in parallel	-	3	220
Channels in series	-	6	1
Fluid capacity	1	20.4	8.7
Mass empty	kg	64.4	2.0

Table 1. Details of cooling panels used
in preliminary experiments

The major advantage of the metal panel is the strength of the material which
reduces the chances of accidental damage to the surface and allows a somewhat higher design operating pressure as compared with the polypropylene
panel. It would be possible to use the metal panel as the evaporator for a
refrigerator but in the present tests chilled water is used as the cooling
medium. The polypropylene panel depends on the supporting frame for its
stiffness but the overall weight is appreciably less than for the metal
panel while the cost is approximately one third.

TEST ROOM

Preliminary experiments have been carried out in a small room specially
adapted to allow high ambient air temperatures and humidities to be maintained up to a temperature of $38^{\circ}C$ and 95 percent relative humidity. The
dimensions of the room are 4.2 m x 3.0 m by 2.7 m high. The walls have
been covered with light-weight perforated metal panels so that the mean
radiant temperature does not lag too far behind changes in the air temperature.
Air is heated and humidified in an external circuit.

Provision has been made for mounting three panels in the room, one on each of
two adjacent walls and one on the ceiling. Each panel is supplied with
water from one of three storage tanks, each maintained at a different constant temperature. This allows each panel to operate with one of three
surface temperatures, namely room air temperature, a minimum temperature
just above freezing ($2^{\circ}C$) and one intermediate temperature.

The test room does not provide the same uniformity of temperature, humidity
and velocity that one would desire from an environmental chamber but does
allow preliminary experiments to be carried out while the panels are being
developed in an environment reasonably representative of actual use.

EXPERIMENTAL RESULTS

The primary objective in designing the panels is to enhance the radiation heat transfer while incurring a minimum convective heat gain. Experiments have therefore been carried out to measure the overall heat transfer rates and hence to estimate the convective component by subtracting that due to radiation. For the low temperatures of interest in this application the thermal emission and absorption coefficients approach unity so that the errors introduced in assuming that the panel behaves as a black body are small compared with uncertainties involved in estimating the convective heat transfer rates. Table 2 gives details of the steady state heat transfer rates, both estimated and measured, for panels mounted vertically on the wall.

	T_S	T_A	Relative Humidity	Radiation h_r	Total $h_c + h_R$	Convection h_c
	$^{\circ}C$	$^{\circ}C$	%	$W/m^2 K$	$W/m^2 K$	$W/m^2 K$
W1	4.0	32.0	39	5.6	11.8	6.2
W2	4.4	31.7	70	5.6	16.6	11.0
W3	4.7	31.7	93	5.6	19.3	13.7
Theory A	4.0	32.0	-	5.6	9.5	3.9
B					11.1	5.5

Table 2. Comparison of surface heat transfer coefficients, measured on panel with values predicted from general correlations given by theory A, McAdams (10) and theory B, Min (11). Experiments W1 to W3 were carried out using a metal panel mounted in a vertical position on the wall of the test room.

Despite the fact that adequate time was allowed for steady-state conditions to become established and that no condensed water was collected from the panel, it is clear that the effective convective component of heat transfer increases by more than a factor of two as the relative humidity rises from 40 percent to 90 percent. The reasons for this are not certain but it seems likely that the cloth acts as a wick in distributing the condensed water and that the natural convection process is substantially modified by the circulation of water vapour. It is of course well known that the heat transfer coefficients associated with condensation are much larger than those found for non-condensing gases and local recirculation of the water between different sections of the panel may account for the observed results.

Some variation has been observed in the surface temperatures on the panel and further experiments are being carried out in order to determine the

significance of these results. It is clear however from these preliminary
experiments that it will be essential to compare the performance of panels
protected by transparent plastic films with the present design, especially
with high humidity conditions.

DISCUSSION

The experimental results presented in this paper show that suitable fabric
materials exist which can be mounted on cold surfaces to prevent the continuous condensation of water vapour from the surrounding air. At the same
time it has been shown that the convective heat transfer between the air
and the surface is significantly increased when the relative humidity of
the air approaches saturation, thus increasing the total energy requirements
of a panel having a given total radiant power. It appears that while a
panel covered with fabric can satisfactorily be operated with low humidities,
the use of a protective film over the panel may be desirable at higher
humidities. The relative importance of radiation and convection heat transfer
in determining the subjective degree of "comfort" is by no means clear.

Fanger (1) has produced a series of correlations showing the conditions under
which comfort is achieved including one in which the relative effects of mean
radiant temperature and air temperature are shown for fixed activity, clothing
and humidity. However it is not clear from the published curves how much
data is actually available for each air temperature and mean radiant temperature, nor whether the subjects are more concerned about being heated up or
cooled down.

A few experiments have been carried out to determine the effect of non-
symmetrical radiation patterns, for example those associated with heated
ceilings. McIntyre and Griffiths (12) carried out tests in which the temperature of the walls in an environmental chamber was maintained at a
different value from that of the ceiling and they commented that the hot
ceiling condition was consistently reported cooler than a uniform condition
having the same air and mean radiant temperatures. No comparable experiments
appear to have been carried out with single surfaces at reduced temperatures.

At the present time there is inadequate data to predict either the subjective
reactions of individuals to cooling panels nor the precise energy requirements
of the panel themselves. Work is proceeding on both these aspects of the
problem but it is still too soon to try and compare the total energy requirements of panels and conventional cooling systems. It will depend particularly
on a full consideration not only of the building but also of the way in which
people use the building. Fast responding systems capable of producing an
acceptable microclimate for the individual may ultimately prove to be more
economical both in terms of energy and finance than conventional systems which
produce prescribed conditions throughout a building, whether it is occupied
or not. One situation to which the cooling panel would seem particularly
well suited is in bedrooms where a panel fixed over the bed would provide a
silent unobtrusive means of cooling an individual. Future needs for energy
economy makes it important to consider unconventional solutions such as this
while at the same time seeking improvements in conventional systems.

Acknowledgements: This work has been supported by the Science Research Council.

REFERENCES

1. Fanger, P.O. 1972. Thermal Comfort. McGraw-Hill Book Company.

2. Langdon, F.J., Humphreys, M.A., and Nicol, J.F. 1973. Thermal Comfort and Moderate Heat Stress. Building Research Establishment Report 2, H.M.S.O. London.

3. A.S.H.R.A.E. 1972. Guide on Fundamentals. American Society of Heating, Refrigerating and Air Conditioning Engineers, New York.

4. Nevins, R.G. 1973. Building Research Establishment Report 2, H.M.S.O. London, p.226.

5. Bedford, T. 1936. The Warmth Factor in Comfort at work. Medical Research Council, London. Industrial Health Research Board Report No. 76.

6. Leopold, C.S. 1951. Design Factors in Panel and Air Cooling Systems. Trans. ASHVE, 57, 61.

7. Morse, R.N. and Kaletzky, E. 1961. A new approach to Radiant Cooling for Human Comfort. J. Inst. Engrs. Australia, 33, 181.

8. Kaletzky, E., Macpherson, R.K., and Morse, R.N. 1963. Effect of Low Temperature Radiant Cooling on Thermal Comfort in a Hot Moist Environment. J. Inst. Engrs. Australia, EM5, 60

9. Kjerulf-Jensen, P. 1975. Radiative Cooling of Work Places at Low Energy Consumption. Symposium on physiological requirements of the Microclimate. Czechoslovak Medical Society, Prague.

10. McAdams, W.H. 1954. Heat Transmission. McGraw-Hill Book Company, 173.

11. Min, T.C., Schutrum, L.F., Parmelee, G.V., Vouris, J.D. 1956. Natural Convection and Radiation in a Panel-heated Room. ASHAE Trans, 62, 337.

12. McIntyre, D.A., and Griffiths, I.D. 1973. Radiant Temperature and Comfort. Symposium on thermal comfort and moderate heat stress, Building Research Establishment Report 2, 113.

PERFORMANCE OF
AN EXPERIMENTAL
REGENERATIVE-EVAPORATIVE
COOLER COMPARED
WITH PREDICTIONS

H. BUCHBERG AND N. LASSNER

Chemical, Nuclear, and Thermal Engineering Department
School of Engineering and Applied Science, University of California
Los Angeles, California, USA 90024

ABSTRACT

An experimental regenerative-evaporative cooling (REC) system has been designed, constructed and tested in the laboratory under simulated environmental conditions. It represents the first phase of an investigation of solar desiccant evaporative cooling processes applied to space cooling with the goal of substantial reduction in electric power usage for air-conditioning.

The test measurements were used to validate and refine an analytical performance model programmed for computing. Performance graphs are generated which permit the determination of supply air temperature and room air temperature and humidity for a specified air flow, latent and sensible heat load and heat exchanger carryover ratio. Predictions of room air temperature (T1) are within 2°F or less and room air humidity ratio (W1) within 0.0008 lb/lb_a or less of observed performance. Supply air temperature (T10) was predicted within 1°F or less. Based on a design capacity of 3.5 tons of cooling at an air flow of 2500 cfm it is estimated that a cooling effect to electrical power usage ratio between 8 and 10 is achievable. The experimental system is capable of providing 3.5 tons of cooling when delivering air at 67.3°F when the outside air is at 100°F with humidity 0.0103 lb/lb_a. For these conditions, 2500 cfm supply air maintains a room temperature of 77.6°F at 70% relative humidity when 1/3 of the load is latent heat.

NOMENCLATURE

CO	heat exchanger carryover ratio (dimensionless fraction)
COP	coefficient of performance (dimensionless)
p	barometric pressure (psia)
p_s	saturation pressure (psia)
QL	latent heat load (Btu/hr)
QS	sensible heat load (Btu/hr)
$\dot{Q}_V$	volumetric flow (ft^3/min or cfm)
RH	relative humidity (dimensionless fraction)
T	temperature (°F)
T_s	saturation temperature (°F)
W	humidity ratio, lbs of water vapor/lb dry air (lb/lb_a)
W_s	saturation humidity ratio (lb/lb_a)

629

Typical Computer Symbols

M7   volumetric flow at location 7 (cfm)

T7   temperature of air-vapor mixture at state 7 (°F)

TS7   saturation temperature corresponding to T7 (°F)

W7   humidity ratio of air-vapor mixture at state 7 (lb/lb$_a$)

WS7   saturation humidity ratio corresponding to T7 (lb/lb$_a$)

Greek Symbols

ε_h heat exchanger effectiveness (dimensionless fraction)

ε_m evaporative cooler effectiveness (dimensionless fraction)

Subscripts

a   dry air   DB   dry bulb

c   critical   WB   wet bulb

s   saturation

1,2,3...   state points or locations

INTRODUCTION

    The performance studies of an experimental regenerative-evaporative cool-
ing (REC) system represents the first phase in the investigation of solar-
desiccant-evaporative processes for space cooling. It is the ultimate purpose
of this investigation to demonstrate a space cooling system which substantially
reduces the electric power requirement compared to conventional vapor-compression cooling. The basic cooling mode is the adiabatic saturation of an air
stream with moisture where sensible enthalpy is transformed to latent heat of
evaporation. This process is recognized as evaporative cooling which apparently was known by the ancients as evidenced in Egyptian frescoes depicting slaves
fanning jars of water [1]*. In more modern times evaporative cooling has been
used with varying degrees of success for air conditioning [1,2,3]. Interest in
evaporative cooling has been rekindled as a result of our need to conserve
energy. A considerable amount of electric power is used to reconstitute the
evaporated working medium in a vapor-compression cooling cycle. Evaporative
cooling processes with water use much less power but at the expense of reject-
ing water vapor to the environment.
    Several evaporative cooling process arrangements may be considered depend-
ing primarily on the humidity of the ambient air environment. In regions of
very low humidity, such as desert areas, cooled humidified air may be blown
directly into the living space using a humidifier popularly known as a "swamp"
cooler. Using the design dry and wet bulb temperatures of Barstow, California
as representative of a high desert region, where $T_{DB} = 104°F$ and $T_{WB} = 73°F$ [4]
an 85% effective swamp cooler would supply air at 78°F with relative humidity
(RH) approximately 80% at the design conditions. For a 36,000 Btu/hr sensible
load and zero latent load a volume flow of 6,700 cfm would maintain a space
temperature of about 83°F at 68% RH. This would be useful but marginal cooling.
The next step in sophistication would be to add a regenerative (matrix) heat
exchanger using humidified and cooled building exhaust air as the coolant for
the heat exchanger to cool fresh air entering the building. For the same cooling load and volumetric air flow, as in the previous example, a space temperature of about 80°F at 48% RH can be maintained at the design condition when the
heat exchanger effectiveness is 80%. By adding a second air humidifier at the
room inlet, the air circulation rate can be substantially reduced with a small

_____
*Numbers in brackets refer to list of references.

lowering of room temperature, but at the expense of some increase in RH. For
example, with an 85% effective evaporative cooling of the building exhaust
air, and with an 80% effective sensible cooling of fresh air with the cool
moist exhaust air stream, a final 67% effective evaporative cooler produces
supply air at 68.5°F. This requires only 3500 cfm air circulation for full
load or approximately one-half of that required by the previous two systems.
The room temperature stabilizes at 78°F with 60% RH. If 1 or 2 evaporative
cooling effects do not permit sufficient lowering of temperature with acceptable levels of air circulation and RH, (due to high ambient air humidity) a
desiccant bed may be added to dry the air. With the addition of a drying agent,
the process becomes a desiccant-evaporative cooling or DEC process.

Figure 1 is a schematic outline of the DEC system under study. Note that
in addition to the 2-stage evaporative cooling and regenerative heat exchange
between exit and inlet air discussed previously, a desiccant bed has been added
as well as a second regenerative heat exchanger to recover enthalpy from the
hot dry air leaving the desiccant after adsorption of water vapor. To make
the process continuous a desorption system must also be added. This requires
a means for raising the temperature of ambient air such as a solar air heater.

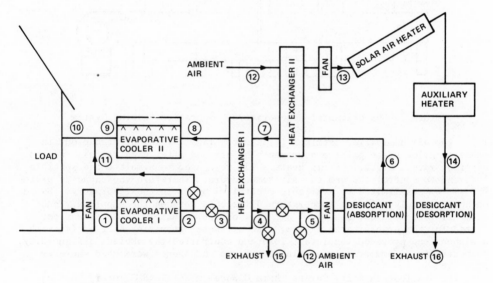

Figure 1. Solar Desiccant-Evaporative Cooling System

It is the purpose of this paper to describe an experimental REC system
representing the first phase in the development of a complete solar DEC system
for space cooling and a computer programmed performance model verified by
actual measurements. The computer generated performance graphs can be used
as an aid in the selection of appropriate designs to meet specified requirements.

EXPERIMENTAL REGENERATIVE EVAPORATIVE COOLING SYSTEM

Description

As shown in the schematic sketch of Fig. 2 and the photograph in Fig. 3,
the experimental REC system comprising one rotary regenerative heat exchanger

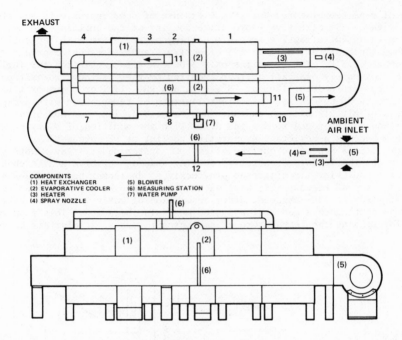

Figure 2. Experimental Regenerative-Evaporative Cooling System

(1)*, two air humidifier matrices (2) and two fans (5) was stretched out in space to allow state point measurements before and after each process and the air flow rate. Electric strip heaters (3) were used to simulate the sensible load and to control the ambient air temperature. Therefore the capacity of the REC system was limited by available electrical power in the laboratory. Based on approximately a 15 deg temperature difference between the supply air and room air temperatures, a nominal air flow of approximately 1200 cfm was permitted for the experimental system. Two air-water atomizing nozzles (4) with a siphon feed provided the latent load and controlled the ambient air humidity, respectively. Commercially available process components were used wherever possible.

The smallest capacity Carnes Therm O'Wheel model T048MS rotary heat exchanger was obtained. The wheel is 48 inches in diameter and the media consists of knitted corrugated aluminum wires folded and skewered into wedge sections about 12 inches deep. The heat exchanger wheel was rated for 2500 cfm and was operated without the purge section. Two air humidifiers were designed using Munter's cooling tower packing, CELdek, described as a cellulose paper impregnated with insoluble anti-rot salts, rigidifying saturants and wetting agents. The packing has a cross-fluted configuration which allows for maximum air and water flow without clogging. Water was recirculated from a collection trough below to a distribution manifold installed above the 12-inch deep packing material with frontal area of approximately 5 ft^2. Two variable speed 1/2

---

* Numbers in parentheses refer to components in Fig. 2.

Figure 3.   Photograph, Experimental REC System

HP fans were installed as shown in Fig. 2. Dampers for air flow control were
placed in the ambient air inlet duct downstream of the fan and in the recirculating duct.

Instrumentation and Test Procedure

The essential measurements made to establish the steady state characteristics of the experimental REC system were wet and dry bulb temperatures at locations 1, 2, 4, 7, 8, 9, and 10 as shown in Fig. 2, pressure difference across
the air humidifiers (2) and both sides of the rotary heat exchanger (1), and
the volumetric air flow in the entrance metering duct at locations 12 and in
the recirculating duct at location 11 when used. Iron-constantan thermocouples
were used for the temperature measurements with a small waterproofed copper
sheath used for the wet bulb which was fed distilled water through an attached
cotton wick. The dry bulb junctions were located upstream of the wet bulbs to
avoid thermal interaction. Temperatures were recorded with a Honeywell strip
chart recorder with an accuracy of about 1°F. Standard magnehelic differential
pressure gauges with an accuracy of about 0.01 inch of water were used for the
pressure difference measurements. A pitot static tube velocity traverse was
made to determine the volumetric air flow rate in a straight duct section downstream from the fan. The metering duct section can be seen in the foreground
of Fig. 3. It is estimated that the air flow was determined with an accuracy
of about 2%.

All measurements were made under steady state conditions which were
achieved within approximately 15 minutes either after startup or after a substantial change in simulated load or simulated ambient air condition. The
independent variables whose values can be set for any test run are T7, W7, M7,
QL, QS and M11. Test runs were made for a range of ambient air conditions
and for different values of sensible and latent loads. One test run was made
with some recirculation air flow M11.

Test Results

The results of 14 test runs are given in Table 1 which covers a range of
test variables as follows:

Table 1. REC System Test Data*

TEST NO.	QS Btu/hr	QL Btu/hr	M7 CFM	T7 °F	W7 lb/lb	T1 °F	W1 lb/lb	T2 °F	W2 lb/lb	T4 °F	W4 lb/lb	T8 °F	W8 lb/lb	T9 °F	W9 lb/lb	T10 °F	W10 lb/lb
1	13192	0	1265	90.5	.0078	70.0	.0107	64.5	.0127	87.0	.0123	67.0	.0080	59.5	.0104	60.0	.0105
2	13192	0	1265	83.0	.0079	70.0	.0110	64.5	.0126	82.5	.0118	66.0	.0082	59.0	.0104	59.5	.0106
3	15908	0	1265	83.0	.0078	73.0	.0111	66.5	.0134	81.0	.0125	67.0	.0083	59.5	.0106	60.5	.0109
4	14589	0	1081	84.0	.0075	75.1	.0115	68.0	.0140	80.5	.0126	68.5	.0081	61.0	.0112	61.5	.0114
5	14920	0	1081	93.0	.0074	76.0	.0120	69.0	.0145	89.0	.0130	70.0	.0082	61.5	.0114	62.3	.0116
6	15440	0	1231	89.0	.0098	77.0	.0127	70.2	.0156	87.0	.0152	71.5	.0102	64.0	.0125	64.5	.0127
7	15470	0	1231	83.0	.0112	79.0	.0137	72.0	.0162	82.0	.0155	72.0	.0116	66.0	.0134	66.8	.0136
8	9051	0	1231	78.0	.0132	75.0	.0154	72.3	.0167	77.5	.0157	72.5	.0135	69.0	.0149	69.5	.0152
9	11594	2313	1231	89.5	.0105	75.0	.0138	70.2	.0156	87.0	.0146	72.0	.0110	65.5	.0132	66.2	.0134
10	9659	6335	1231	89.5	.0105	73.0	.0144	70.5	.0157	87.0	.0152	71.8	.0110	65.2	.0130	66.0	.0133
11	12878	5759	1231	89.5	.0105	76.0	.0141	71.0	.0160	87.0	.0152	72.0	.0110	65.2	.0130	66.0	.0132
12	14243	0	1024	90.2	.0110	79.0	.0143	72.5	.0166	87.2	.0158	73.2	.0114	66.2	.0135	67.8	.0139
13	10065	0	955	93.0	.0117	78.5	.0154	73.5	.0171	91.5	.0142	75.0	.0124	68.5	.0148	69.0	.0150
14	4470	0	805	92.0	.0122	74.1	.0152	71.5	.0163	88.0	.0142	72.5	.0130	68.1	.0146	69.0	.0148

Table 2. REC System Predicted Performance (Refined Model)*

TEST NO.	QS Btu/hr	QL Btu/hr	M7 CFM	T7 °F	W7 lb/lb	T1 °F	W1 lb/lb	T2 °F	W2 lb/lb	T4 °F	W4 lb/lb	T8 °F	W8 lb/lb	T9 °F	W9 lb/lb	T10 °F	W10 lb/lb
1	13192	0	1265	90.5	.0078	69.5	.0107	63.9	.0126	87.8	.0123	66.7	.0081	59.6	.0107	59.6	.0107
2	13192	0	1265	83.0	.0079	69.2	.0106	63.9	.0126	81.5	.0122	66.2	.0083	59.5	.0106	59.5	.0106
3	15908	0	1265	83.0	.0078	72.0	.0110	65.6	.0133	81.6	.0128	69.6	.0083	60.4	.0110	60.4	.0110
4	14589	0	1081	84.0	.0078	73.2	.0112	66.0	.0136	82.7	.0130	68.0	.0083	60.5	.0112	60.5	.0112
5	14920	0	1081	93.0	.0074	74.2	.0114	67.0	.0138	91.0	.0133	71.0	.0083	61.5	.0115	61.5	.0115
6	15540	0	1231	89.0	.0098	76.3	.0128	69.6	.0153	88.0	.0148	71.6	.0103	64.6	.0128	64.6	.0128
7	15470	0	1231	83.0	.0112	78.4	.0140	71.5	.0164	81.0	.0160	72.8	.0118	66.8	.0140	66.8	.0140
8	9051	0	1231	78.0	.0132	75.5	.0150	71.4	.0165	74.3	.0160	72.5	.0135	68.7	.0150	68.7	.0150
9	11594	2313	1231	89.5	.0105	74.5	.0138	69.7	.0155	88.4	.0150	72.7	.0110	65.5	.0134	65.5	.0134
10	9659	6335	1231	89.5	.0105	73.2	.0146	69.9	.0158	88.4	.0153	72.5	.0112	66.0	.0135	66.0	.0135
11	12878	5759	1231	89.5	.0105	76.2	.0145	71.3	.0165	88.5	.0158	73.6	.0110	66.5	.0137	66.5	.0137
12	14243	0	1024	90.2	.0110	78.8	.0143	72.0	.0168	88.7	.0165	74.0	.0116	67.0	.0140	67.5	.0143
13	10065	0	955	93.0	.0117	77.1	.0147	71.5	.0165	91.0	.0162	74.5	.0125	68.0	.0146	68.0	.0146
14	4470	0	805	92.0	.0122	72.4	.0144	69.2	.0155	90.7	.0150	74.5	.0126	67.3	.0144	67.3	.0144

*Recirculation air flow M11 is zero except for Test No. 12, M11 = 143 cfm. Barometric pressure for all tests = 14.5 psia. Carryover (CO) varied from 0.05 to 0.15, decreasing with increasing flow (M7).

Outdoor ambient temperature (T7)   78 to 93 deg F
Humidity ratio (W7)   0.0075 to 0.0132 lb_w/lb_{air}
Sensible load (QS)   4470 to 15470 Btu/hr
Latent load (QL)   0 to 5760 Btu/hr
Make-up air flow (M7)   805 to 1265 cfm

The data indicate that supply air temperature is influenced most by the ambient air humidity and only minimally by ambient air temperature changes. For tests 6, 7 and 8, at constant air flow of 1231 cfm, even though the ambient air temperature decreased from 89° to 78°F, the supply air temperature increased from 64.5° to 69.5°F as the ambient humidity increased from 0.0098 to 0.0132 lb/lb_a. Tests 1, 2 and 3 at 1265 cfm and tests 4 and 5 at 1031 cfm show no significant effect on supply air temperature for substantial changes in ambient air temperature. As expected, tests 9, 10 and 11 show that the conditioned space temperature for a particular air flow is sensitive to the sensible load as well as ambient air humidity which controls supply air temperature and the humidity varies with the latent load for a fixed ambient condition.

COMPUTER MODEL OF REC SYSTEM PERFORMANCE

Description

The processes involved in the REC system are all well known and adequately described in many text books and papers [4,5]. The performance analysis used in this study is similar to that presented by Dunkle and Norris [5]. The processes involved and computational program are briefly described.

Thermodynamic Relations for Moist Air. For the purposes of this study it is assumed that a moist-air system behaves in accordance with an ideal-gas mixture and Dalton's rule. Then,

$$W_s = 0.6620[p_s/(p-p_s)] \tag{1}$$

The empirical relation of Keenan and Keys [6] is used to determine the value of saturated water-vapor pressure p_s as a function of saturation temperature T_s.

$$\log_{10} \frac{3206.18}{p_s} = \frac{z}{1165.09-z} \left[\frac{3.244+3.260 \times 10^{-3}z + 2.007 \times 10^{-9}z^3}{1+1.215 \times 10^{-3}z} \right] \tag{2}$$

$$z = T_c - T_s = 705.40 - T_s \tag{3}$$

The first analysis assumes that a constant thermodynamic wet bulb line can be used to represent the evaporative cooling process on a psychrometric chart, where

$$W_s - W = \frac{(0.240+0.43W_s)(T-T_s)}{1094.2+0.43T-T_s} \tag{4}$$

Component Performance Data. Manufacturers' published performance data for the humidifier packing material and for the rotary regenerative heat exchanger were used in calculating the mass exchanger effectiveness ε_m and the heat exchanger effectiveness ε_h as functions of volumetric air flow Q_V. Evaporative cooler effectiveness is defined as

$$\varepsilon_{m,1} \doteq \frac{W_2-W_1}{W_{s,1}-W_1} \quad \text{and} \quad \varepsilon_{m,2} \doteq \frac{W_9-W_8}{W_{s,8}-W_8} \tag{5}$$

where

$$\varepsilon_{m,1} = 1.01 - 6.70 \times 10^{-5} \dot{Q}_V + 8.25 \times 10^{-9} \dot{Q}_V^2 = \varepsilon_{m,2} = \varepsilon_m \qquad (6)$$

for a 5.69 ft^2 frontal area. The value of ε_m was assumed to be the same for both heat and mass exchange. Regenerative heat exchanger effectiveness is defined as

$$\varepsilon_h = \frac{T_4 - T_3}{T_7 - T_3} \quad \text{or} \quad \frac{T_7 - T_8}{T_7 - T_3} \qquad (7)$$

where

$$\varepsilon_h = -4.90 \times 10^{-5} \dot{Q}_V + 9.74 \times 10^{-1} \qquad (8)$$

for a 5.71 ft^2 frontal area.

Computational Procedure, Parameters and Relations (using computer nomenclature). The inputs are:

1. Sensible heat load QS in Btu/hr which accounts for heat transfer through the building structure, infiltration and internal heat sources.

2. Latent heat load QL in Btu/hr which accounts for the internal moisture sources and infiltration.

3. Fresh air supply M7 and recirculation air M11 in cmf. The room air supply M10 is M7 + M11.

4. Barometric pressure p in psia.

5. Carryover CO or fraction of mass flow carried over from countercurrent air stream compared to net flow through the exchanger.

To begin a computation series, initial desirable values are chosen for T1 and W1. From state 1 to state 2 the room exhaust air passes through the first evaporative cooler, then

$$T2 = T1 + \varepsilon_m(TS1 - T1) \qquad (9)$$

and,

$$W2 = W1 + \varepsilon_m(WS1 - W1) \qquad (10)$$

Cooler effectiveness ε_m is given by Eq. (6). For the specified values of T1 and W1, the corresponding values of TS1 and WS1 that satisfy equations 1, 2, 3 and 4 are calculated by subroutine INTER(T,W,TS,WS).

States 2 and 3 are equivalent. From state 3 to 4 the air undergoes an increase in enthalpy as it passes through the heat exchanger matrix due to sensible heat transfer and a slight decrease in humidity due to adiabatic mixing with air carried over from the countercurrent air stream. Then

$$T4 = T3 + \varepsilon_h(T7 - T3) \qquad (11)$$

$$W4 = W3 \qquad (12)$$

and ε_h is given by Eq. (8). Temperature T4 and humidity W4 are modified by the carryover fraction CO using the approximate relations,

$$T4' = (1 - CO)T4 + (CO)T7 \qquad (13)$$

$$W4' = (1 - CO)W3 + (CO)W7 \qquad (14)$$

The ambient air condition at state 7 is an input parameter; T7 was varied from an initial value of 80 to 120°F in 5° steps and W7 from 0.0057 to 0.0214 pounds water vapor per pound of dry air (lb/lb$_a$) in 0.0007 lb/lb$_a$ steps.

In going from state 7 to state 8 the process is similar to that from 3 to 4, resulting in

$$T8 = T7 + \varepsilon_h (T7-T3) \tag{15}$$

and

$$W8 = W7 \tag{16}$$

but, T8 and W8 are modified by carryover,

$$T8' = (1-CO)T8 + (CO)T3 \tag{17}$$

$$W8' = (1-CO)W7 + (CO)W3 \tag{18}$$

The air is then cooled further by passing it through the second evaporative cooler, giving

$$T9 = T8 - \varepsilon_m (T8-TS8) \tag{19}$$

$$W9 = W8 - \varepsilon_m (W8-WS8) \tag{20}$$

where ε_m is obtained from Eq. (6) and TS8 and WS8 are calculated by INTER(T,W, TS,WS). If M11 is zero, state 10 is equivalent to state 9. If M11 is not zero and adiabatic mixing occurs between air at state 2 and air at state 9,

$$T10 + (T9)(M7)/(M7+M11) + (T2)(M11)/(M7+M11) \tag{21}$$

$$W10 = (W9)(M7)/(M7+M11) + (W2)(M11)/(M7+M11) \tag{22}$$

Finally, for the specified values of QS and QL, T1 and W1 are calculated from,

$$T1 = T10 + QS/[(M7+M11)(.073+.04W10)(60)(.24+.43W10)] \tag{23}$$

$$W1 = W10 + QL/[(M7+M11)(.073+.04W10)(60)(1062.2+.43T10)] \tag{24}$$

where

$$(.073+.04W10) = \text{density of air-vapor mixture}$$

$$(.24+.43W10) = \text{specific heat capacity of air-vapor mixture}$$

$$(1062.2+.43T10) = \text{specific enthalpy of water vapor}$$

The computed values of T1 and W1 given by Eqs. (23) and (24) are compared to the initial values. If the differences are greater than the absolute values of 0.05°F for the temperature and 0.000143 lb/lb$_a$ for the humidity, the calculations are repeated with the new values of T1 and W1. The iterative process continues until the initial and final values approach within the prescribed difference.

The computer program output is available in tabular and graphical format. For specified input values of QS, QL, M7, M11, p and CO, all of the state point temperatures and humidities are tabulated for each ambient air condition represented by values of T7 and W7. The performance of the system was also mapped using the 936 Calcomp Plotter as shown in Fig. 4. For each set of input values, the supply air temperature T10 and the room temperature T1 is plotted as a function of ambient air humidity W7 for the different values of T7.

Constant room air humidity lines are superimposed on the room air temperature plot.

COMPUTER MODEL VERIFICATION AND IMPROVEMENT

A comparison of predicted results with the experimental data indicated that the predicted inlet and exit room air temperatures and humidities were consistently lower than the measured values. Inlet or supply air temperature predictions ran about 3°F lower than the measured values. Three possible sources for the difference between actual and predicted performance were investigated.

1. An estimate of heat gains through uninsulated duct walls indicated that its effect was negligible, accounting for only a fraction of 1°F temperature rise.

2. The measured heat exchanger effectiveness agreed with the manufacturer's data within 5%.

3. The actual evaporative cooler processes did not correspond to a constant thermodynamic wet bulb process. For all of the tests the humidification process line had a steeper negative slope on a psychrometric chart as shown

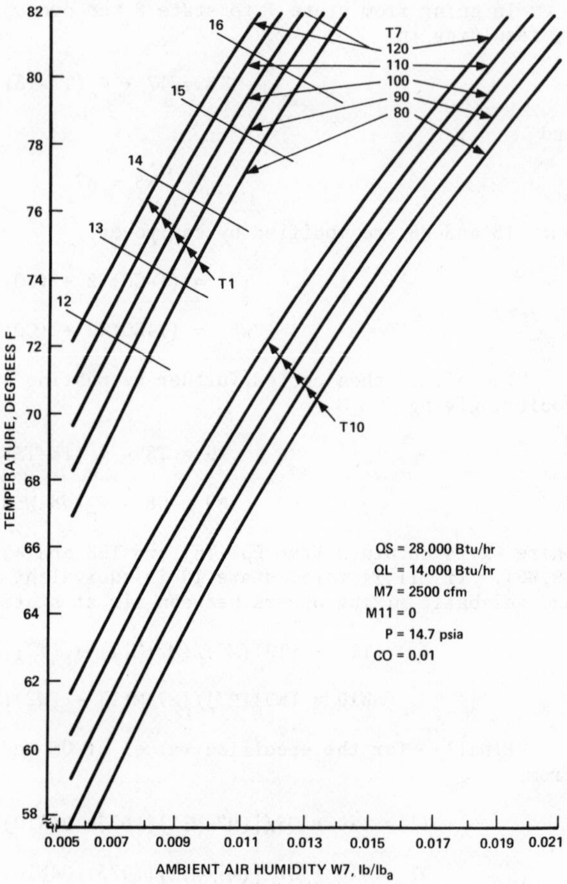

Figure 4.   Performance Map of Experimental REC System (computer output)

in Fig. 5 for test number 1. The computer program was consequently modified by replacing Eq. (4) with a process line representing the average performance of 14 tests, where

$$W = W1 + 3.60 \times 10^{-4}(T1-T) \tag{25}$$

The values of T2, W2 (Eqs. (9) and (10)) and T9, W9 (Eqs. (19) and (20)) were calculated using modified saturation values TS' and WS' based on the intersection of the empirical process line with the saturation curve.

The predicted performance of the experimental REC system for conditions equivalent to the tests shown in Table 1 are given in Table 2. Note that the predicted values of room air inlet and exit temperatures compare very well with the measurements. A comparison of measured and predicted values of T1 and W1 is given in Fig. 6. It should be noted that the determination of the heat exchanger carryover fraction CO using the steady state measurements of wet and dry bulb temperatures was somewhat difficult because the humidity ratio differences were relatively small. However, the predicted performance based on the estimated values of CO agreed quite well with the test results.

DISCUSSION AND CONCLUSIONS

The experimental system was designed for 2500 cfm with a total cooling capacity of approximately 3.5 tons of refrigeration. After demonstrating the validity of the computer performance model for more than 14 tests, computational studies were conducted to determine the effects of climatic variables, design and operating variables and load. Similar to what was concluded from test observations, outdoor ambient air humidity generally has the strongest influence on supply air temperature. For the same load a variation in air flow of as much as 2.5 showed only a small effect on air temperature over a wide range of environmental conditions. The interaction between supply air flow and load to influence the room condition is of course obvious from an enthalpy balance. It was also demonstrated that carryover degrades the performance of the system. By reducing the value of CO from 0.10 (the average value observed during tests when operating without a heat exchanger purge) to 0.01 (achievable with purging) a lowering of supply air temperature by 1.5 to 2°F was obtained.

Because of space limitations, only one typical performance map (Fig. 4) is shown for the experimental system operating at full load consisting of QS = 28,000 Btu/hr and QL = 14,000 Btu/hr. For purposes of illustration the performance of the system operating in 5 different climate zones in Southern California was determined using Fig. 4. The one percent design dry and wet bulb temperatures [4] for important cities in each zone were used as the reference climate for the zone. The results given in Table 3 indicate the advantage and limitations of the REC system.

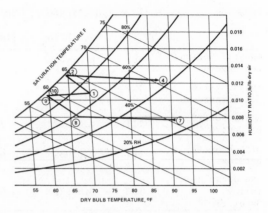

Figure 5.   Experimental Cycle Performance (Test #1, Table 1)

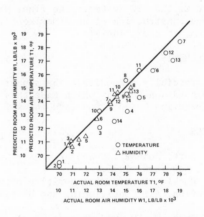

Figure 6.   Comparison of Experimental and Predicted Room Temperatures and Humidity (numbers refer to tests in Table 1)

It appears that the system would be useful in the high desert and valley regions with climate similar to Barstow and Bakersfield, somewhat marginal for the coastal and inland valley with climate similar to Los Angeles and San Bernardino and unsuited for the relatively humid and hot lower desert region with climate similar to El Centro. It is clear that for the more humid hot regions, evaporative cooling can only be used in conjunction with a drying stage which might appropriately be regenerated with solar heated ambient air.

Table 3. REC System Performance in Various Design Climates

Location	1% Design Climate		Supply Air	Room Condition		
	T7, °F	W7, lb/lb$_a$	T10, °F	T1, °F	W1, lb/lb$_a$	RH %
Coastal Region Los Angeles	94	0.01185	69	79	0.0156	71
Inland Valley San Bernardino	101	0.01275	72	82	0.0170	71
Lower Desert El Centro	111	0.01607	77	87	0.0203	71
High Desert Barstow	104	0.01035	68	78	0.0147	71
San Joaquin Valley Bakersfield	103	0.00971	67	77	0.0143	71

Another performance factor of interest is the ratio of cooling effect to the electric power needed to achieve it (COP) including fan power, power to recirculate water in the humidifiers and power to rotate the heat exchanger wheel. Based on a design capacity of 3.5 tons of cooling at an air flow of 2500 cfm it is estimated that COP's of 8-10 are achievable. The estimated fan power includes pressure losses due to the rotary heat exchanger, humidifiers, ducting and an air filter.

We have now completed a desiccant bed performance model and are engaged in computer aided design studies of an experimental unit which will be constructed and added to the REC system. The final phase of the investigation will be the design and construction of an experimental solar air heater and recuperator for a desiccant regeneration stage.

ACKNOWLEDGMENTS

We gratefully acknowledge the support and encouragement of the Southern California Edison Company which made this investigation possible. The UCLA Campus Computing Network was used for all computer computations.

REFERENCES

    1. Watt, John R., Evaporative Air Conditioning, The Industrial Press, New York, New York, 1973.
    2. Dunkle, R. V., "Regenerative Evaporative Cooling Systems", Australian Refrigeration, Air Conditioning and Heating, Vol. 20, No. 1, pages 13-23, Jan. 1966.
    3. Rush, W. F., and Macriss, R. A., "Munters Environmental Control System", Appliance Engineer, Vol. 3, No. 3, 1969.
    4. American Society of Heating, Refrigeration and Air Conditioning Engineers, Handbook of Fundamentals, The Society, New York, New York, 1972.
    5. Dunkle, R. V., and Norris, D. J., "General Analysis of Regenerative Evaporative Cooling Systems, Proceedings of the XIIth International Congress of Refrigeration, Madrid, August 30-September 6, 1967.
    6. Keenan, J. H., and Keyes, F. G., Thermodynamic Properties of Steam, John Wiley and Sons, Inc., New York, New York, 1950, p. 14.

AN OPEN ABSORPTION SYSTEM UTILIZING SOLAR ENERGY FOR AIR CONDITIONING

GERSHON GROSSMAN AND ILIAU SHWARTS

Department of Mechanical Engineering
Technion — Israel Institute of Technology
Haifa, Israel

ABSTRACT

The possibility of providing the cooling needs of buildings by means of solar energy has a large potential for energy conservation. A number of solar air conditioning systems have been experimented with mostly relying on solar heat to power a closed absorption or a steam-ejector type cooling device. These systems normally have a low coefficient of performance when used with a low temperature heat source such as flat plate solar collectors.

The present work is a study of an air conditioning system based on de-hydration of the air in an open absorption system where the desiccant solution is regenerated by solar energy. Ambient air is dehydrated by means of the desiccant. The dry air is cooled back to ambient temperature by indirect heat exchange, then cooled further by combination of evaporative and mechanical cooling. The advantages of the system are that the energy to power the system can be supplied at temperatures not higher than 160°F, which are easily obtained with solar energy, and that the entire latent heat can be removed at temperatures of 85°F, compared with latent heat removal temperature of 34-41°F in mechanical air-conditioning.

INTRODUCTION

The possibility of providing the cooling needs of buildings by means of solar energy has attracted considerable interest over the past two decades. A number of working systems has been built and experimented with, on various scales [1-5]. While there have been several suggestions to rely on natural air conditioning [6], the majority of the schemes proposed were based on utilizing solar heat to power an absorption or a vapor-ejector type cooling system. Alternatively, it has been suggested to use mechanical refrigeration devices propelled by a solar heat engine.

The above systems, although based on well-known thermodynamic cycles, are highly limited in performance by the temperature of the heat source. In the case of the solar driven mechanical device, this determines the thermal efficiency of the heat engine; in the absorption and the ejector cycles the heat source temperature affects the overall system. The maximum possible coefficient of performance of a heat-powered cooling device, based on the ideal Carnot cycle, is given by:

$$COP = \frac{T_E}{T_G} \frac{T_G - T_C}{T_C - T_E} \tag{1}$$

Where T_E, T_C and T_G are the heat removal, the heat reflection and the solar heat supply temperatures, respectively. When common flat-plate collectors are used to intercept the solar radiation the heat is obtained at relatively low temperatures. With concentrating collectors, higher temperatures can be reached but to date, these collectors are expensive and complex to operate due to their need to track the sun. The situation is made worse by the fact that the heat removal in the evaporator and condenser and the solar heat supply are done through indirect contact heat exchangers. This requires T_E to be lowered, raises the effective T_C and lowers T_G each by about 5^OF, which reduces the COP considerably.

Many of the above difficulties may be overcome by the use of an open absorption system for solar air conditioning. The basic idea consists of de-hydrating air by means of chemical absorption using a desiccant. Water is then evaporated into the dry air in a controlled manner to lower its tempera-ture. The saturated desiccant is regenerated by solar energy. This simple concept may be applied with some additional sophistication to give an effec-tive operating device.

While the idea of open absorption is not new, little work has been done on it in comparison with other types of solar cooling. One of the pioneers in the area was Lof[7], who early in 1955 studied an open absorption system with liquid triethylene glycol as the desiccant. Dunkle[8] in 1965 suggested a related method, based on a solid rather than liquid desiccant (e.g. silica gel). The use of a solid desiccant had been proposed earlier by Dannies[9], in a passive, non-sophisticated open absorption cycle. Kapur [10] in 1960 compared the open absorption cycle with other types of cooling processes - mechanical, closed absorption and ejector - and found it most applicable for solar air conditioning in tropical climates such as India.

The present work is a feasibility study on the use of the open absorption cycle for solar air conditioning, performed as a first step toward the con-struction of a working system. The study is based in part on the experience accumulated thus far by others and makes use of the inherent advantages of the cycle. The basic process is discussed and various additions and alternations considered to obtain an optimal process from the thermodynamic and technolo-gical point of view.

The Open Absorption Cycle

The principle of the open absorption system is similar to most other cooling cycles, where water serves as the regrigerant. The principal gain here is that most of the heat transfer functions are done by direct contact, thus eliminating several temperature differentials which are deleterious to the COP. Another important feature is that the latent part of the heat load is removed at ambient temperature rather than at the considerably lower dew point. This makes the concept particularly attractive for humid climates.

The open absorption cycle may be implemented in a number of different schemes, differing from each other by the addition of sub-systems designed to increase the efficiency of the process. Let us consider them starting with the most basic one:

Scheme I

The basic open absorption scheme is illustrated in figure 1a and on the psy-chometric chart of figure 1b. The air to be treated (state 2 - combining re-circulated room air with a certain percentage of fresh outdoors air) enters the dehumidifier D and leaves at a reduced water content and higher tempera-ture (state 5). It is now passed through the heat exchanger E1 and cooled by indirect contact with ambient air or with water from a cooling tower. The air, now at state 6, enters the evaporative cooler C where its temperature is lo-wered by increasing its humidity in a controlled manner. Leaving at state 9,

it is introduced into the room. Meanwhile, the desiccant which enters the
dehumidifier at a dry and cool state 31, absorbs humidity and leaves at a
higher temperature and water content (state 32). It is then passed through
the regenerator R where it is heated directly or indirectly by solar heat.*
With its water vapor pressure increased, the hot desiccant transfers part of
its water content to air(state 0)brought in from outside, and leaves the regenerator to reenter the Dehumidifier D.

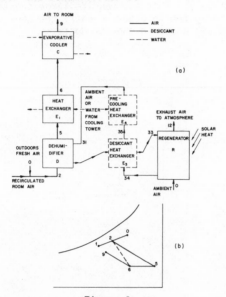

Figure 1

This basic scheme,represented in figure 1a by a block diagram is the
cycle discussed by Dannies [9] for his air conditioned house. This is also
the cycle discussed by Kapur [10] in his comparative study. The system used
by Lof [7] is essentially the basic scheme I, with one important improvement:
It adds a heat exchanger designed to transfer heat from the hot desiccant
leaving the regenerator to the cooler desiccant flowing into it. This heat
exchanger (E3) is shown in dotted lines in figure 1a. Thus, the needs to
heat the desiccant in the regenerator and to cool it in or before it enters
the dehumidifier are both reduced. An absorbent heat exchanger is quite
common in commercial closed absorption system and provides a substantial
energy saving. Its use is made possible here by operating with a liquid
desiccant and points toward an important advantage of the liquid desiccant
over the solid. In Lof's system, when compared with the block diagram of
figure 1a, the regenerator R consists of a stripping column and a solar
air heater; the dehumidifier D and the heat exchanger E1 are both combined
into one unit - the absorber. The air thus goes from state 2 to state 6 in
one step, as indicated by the dotted line and the psychometric chart of
figure 1b.

---

* The two heat exchangers indicated in dotted lines are optional, and should
  be ignored for the purpose of the present description.

Scheme II

One of the shortcomings of Scheme I is in that the dry air leaving the dehumidifier is cooled before entering the evaporative cooler to a temperature
not lower, and perhaps somewhat higher than the ambient (state 6). This
limits the low temperature which can be reached at state 9 and the enthalpy
of the air introduced into the room is consequently not much lower than that
of the air in the room (State 1). This may require large quantities of air
to be circulated through the system in order to carry the cooling load, requiring large fans and creating noise in the air ducts.

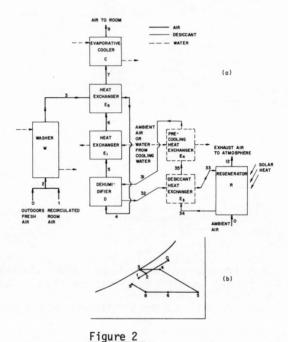

Figure 2

An improvement to scheme I which overcomes this problem is implemented
in Scheme II, illustrated in figure 2a and on the psychometric chart of
figure 2b. The air to be treated (state 2) is saturated with water in the
washer W and its temperature reduced to the adiabatic saturated temperature
(state 3). The air is then passed through the heat exchanger E2 where it
absorbs heat (by indirect contact) from the dry air which had passed through
the dehumidifier D and the heat exchanger E1. Leaving E2 at state 4, the
air enters the dehumidifier and follows pretty much the same path as in scheme
I, with the additional cooling in the heat exchanger E2. The desiccant cycle
here remains unchanged.

Scheme II as has been described here is the cycle used by Dunkle [8]
is his solid desiccant open absorption system. the washer in Dunkle's system
consists of water sprays in an air duct, and the heat exchanger E2 is a rotary bed of solid particles. The improvement over scheme I is obtained by first
converting as much as possible of the sensible heat contained in the air to
latent heat, which can be handled by the desiccant and extracted at ambient
temperature.

Apart from the thermodynamic improvement, the additions to the system made in scheme II have an important practical aspect: Passing the air through the washer W before it comes into contact with the desiccant removes dust and dirt particles, which could contaminate the desiccant. Considering the quantities of air which are to be handled by this open system, it is impossible to operate for any length of time without first removing the dirt from the air. An added benefit is that both the washer and the evaporative cooler operate as cooling towers, and the excess of water leaving them can supply part of the cooling water needed for the desiccant and dry air heat exchangers.

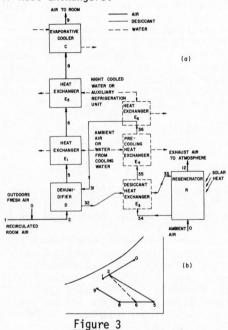

Figure 3

Scheme III

Furhter improvement in the operation of the system described in scheme I may be obtained by utilizing night cooled water or an auxiliary refrigeration u- nit, which also provides a back-up system to be used when not enough solar energy is available. The scheme is illustrated in figure 3a and on the psy- chometric chart of fig. 3b. The dry air leaving the heat exchanger E1 at state 6 is cooled in the heat exchanger E5 by night cooled water or by an auxiliary mechanical refrigeration unit to a temperature lower than ambient by several degrees (state 8). This reduces the enthalpy of the air entering the room and allows lower temperatures to be reached at the exit from the evaporative cooler (state 9). Night-cooled water can also be used to pre- cool the solution entering the dehumidifier in the heat exchanger E6, as shown.

Other schemes

Schemes II and III may be combined together where night cooled water is available to increase the system's COP. Ways of operating the cycle in two stages may also be investigated. At this point, the system seems to become too complex, and the additional gain in COP is small.

Finally, it should be noted that the open absorpber cycle provides
a neat way for storing refrigeration capacity. With enough desiccant in
the dehydrated state (31) kept in store, the system may be operated
during parts of the evening and night when solar energy is not available
for refrigeration. This, of course, can only be done with a cheap desiccant, and other methods of storage may prove to be more economical.

Choice of desiccant

The desiccant to be used in the open absorption system must posses
several   qualities.   Since it comes in direct contact with the air
supplied to the room, it must be non-toxic, non-flammable and odorless.
It should be easily regenerated with solar energy, and hence have a low
vapor pressure at ambient temperature and   a high one at, say, 80°C.
Finally, for economic operation it must be inexpensive and non-corrosive.
      The desiccants available which satisfy these requirements operate
in two different mechanisms: solid desiccants usually adsorb the water
on their surface; liquid desiccants usually absorb the water. In the
first category, we may list materials such as silica gel,activated alu-
mina, activated carbon, molecular sieves, and solid calcium chloride.
The second category include solutions in water of lithium chloride,
lithium bromide, calcium chloride, ethylene glycol and propylene glycol.

Coefficient of performance

The efficient performance of the open absorption system depends on the
individual components. The central problem in the construction of the
system is proper evaluation and design of the dehumidifier and regenerator units.
      The efficiency of the overall system strongly depends on the degree
of dehumidification of the air. The dehumidifier is a mass-heat transfer
apparatus where the treated air comes in direct contact with the desiccant. The performance thus depends on the particular desiccant used.
A computer model has been developed in the present study which determines
the temperatures and water content in both the air and desiccant leaving
the dehumidifier as functions of these parameters at the inlet and the
relative flow rates. The calculations are based on energy and mass conservation, transfer rates of heat and water mass and thermodynamic properties of the two fluids. Similar calculations with the same program
have been performed for the regenerator.

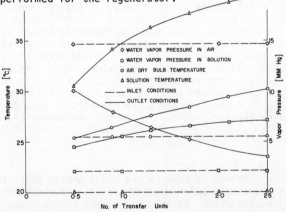

Figure 4 shows typical results of the calculations preformed for a de-humidifier operating with Lithium-bromide solution as the desiccant. Tempera-tures and water vapor pressure for the air and the solution are plotted as a function of the number of transfer units. It is assumed here for simplicity that the number of heat transfer units is equal to the mass transfer units. Typical values have been selected for the temperature and vapor pressures of the air and the solution at the inlet, which are kept constant. Figure 4 shows that an increase of the number of transfer units from 0.5 to 2.0 permits to achieve a decrease in water vapor pressure in the air from 10.4 to 4.5 mm Hg. This corresponds to a decrease in relative humidity from 42% to 15%.

Extensive calculations of this kind have been carried out in order to determine the design parameters of both the dehumidifier and regenerator. The results indicate that these sub-systems should be designed to have 2.0-2.5 transfer units and a solution-to-air mass flow ratio of 0.7 to 0.9. The dehumidifier will perform efficiently under solution inlet temperatures of 25-30°C and LiBr concentration of 55-58%. Temperatures above 70°C are re-quired for regeneration. These temperatures are easily obtained by flat plate solar collectors in the summer.

REFERENCES

1. A.M. Zarem and D.D. Erway: Introduction to the Utilization of Solar Energy (McGraw-Hill, New York, 1963).

2. F. Kreith: Solar Heating and Cooling, McGraw Hill, 1975.

3. J.A. Duffie and W.A. Beckman: Solar Energy Thermal Processes, Wiley (1974).

4. R.K. Swartman, Vinh Ha, and A.J. Newton: Review of Solar Powered Refrigeration (Paper No. 73-WA/Sol-6, ASME, 1973).

5. E.A. Farber: The Direct Use of Solar Energy to Operate Refrigeration and Air Conditioning Systems (Tech. Progress Report No. 15, Florida Engineering and Experimental Station, University of Florida, Gainsville, Nov. 1965.

6. H.R. Hay and J.I. Yellott: Natural Air Conditioning with Roof Ponds and Movable Insulation (ASHRAE Transactions, Vol. 75, Part 1, 1969, p. 165).

7. Lof, G.O.G., Solar Energy Research, Madison, University of Wisconsin Press, 33, 1955. "House Heating and Cooling with Solar Energy".

8. Dunkle, R.V., Mech. and Chem. Engr. Trans., Inst. Engrs., Australia, MCI, 73 (May 1965). "A Method of Solar Air Conditioning".

9. J.H. Dannies: Solar Air Conditioning and Solar Refrigeration (Solar Energy, 3, 1958, 34-39).

10. J.C.Kapur: A Report on the Utilization of Solar Energy for Refrigeration and Air Conditioning Applications, (Solar Energy, 4, 1960, 39-47).

11. Baum, V.A., Kakabaev, A., Khandurdyev, A., Klychiaeva, O., and Rakhmanov, A., Paper presented at International Solar Energy Congress, Paris (1973). "Utilisation de L'energie Solaire Dans les Conditions Particulieres Des Regions A Climat Torride er Aride pur la Climatisation en Ete".

HEAT FLOW INTO THE GROUND UNDER A HOUSE

R. W. R. MUNCEY AND J. W. SPENCER

CSIRO Division of Building Research
Highett, Victoria 3190, Australia

ABSTRACT

        In calculating non-steady heat flow within buildings, the assumption of one-dimensional heat flow is almost universal. However, this assumption is not acceptable for heat flow at the ground, especially for detached domestic buildings. This paper presents a mathematical model which idealises such a situation and determines a one-dimensional equivalent.

        A temperature is represented by a Fourier series in two dimensions, suitably truncated to give a practical value of rate of change at the perimeter and under these conditions the average heat flow over the surface can be calculated. Conversely a similar pattern of heat flow permits calculation to be made of the average temperature. In practical situations there is always an air film resistance above the ground and a wall around the perimeter. The theory includes these effects.

        Under steady state conditions the ratio of heat flow to temperature can be used to estimate the U-value of the ground, which is shown to agree with experiment. As has been found empirically in the past, an approximate relationship exists between slab perimeter and ground thermal resistance.

        Evaluation of the ratio of heat flow to temperature at various frequencies enables the complex admittance parameter used in the matrix method to be estimated. It is shown that simple resistance-capacitance elements can represent the behaviour of the ground to sufficient accuracy. A method is given for assessing the values of the elements to represent widely differing types of ground.

LIST OF SYMBOLS

A   area of slab (m^2)

c   volumetric heat capacitance of ground (J/m^3K)

C   thermal capacitance for a particular thickness of ground (J/m^2K)

e   horizontal distance corresponding to unit temperature change at the tangent slope at "temperature" of 0.5 (m)

f,g   integers describing the spacing apart of slabs in the x and y directions

F   dimensionless ratio [ground thermal resistance for a given area/ $(0.25 \text{ perimeter})^2$] / [ground thermal resistance for the square of equal perimeter]

H   air film resistance to heat flow (m^2K/W)

j $\sqrt{-1}$

ℓ depth of ground (m)

L   large value approached by ℓ (m)

m,n   integers

M   limit of summation of series in the x direction

N   limit of summation of series in the y direction

P_{11}, P_{12}
P_{21}, P_{22} elements in the transfer matrix for one-dimenstional heat transfer

Q   Heat flow pattern at the surface of a slab corresponding to a temperature pattern V.

Q_1 heat flow per unit area into a slab under cyclic conditions with unit temperature above the slab, zero temperature elsewhere (W/m^2)

Q_2 heat flow per unit area into a slab under cyclic conditions with unit temperature over the whole area (W/m^2)

Q_3 heat flow per unit area into a slab under steady state conditions with a constant temperature of 1 - 1/fg above the slab and -1/fg elsewhere (W/m^2)

R   ground thermal resistance (m^2K/W)

R_s ground thermal resistance for a square slab of 10 m side with a ground thermal conductivity of 1 W/mK (m^2K/W)

t   time (sec)

T   constant temperature amplitude for slab temperature pattern V ($^{\circ}$C)

T_1, T_2 temperatures at opposite faces of a slab in considering one-dimensional heat transfer ($^{\circ}$C)

u,v   half-length and half-width of slab in the x and y directions (m)

V   temperature pattern assumed over slab

W   heat flow amplitude for heat flow pattern Q (W/m^2)

W_1, W_2 heat flows at opposite faces of a slab in considering one-dimensional heat transfer (W/m^2)

X   air temperature pattern assumed over slab

x,y,z   rectangular coordinates

Z   constant temperature amplitude for air temperature pattern X

α, β constants

γ defined by $\gamma^2 = -\left((\pi m/fu)^2 + (\pi n/gv)^2 + j\omega c/\lambda\right)$.

λ ground thermal conductivity (W/m K)

ω angular frequency (rad/s)

INTRODUCTION

With increasing sophistication in the maintenance of comfort conditions in buildings and the developing fears for future world energy supplies, the calculations of temperatures and heat flows in buildings are being carefully examined. In a review as long ago as 1962, Stephenson characterised the basic assumptions of calculation models as follows:

1.   heat flow is one-dimensional;

2.   the differential equation describing the heat flow is linear;

3.   it is possible to combine the outside air temperature, wind velocity, and radiation intensity into a single factor called the sol-air temperature;

4.   the heat transfer by convection and radiation at the inside surface can be represented by a combined film conductance which couples the inside surface with the inside air.

One heat path that has proved difficult to model is the exchange of heat between the ground and a building placed theron because the heat flows within the ground cannot be considered one-dimensional. Information concerning the equivalent thermal transmission of ground is notably scarce with the pioneering work of Billington (1951) still being quoted approvingly in a BRS note (1972). Other works and studies are from Vuorelainen (1960, 1963) and Shelton (1975).

The references given tend to show that for the steady state situation, the U-value is related to the perimeter of the building or slab but no simple rules are available to set the U-value for given dimensions and thermal properties of the ground. Means of representing the ground appropriate to the harmonic and non-steady cases are also required. This paper gives results from which U-values may be assessed and representations found for the ground for use in one-dimensional calculations.

MATHEMATICAL MODEL

Consider the rectangular array (Figure 1) of rectangular (2u x 2v) slabs, notionally of zero thickness, on the surface z=0 of a semi-infinite solid z>0. Suppose that the temperature V of this surface is constrained to be

$$V = T \cos(\pi mx/fu) \cos(\pi ny/gv) \exp(j\omega t), \tag{1}$$

where ω = angular frequency, t = time, $j = \sqrt{-1}$, m and n are integers, and T(m,n) is a constant.

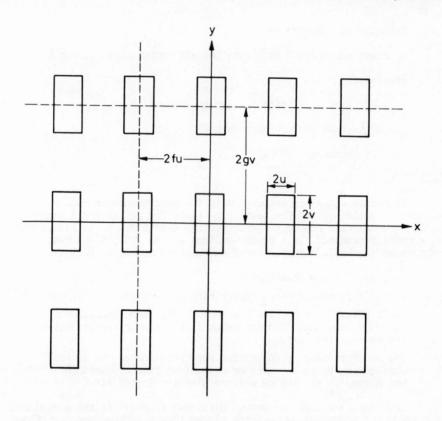

Figure 1.   Array of rectangular slabs on the surface of a semi-infinite
             solid.

        Then, Muncey (1967) has shown that the heat flow Q across the surface
is given by

$$Q = W \cos(\pi m x / f u) \cos(\pi n y / g v) \exp(j\omega t). \tag{2}$$

        Any given temperature distribution over the whole surface which has
the symmetry of the slab pattern can be represented as a sum of terms such as
V with various m,n and appropriate T(m,n). Since the temperatures are
symmetrical about the centre of each slab for directions parallel to the axes

only cosine terms are required. The special case of V with $m = n = \omega = 0$ will be excluded from further discussion as it corresponds to the steady-state problem in which the temperature of the whole solid approaches T in a known manner.

Muncey (1967) showed further that the temperature at and the heat flow across the surface $z = \ell$ are given by expressions of the forms (1) and (2) with

$$T_{z=\ell} = T \cos\gamma\ell - W (\sin\gamma\ell/\gamma\lambda)$$

$$W_{z=\ell} = T (\gamma\lambda\sin\gamma\ell) + W \cos\gamma\ell, \tag{3}$$

where λ = thermal conductivity,

$$\gamma = \gamma(m,n) = -j[(\pi m/fu)^2 + (\pi n/gv)^2 + j\omega c/\lambda]^{\frac{1}{2}}, \tag{4}$$

and c = volumetric heat capacity.

For a given m, n, ω, if either $T_{z=\ell}$ or $W_{z=\ell}$ represents a known boundary condition then equations (3) can be used to determine W. In the special case considered here both heat flow and temperature must tend to zero at large ℓ, = L say. While L is finite, the only solution is the trivial $T = W = 0$ but as L becomes infinite, the solution

$$W(m,n) = j\gamma(m,n)\lambda T(m,n) \tag{5}$$

becomes available. The physical situation of interest is that of an isolated slab at a uniform temperature of unity and the rest of the area at zero. This is reasonably approximated to by an array as in Figure 1 provided f and g are adequately large, i.e. the interaction between the heat flows from different slabs can be neglected. This periodic temperature distribution can then be expressed as a Fourier series of terms such as (1) with appropriate values of $T(m,n)$. As the number of terms needed will increase with f and g, some compromise is needed for f and g; $f = g = 16$ has been found to give adequate accuracy and all numerical calculations have been made with these values.

If an infinite Fourier series is used to define the temperature, the corresponding series for Q is divergent. This corresponds to the physical phenomenon that a step-wise temperature change implies infinite heat flows. In practice, this will not occur because

(a)   the temperature distribution at the edges of the slab is not an ideal square wave, but tapers over a distance of the order of the thickness of a wall, and

(b)   the "constant" temperature will occur not at the surface but at some point in the air above and there will be an air film resistance to heat flow.

The edge taper can be simulated by truncating the series at finite values M, N. If the "edge distance", e, is defined as the horizontal distance for unit change in temperature at the edge of the slab, then it is shown in Appendix 1 that

$$e = fu/M = gv/N$$

and these equations serve to define M, N. Figure 2 shows the form of temperature near the edge of a 10 m slab when the series is truncated at 800 terms and e is then 0.1 m.

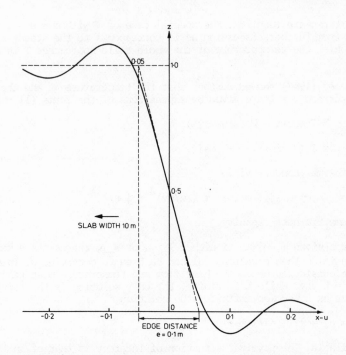

Figure 2.   Temperature or heat flow profile at slab edge. The curve is
            calculated from $z = \sum\limits_{m=1}^{800} (2/\pi m) \, \sin(\pi m/f) \, \cos(\pi mx/fu)$
            with $f = 16$, $u = 5$ for values of x near $x = u$. It corresponds
            to an edge distance of $fu/800 = 0.1$ m.

For an air film resistance H over the surface of the solid, the air
temperature X will be given for each spatial and temporal harmonic by

$X = V + HQ$

$\quad = Z \cos(\pi mx/fu) \, \cos(\pi ny/gv) \, \exp(j\omega t)$ \hfill (7)

where on dividing by common harmonic factors

$Z = T + HW$ \hfill (8)

The combination of equations (5) and (7) leads to

$W = j\gamma\lambda Z/(1 + j\gamma\lambda H)$ \hfill (9)

The problem is now to choose a series of values of $Z(m,n)$ so that X is unity
above the slab and zero elsewhere, with a harmonic series suitably truncated
to allow edge effects. This leads to

$$Z(m,n) \quad = \quad \begin{cases} 1/fg & m = n = 0 \\ (2/\pi mg) \, \sin(\pi m/f) & m \neq 0,\ n = 0 \\ (2/\pi nf) \, \sin(\pi n/g) & m = 0,\ n \neq 0 \\ (4mn/\pi^2) \, \sin(\pi m/f) \, \sin(\pi n/g) & m \neq 0,\ n \neq 0 \end{cases} \quad (10)$$

and

$$W(m,n) = \frac{[(\pi m/fu)^2 + (\pi n/gv)^2 + j\omega c/\lambda]^{\frac{1}{2}} \lambda \, Z(m,n)}{1 \pm [(\pi m/fu)^2 + (\pi n/gv)^2 + j\omega c/\lambda]^{\frac{1}{2}} \lambda H} \tag{11}$$

(m, n, ω not all zero.)

One-Dimensional Equivalent

It has been shown (van Gorcum 1950, Muncey, 1953) that the (one-dimensional) heat flow through an infinite sheet of material of constant thickness satisfies the equations

$$T_2 = P_{11} T_1 + P_{12} W_1 \tag{12a}$$
$$W_2 = P_{21} T_1 + P_{22} W_2, \tag{12b}$$

where $T_1 \exp (j\omega t)$ and $T_2 \exp (j\omega t)$ are the temperatures at and $W_1 \exp (j\omega t)$ and $W_2 \exp (j\omega t)$ are the heat flows across the two "surfaces" of the sheet and the P_{ij}^2 (ω) are the coefficients in equations (3) with m = n = 0, i.e. with

$$\gamma = \gamma(0,0) = -j \, (j\omega c/\lambda)^{\frac{1}{2}} \tag{13}$$

In order to incorporate flow-to-ground into the usual thermal equations, values of P_{ij} must be determined such that the heat flow to the ground is described by equation (12a) when T_1, W_1 are defined as the temperature above and heat flow across slab and T_2 is the temperature over the area between slabs. This can be done by using equations (10) and (11), or simple modifications of these, to calculate heat flows for some special cases, namely:
1.   temperature $\exp (j\omega t)$ above the slab, zero elsewhere;
2.   temperature $\exp (j\omega t)$ everywhere, (m = n = 0);
3.   temperature 1-1/fg above the slab, -1/fg elsewhere
             (i.e. steady state, ω = 0).
If the heat flows as calculated for these three cases are $Q_1 \exp (j\omega t)$, $Q_2 \exp (j\omega t)$, and Q_3, substitution into equation (12a) yields:

$$0 = P_{11}(\omega). \; 1 + P_{12}(\omega).Q_1,$$
$$0 = P_{11}(\omega). \; 1 + P_{12}(\omega).Q_2, \tag{14}$$
$$-1/fg = P_{11}(0). \; (1-1/fg) + P_{12}(0).Q_3.$$

Further, for the steady state case 3, $\gamma = 0$ and equation (3) then shows that $P_{11}(0) = 1$.

Thence, simple manipulation allows evaluation of the thermal admittance P_{11}/P_{12}, and thermal transfer factor $1/P_{12}$ as

$$P_{11}(0)/P_{12}(0) = 1/P_{12}(0) = -Q_3 \tag{15a}$$
$$P_{11}(\omega)/P_{12}(\omega) = \qquad\qquad -Q_1 \tag{15b}$$
$$1/P_{12}(\omega) = Q_2 - Q_1 \tag{15c}$$

Equation (15a) implies that the slab-film combination has a U-value of Q_3 or a resistance of $1/Q_3$, i.e. the resistance of the slab alone is $(1/Q_3)$- H.

Values of Slab Resistances

The resistance R_s ($m^2 K/W$) of a square slab, 10 m side, i.e. perimeter p = 40 m, on ground of conductivity $\lambda = 1.0$ W/mK has been calculated for various values of film resistance H and edge distance e. These are shown in Figure 3.

These results can be immediately extended to other square slabs and other values of conductivity by using the result (see Appendix 2) that

$$R(\beta\lambda,\alpha P, e,H) = (\alpha/\beta)R_s(\lambda,P,e/\alpha,\beta H/\alpha) \tag{16}$$

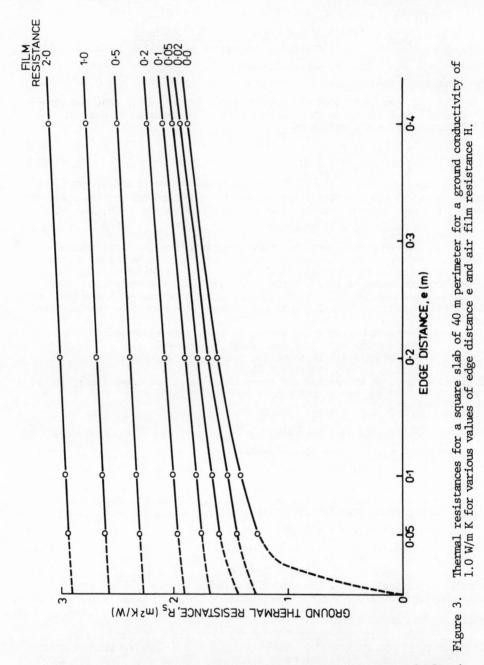

Figure 3. Thermal resistances for a square slab of 40 m perimeter for a ground conductivity of 1.0 W/m K for various values of edge distance e and air film resistance H.

where $\lambda = 1.0$ and $P = 40$ are the standard values for which R_s was calculated.

For rectangular slabs, resistance values have also been calculated. These are plotted in Figure 4 as the ratio of the resistance of the rectangle to that of a square of equal perimeter against the ratio of area to $(\text{perimeter}/4)^2$.

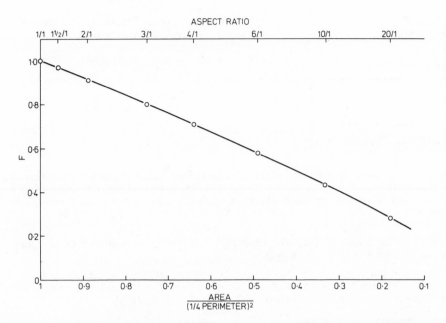

Figure 4. Effect of shape on ground thermal resistance. F is defined as

$$F = \frac{\text{ground thermal resistance for given (area/} \tfrac{1}{4} \text{ perimeter)}^2}{\text{ground thermal resistance for the square of equal perimeter}}$$

The calculations have been further extended to the shapes shown in Table 1 and all, except for open square with small opening, gave resistance ratios very close to those shown in Figure 4 for comparable ratios of area to (perimeter/4)^2.

Shape	Area, A m^2	Perimeter, P m	$\dfrac{A}{(\tfrac{1}{4}P)^2}$	F* Calculated	F* From eqn 16 and Figs 3 and 4
A	75	40	0.75	0.79	0.80
B	100	50	0.64	0.72	0.71
C	125	60	0.56	0.65	0.63
D	125	60	0.56	0.66	0.63
E	200	80	0.50	0.80	0.59
F	125	60	0.56	0.66	0.63
G	300	120	0.33	0.43	0.44

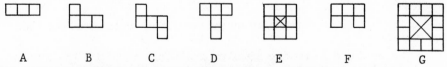

A   B   C   D   E   F   G

$$* \quad F = \frac{\text{Thermal resistance of shape shown}}{\text{Thermal resistance of square of equal perimeter}}$$

Table 1.   Comparison of F values for various shapes for an edge distance, e, of 0.2m by direct calculation and from curves of Figs 3 and 4.

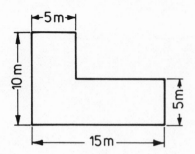

The procedure for calculating the thermal resistance for a slab of general rectilinear shape is set out below, as it relates to the slab indicated in Figure 5.

Figure 5.   Example slab used for showing calculation method.

(a)   Determine R_s for the required conditions using Figure 3. For the slab shown the perimeter $P \cong 50$ m so $\alpha = 1.25$, and for a ground conductivity of 0.8 W/m K, $\beta_2 = 0.8$. Assuming an edge distance of 0.2 m and an air film resistance of 0.5 m²K/W the value of R_s (1,40,0.2/1.25, 0.8x0.5/1.25) is found from Figure 3 to be 2.16.

(b)   Using equation (16) the thermal resistance for a square of the same perimeter is $(\alpha/\beta) R_s = 1.56x2.16 = 3.37$.

(c)   For the slab shown the area A is 100 m², so that $A/(\tfrac{1}{4}P)^2 = 0.64$. Reference to Figure 4 then gives F = 0.72 and hence the thermal resistance of the ground for the case illustrated is 3.37x0.72 = 2.4 m²K/W.

MODEL FOR THERMAL CALCULATION

The above treatment has evaluated the resistance of the ground relative to slabs of various shapes and dimensions. If, in fact, a concrete slab or a suspended floor be installed "above" the slab areas treated in the calculation, the resistance of these additional structures is to be added in series. It is recommended that, in the general case, an edge distance of 0.2 m be treated as normal, it being noted that this corresponds to the thickness of an average wall.

The authors calculated the response of the slab-on-ground to temperatures cyclical with respect to time as has been previously indicated. The frequencies chosen lay between 1 rad in 768 hours to 1 rad in 12 hours in a geometric progression of ratio 2:1. A number of slabs were evaluated of which a typical result is shown in Figure 6. Therein the values of P_{11}/P_{12} and $1/P_{12}$ as defined earlier are shown using crosses when derived from a constant heat flow distribution and circles when from a constant temperature distribution - the air-film effect being included. It is typical of such values that the transfer parameters at low frequencies have a negative real and positive imaginary part and at higher frequencies the plot "spirals" to zero. The admittance parameter (P_{11}/P_{12}) has negative real and imaginary parts and the plot has approximately a semi-elliptical shape, equal to the transfer parameter at steady state and minus the reciprocal of the film resistance at high frequencies.

The simpler representation of one resistance distributed capacity section was derived:-

1.   by setting a resistance equal to that required to give the correct steady state value

2.   by setting the distributed capacitance equal to that of the ground in the given instance having the area and resistance of the particular slab

This technique leads in all cases tried to a good representation for the admittance parameter but gives only a fair representation of the transfer parameter. A notably improved representation can be achieved using two sections in parallel - one with a resistance and capacitance 2.5 and 2.0 times that given above and the second 1.67 and 0.02 those above. The latter should be used in situations where a good representation of the transfer parameter is required.

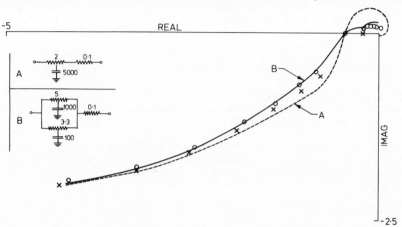

Figure 6. Comparison of admittance (P_{11}/P_{12}) and transfer ($1/P_{12}$) parameters for 10 m x 10 m slab calculated from Fourier series and from practical representations A and B. In the practical representations A and B the capacitors are shown attached to the resistors to indicate that the capacitance is distributed rather than lumped.

REFERENCES

1. Billington, N.S. (1951). Heat loss through solid ground floors. J. Inst. Heat. Vent. Eng., 19(195), 351-372.

    Great Britain, Building Research Station 1972. Heat losses through ground floors. BRS Digest No. 145.

2. Muncey, R.W.R. (1953). The calculation of temperatures inside buildings having variable external conditions. Aust. J. Appl. Sci. 4(2), 189-196.

3. Muncey, R.W.R. (1967). The conduction of fluctuating heat flow. App. Sci. Res. 18, 9-14.

4. Shelton, J. (1975). Underground storage of heat in solar heating systems. Solar Energy, 17(2), 137-143.

5. Stephenson, D.G. (1962). Methods of determining non-steady-state heat flow through walls and roofs of buildings. The I.H.V.E. Journal, 30, 64-73.

6. Van Gorcum, A.H. (1950). Theoretical considerations on the conduction of fluctuating heat flow. Appl. Sci. Res. A2:272.

7. Vuorelainen, O. (1960). The temperatures under houses erected immediately on the ground and heat losses from their foundation slab. The State Institute for Technical Research, Finland, Publication 55.

8. Vuorelainen, O. (1963). A practical method for calculation of the heat losses into the ground from buildings erected immediately on the ground. The State Institute for Technical Research, Finland, Publication 76.

APPENDIX 1.   EDGE SLOPE

Let M be the upper limit of summation for m. In the x direction, the series used for T is

$$T(x) = 1/f + \sum_{m=1}^{M} [(\sin \pi m/f)/(\pi m/f)] \cos(\pi m x/fu)$$

The edge slope is given by

$$\left| dT(x)/dx \right|_{x=u} = \sum_{m=1}^{M} (2/fu) \sin^2(\pi m/f)$$

For unit temperature increment,

$$1/\delta x = (2/fu) \sum_{m=1}^{M} \sin^2(\pi m/f) = M/fu$$

since each f terms contribute $f/2$ and provided $M \equiv 0 \pmod{f}$, though this is not a critical restriction when M is large. So the number of terms to be used in the series for an edge distance e is given by

$$M = fu/e$$

and a similar expression $N = gv/e$ holds in the y direction.

APPENDIX 2.   DERIVATION OF EQUATION (16)

Equation (9) gives $W = j\gamma\lambda Z/(1 + j\gamma\lambda H)$. For the steady state case ($\omega = 0$) both W and $j\gamma$ are real. The ground thermal resistance is then given by

$$R = (1/W) - H = (1/\lambda)(1/j\gamma + \lambda H)/Z - H$$

If λ be changed to $\beta\lambda$ and H to H/β where β is constant, then the resulting thermal resistance is R/β. If the perimeter y be changed to αP, where α is a constant, then γ becomes γ/α provided that the series is maintained at the same length, i.e. P/e is constant (Appendix 1) so that e must become αe. The resulting thermal resistance is then αR.

If R_s is the ground thermal resistance for the condition $P = 40$ and $\lambda = 1$, then the thermal resistance R when the perimeter changes from P to αP and the conductivity changes from λ to $\beta\lambda$ is given by

$$R(\beta\lambda, \alpha P, e, H) = (\alpha/\beta) . R_s(\lambda, P, e/\alpha, \beta H/\alpha).$$

HEAT STORAGE AND TRANSIENT TEMPERATURE DISTRIBUTION IN THREE-DIMENSIONAL COMPOSITE SOLID SYSTEMS

JÜRGEN PLEISS

G. Bauknecht GmbH
Stuttgart, Federal Republic of Germany

ERICH HAHNE

Institut für Thermodynamik und Wärmetechnik
Universität Stuttgart
7 Stuttgart 1, Seidenstrasse 36
Federal Republic of Germany

ABSTRACT

Conventional calculations of heat losses and temperature fields are based on one-dimensional heat flow. This is realistic for parts of a wall area. For edge and corner sections, however, the heat flow is two- and three-dimensional. Solutions to such problems are based on Newman's Rule (multiplicative superposition) and can be found in literature, provided that the systems are isotropic and homogeneous. In many cases only the solution for transient temperature fields is given. It is shown here that Newman's Rule can successfully be applied for the prediction of both transient temperature fields and heat losses of both homogeneous and composite solid systems. The application to composite systems requires the introduction of an equivalent Biot-and Fourier-number. With these equivalent numbers existing diagrams for one-dimensional temperature fields can be used. The method was checked by numerical calculation with good agreement. For a practical example experimental measurements were taken on a storage radiator and compared to the numerical calculation.

NOMENCLATURE

c	specific heat
h	heat transfer coefficient
k	thermal conductivity
Q	heat
T	temperature
T*	new temperature at the time $t + \Delta t$
t	time
Δt	timestep
W	heat source intensity per volume
x, y, z	coordinates
ϱ	specific density
ϑ	temperature difference (excess temperature)

Subscripts

o   initial

∞   surroundings

INTRODUCTION

Transient temperature fields and transient heat contents in infinitely large flat plates, infinitely long cylinders, in spheres and in semi-infinite walls representing one-dimensional problems are excessively treated in literature. Mathematical results are given in tables and diagrams.

Two-dimensional problems such as the temperature distribution in bars or finite cylinders can be solved by applying "Newman's Rule": This is the multiplicative superposition of two solutions of the respective one-dimensional problem; e.g. for a finite cylinder, the solution of the infinitely long cylinder and the solution of the infinitely large plate. The thickness of this plate equals the finite length of the cylinder under consideration. This method was experimentally verified by Budrin and Krassowski /1/.

The method of multiplicative superposition can also be applied to transient three-dimensional temperature distributions. For a cube, the following relations are valid:

$$\left(\frac{\vartheta}{\vartheta_0}\right)_{\substack{Center \\ of\ cube}} = \left(\frac{\vartheta}{\vartheta_0}\right)^3_{\substack{middle \\ of\ plate}} \tag{1}$$

$$\left(\frac{\vartheta}{\vartheta_0}\right)_{\substack{middle \\ of\ cube\ surface}} = \left(\frac{\vartheta}{\vartheta_0}\right)^2_{\substack{middle \\ of\ plate}} \cdot \left(\frac{\vartheta}{\vartheta_0}\right)_{\substack{surface \\ of\ plate}} \tag{2}$$

$$\left(\frac{\vartheta}{\vartheta_0}\right)_{\substack{middle \\ of\ cube\ edge}} = \left(\frac{\vartheta}{\vartheta_0}\right)_{\substack{middle \\ of\ plate}} \cdot \left(\frac{\vartheta}{\vartheta_0}\right)^2_{\substack{surface \\ of\ plate}} \tag{3}$$

$$\left(\frac{\vartheta}{\vartheta_0}\right)_{\substack{corner \\ of\ cube}} = \left(\frac{\vartheta}{\vartheta_0}\right)^3_{\substack{surface \\ of\ plate}} \tag{4}$$

All results are based on the solutions of one-dimensional fields and consequently apply only to conditions assumed for those. These conditions are: a homogeneous and isotropic solid material, temperature independent properties, no heat sources or sinks within the material and constant and uniform boundary conditions.

It was the aim of this investigation to develop a computer program for the calculations of transient temperature fields without these restrictive conditions. On the basis of these calculations a method was developed which allows the application of Newman's Rule to non-homogeneous materials and to the transient behaviour of heat contents of the materials. For a comparison, the theoretical results were checked by means of experiments performed on a storage heater.

COMPUTER PROGRAM FOR TEMPERATURE FIELDS

Based on the heat conduction equation (simplified for temperature independent properties)

$$\frac{\partial T}{\partial t} = \frac{k}{c\varsigma}\left(\frac{\partial^2 T}{\partial x^2} + \frac{\partial^2 T}{\partial y^2} + \frac{\partial^2 T}{\partial z^2}\right) + \frac{W}{c\varsigma} \tag{5}$$

the finite difference approximation is applied, which yields for each grid point:

$$\frac{T^*(x,y,z)-T(x,y,z)}{\Delta t} = \frac{k}{c\varsigma}\left[\frac{T^*(x-\Delta x,y,z)+T^*(x+\Delta x,y,z)-2T^*(x,y,z)}{\Delta x^2}\right.$$

$$+ \frac{T^*(x,y-\Delta y,z)+T^*(x,y+\Delta y,z)-2T^*(x,y,z)}{\Delta y^2}$$

$$\left.+ \frac{T^*(x,y,z-\Delta z)+T^*(x,y,z+\Delta z)-2T^*(x,y,z)}{\Delta z^2}\right]$$

$$+ \frac{W}{c\varsigma} \tag{6}$$

The resulting system of equations has to be solved simultaneously. In equation (6) only unknown temperatures T* for the time t+ Δt appear on the right side. This fact calls for an "implicit" solution method: The SIP-algorithm (strongly implicit procedure) by Weinstein, Stone and Kwan /2/ was selected as appropriate. The advantage of the chosen implicit method is the stability for all values of Δt and consequently better calculating accuracy.
The iterative procedure was developed to solve systems of three-dimensional elliptic and parabolic differential equations.
The procedure is fast, is insensitive to rounding errors and renders solutions for difficult problems, where other procedures, e.g. the alternating direction technique fail.

CALCULATING PROCEDURE

A grid is introduced in the body under consideration in such
a way that grid lines coincide with the outer boundaries of the
body. The procedure starts from a given temperature distribution
with a temperature T(x,y,z) assigned to each nodal point. From a
heat balance for each nodal point a residuum is obtained from
which the temperature at a time t+ Δt is calculated. The heat balance is repeatedly applied for each nodal point until the residuum falls below a prescribed value (truncation error limit
ΔT = 0,01 K). When this is achieved for all nodal points the temperature field T*(x,y,z) at the later time t+ Δt is reached.
This procedure is practised for consecutive time steps. In order
to accelerate the numerical calculation the method of so-called
"overrelaxation" is used.
After every time step the thermal properties k and c are
corrected corresponding to the temperature of each nodal point.
This is necessary when large temperature differences occur in the
body for which constant properties cannot be assumed. Heat transfer coefficients are corrected for both a local and a time dependent variation of temperature. This is important when free convection heat transfer is considered in combination with thermal
radiation. With this calculation procedure it is possible to calculate the transient temperature distribution and heat contents
in one-, two- and three-dimensional bodies in Cartesian, cylindrical or spherical coordinate systems for any arrangement of the
grid.
Using the newly developed computer program the transient temperature distribution within homogeneous, isotropic cubes, short
cylinders and spheres was calculated. For a check the computed
temperatures in the center and at the surface of the various bodies were compared to those temperatures found by the application
of Newman's Rule from diagrams for one-dimensional heat conduction.
The agreement was extremely good. For a further test the numerical
predictions were examined against an experimental investigation.

EXPERIMENTAL COMPARISON

In a commercially available storage radiator* the transient
temperature distributions during various cycles of charging and
discharging were measured. These measurements were compared to
predictions made by the newly developed computer program.
In figure 1 the schematic construction of the radiator is
shown. The core of the storage material is heated by electric resistance wires. It is mantled on all sides by an insulating layer
with an air-gap on top. Metal plates form the outer surface. From
these, heat is dissipated by convection and radiation. In z-direc-
tion the radiator is subdivided into 9 equally sized layers. For
the calculation 400 nodal points were taken, 40 of those are
shown in the figure as circles.

* Storage radiator S 150 E
  G. Bauknecht GmbH, Stuttgart, F.R.G.

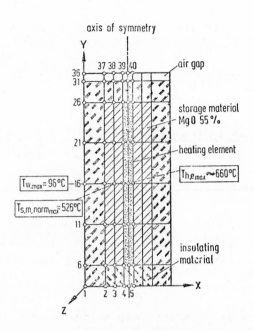

Fig. 1 Cross-Section of Storage Radiator

The mean temperature as measured at the end of a charging period in the center of the heating element is 660°C, in the center of the storage material the temperature is 526°C and at the outside surface 96°C.

TRANSIENT HEAT CONTENTS

With the results obtained numerically for the transient heat contents of a three-dimensional body it was examined whether Newman's Rule is also applicable for this case. The ratio of instant heat contents and the original heat contents Q/Q_0 for an infinite flat plate which is made up of small elements can be obtained from

$$\left(\frac{Q}{Q_0}\right)_{\substack{flat \\ plate}} = \frac{\sum_{i=1}^{n}\left[c\varrho\Delta x\Delta y\Delta z(T_x - T_\infty)\right]_{i\,flat\,plate}}{\sum_{i=1}^{n}\left[c\varrho\Delta x\Delta y\Delta z(T_{x_0} - T_\infty)\right]_{i\,flat\,plate}} \qquad (7)$$

Matrix multiplication yields for a cube of uniform initial excess temperature ϑ_0

$$\left(\frac{Q}{Q_0}\right)_{cube} = \left(\frac{Q}{Q_0}\right)^3_{flat\,plate} \tag{8}$$

since

$$\left(\frac{Q}{Q_0}\right)_{cube} = \sum_{i=1}^{n} \frac{[\Delta x^3 (\vartheta/\vartheta_0)_{cube}]_i}{(2X)^3} = \sum_{i=1}^{n} \frac{[\Delta x (\vartheta/\vartheta_0)_{flat\,plate}]_i^3}{(2X)^3}$$

and for a short cylinder

$$\left(\frac{Q}{Q_0}\right)_{\substack{short\\cylinder}} = \left(\frac{Q}{Q_0}\right)_{\substack{long\\cylinder}} \cdot \left(\frac{Q}{Q_0}\right)_{\substack{flat\\plate}} \tag{9}$$

The comparison of the results obtained by the numerical method and Newman's Rule showed very good agreement.

COMPOSITE SYSTEMS

The good agreement found between the temperatures and heat contents calculated by the numerical method and those obtained from the application of Newman's Rule for three-dimensional homogeneous and isotropic bodies raised the question whether there is the possibility of applying Newman's Rule also to composite media. If so, predictions of the transient processes within bodies composed of different materials would be considerably facilitated, since diagrams for one-dimensional transient changes in homogeneous bodies are widely available /3,4/.

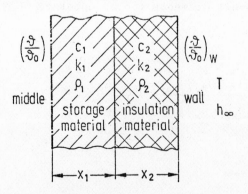

Fig. 2 Two layer composite
        system

The investigation was performed on a system composed of two materials: a storage material on the inside and an insulating material on the outside of the system. These materials differed remarkably in their thermal properties. A schematic setup is presented in figure 2.

The result of the calculations was, that one-dimensional solutions, which can be used for the application of Newman's Rule, will be obtained for a two layer system, when "equivalent" quantities are introduced. Thus the equivalent Fourier number becomes:

$$Fo_{equ} = \frac{k_{equ}\, t}{(c\varrho)_{equ}\, X^2} \qquad (10)$$

and the equivalent Biot number:

$$Bi_{equ} = \frac{hX}{k_{equ}} \qquad (11)$$

with

$$X = x_1 + x_2 \qquad (12)$$

and

$$(Xc\varrho)_{equ} = (xc\varrho)_1 + (xc\varrho)_2 \qquad (13)$$

As expected the value of the heat transfer coefficient h is not affected.

For the equivalent thermal conductivity, however, a simple arithmetic relation could not be obtained. It is obvious that this equivalent thermal conductivity must be a function of the thermal conductivity of the different materials and their thicknesses:

$$k_{equ} = k_{equ}(k_1, k_2, {}^{x_1}/_{x_2}) \qquad (14)$$

The equivalent thermal conductivity was determined in the following way: In the diagrams in literature /3/ the ratio of excess temperatures ϑ/ϑ_0 and dissipated heat Q/Q_0 are plotted as functions of Fo with 1/Bi as a parameter. For our purposes the diagrams for flat plates were considered and values Fo = 0.5(1/Bi) and Fo = 1/Bi were chosen. With the data ϑ/ϑ_0 and Q/Q_0 taken from the diagrams for these Fourier numbers, figure 3 was drawn as an auxiliary chart.

Equations (10) and (11) give

$$Fo_{equ} = \left(\frac{1}{Bi}\right)\left(\frac{h}{(c\varrho)_{equ}\, X^3}\right) t \qquad (15)$$

The quantities in the second bracket can be chosen in a way that the value of the bracket becomes 0.5, and correspondence is achieved between equation (15) and the Fourier numbers as they were chosen for the construction of figure 3, after one or two hours of cooling. The temperatures in the middle of our system (fig. 2) and on the wall, as well as the dissipated heat were calculated by the numerical method with the quantities as chosen to make the second bracket in equation (15) equal 0.5. The resulting temperature ratios and heat ratios were inserted into the respec-

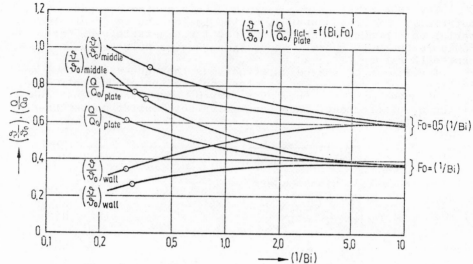

Fig. 3 Auxiliary chart for the determination
of k_{equ}

tive curves of fig. 3. From the value 1/Bi on the abscissa of
fig. 3 from equation (11) the equivalent thermal conductivity is
obtained as:

$$k_{equ} = \frac{1}{Bi} h X \qquad (16)$$

The numerical calculations were performed with six different
ratios of k_1/k_2 and three different layer thicknesses x_1/x_2.
The results are plotted in fig. 4.

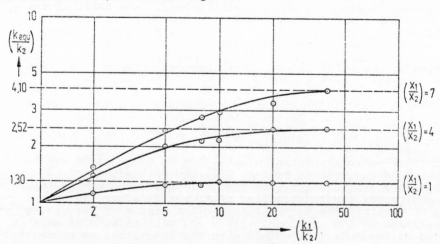

Fig. 4 Equivalent thermal conductivity for a
two layer system

A strong influence of the insulating material is observed.
For a given thermal conductivity k_2 the equivalent thermal conductivity approaches a constant value when a certain ratio of
k_1/k_2 is exceeded. For a two layer system with equal layer thickness e.g. the equivalent thermal conductivity ratio assumes the
value k_{equ}/k_2 = 1.3 regardless whether the thermal conductivity
of the storage material is 10 times a 100 times the conductivity of the insulating material. The equivalent thermal conductivity is only 1.3 times the thermal conductivity of the insulating material. In the case of a storage radiator or any other
system which dissipates heat from a good conductor through an
insulating layer this means that the heat transport is primarily
controlled by the insulating material and there is no use of increasing the conductivity of the well conducting layer.
    The equivalent thermal conductivity ratio encreases only
moderately with an increasing ratio of x_1/x_2. For house heating
this means that an increase in insulation thickness for instance
from a value x_2 = $x_1/4$ to x_2 = x_1 reduces k_{equ}/k_2 = 2.52 to
k_{equ}/k_2 =1.3.
    Introducing these values x_2 and k_{equ} into the Fourier and
Biot numbers it can be found in the diagrams in literature that
a four-fold increase in insulation thickness reduces the transient heat losses by about 40 %.
    The transient temperatures and heat losses as obtained with
the equivalent quantities from diagrams were also compared to data
calculated numerically. Very good agreement was found.
    Finally such a comparison was performed for cubes and short
cylinders and it was observed that data obtained from Newman's
Rule with solutions for the one-dimensional field based on equivalent quantities, agree well with those calculated numerically.

REFERENCES

/1/   Budrin, D.W. and B.A. Krassowski: Studies of the Uralian
       Industrial Institute (in Russian) Delivery XVII (1941)

/2/   Weinstein, H.G., H.L.Stone and T.V. Kwan: Iterative Proce-
       dure for Solution of Systems of Parabolic and Elliptic
       Equations in Three Dimensions
       I & E Fundamentals, Vol. 8, No.2, May 1969, pp.281-287

/3/   Grigull, U.: Temperaturausgleich in einfachen Körpern-
       Ebene Platte, Zylinder, Kugel, halbunendlicher Körper
       Springer-Verlag, Berlin/Göttingen/Heidelberg, 1964

/4/   Gröber, Erk, Grigull: Grundgesetze der Wärmeübertragung
       3. neubearbeitete Auflage, 3. verbesserter und erweiterter
       Neudruck
       Springer-Verlag, Berlin/Göttingen/ Heidelberg, 1963

STUDY OF TEMPERATURE CONDITIONS OF HEATED FLOOR STRUCTURES AND BUILDING FOUNDATION SOILS WITH NEGATIVE TEMPERATURE OF INTERNAL MEDIA

A. G. GINDOYAN AND V. Ya. GRUSHKO

TSNIIpromzdaniy, Moscow, USSR

ABSTRACT

Herein, the mathematical model of heat and mass transfer processes in the floor structures and building foundation soils with
negative temperatures of internal atmosphere has been studied.
The thermal interrelation of a building and foundation soil (non-
stationary case) has been considered. The zones with steady and
non-steady temperatures mode of floors and foundation soils have
been determined. The variation task has been set to determine
the power of a soil heating system and the solution of this task
is given for a portion of an area with a heat conducting insert
in the shape of column.

NOMENCLATURE

t_K temperature, °K (°C)

θ_K dampness potential, °M

τ time, S (h)

λ_{q_K} coefficient pf thermal conductivity, W/m K (kcal/m h °C)

λ_{m_K} coefficient of mass conductivity, kg/m s °M (kg/ m h °M)

C_{q_K} specific heat, J/kg K (kcal/kg °C)

C_{m_K} specific mass, kg/kg°M

δ_K^I thermal gradient coefficient related to mass transfer potential difference, °M/°K (°M/°C)

ε_K phase conversion number ($\varepsilon_{K\Lambda}$ - water into ice)

ζ specific phase conversion heat, J/kg (kcal/kg); (ζ_Λ - specific heat of ice thawing)

K number of a layer of the floor structural material

n number of layers in the floor structure

$n+1$ number of a foundation soil layer

a_{q_K} coefficient of temperature conductivity, m^2/S (m^2/h)

a_{m_K} coefficient of potential conductivity, m^2/S (m^2/h)

α_B coefficient of heat transfer between the floor and air inside the building, $w/m^2 K (kcal/m^2 h°C)$

θ_g soil dampness potential at the level of ground water, $^\circ M$

θ_δ air dampness potential inside the building, $^\circ M$

T_1 air temperature inside the building, $^\circ K$ ($^\circ C$)

T_2 soil temperature at the ground water level, $^\circ K$ ($^\circ C$)

T_0 soil temperature at a critical point, $^\circ K$ ($^\circ C$)

γ_0 unit weight of dry material, kg/m^3

L building length, m

B building width, m

ℓ_0 depth of heat sources from the floor surface, m

ℓ_2 depth of the plane of contact between the heating plate and foundation soil, m

β_δ coefficient of mass transfer between the floor and internal air, kg/m^2 S $^\circ M$ ($kg/m^2 h$ $^\circ M$)

u_κ weight dampness of material (kg/kg)

t_n yearly average temperature of the soil surface, $^\circ K$ ($^\circ C$)

t_A annual temperature variation amplitude of the soil surface, $^\circ K$ ($^\circ C$)

$\omega = \frac{2\pi}{T}$ frequency of annual variations (T = 1 year, period of annual variations, $S(h)$)

H_M depth of season soil freezing, m

T_c soil temperature at the level of zero annual amplitudes $^\circ K$ ($^\circ C$)

ε phase displacement of annual temperature variation

$\mathscr{v} = (t - T_c)/(T_0 - T_c)$ relative temperature

$\mathscr{x} = 2x/B, \xi = 2z/B$ relative coordinates

$F_0 = 4a\tau/B^2$ Fourier criterion

$P_d = \omega B^2/4a$ Predvoditelev criterion

$\mathscr{v}_a = T_a/(T_0 - T_c)$ relative conditional amplitude of soil surface temperature variation

F area of the building cold profile, m^2

Q heating system capacity, $W(kcal/h)$

$\delta(z)$ Dirac function

$\chi(z)$ Heaviside unit function

Provision for durability and reliable peformance of buildings with negative temperature of internal media (industrial cold stores, liquified gas storage, etc.) requires special measures to protect foundation soils against freezing and frost heaving. The lack of these measures or their poor efficiency result in freezing and frost heaving of foundation soils, which in turn causes strains and rather often collapse of building structures.

With the building width in the order of 50 m the freezing depth reaches 10 to 15 m. Rehabilitation of the building and recovery of the base temperature conditions necessitate temporary interruption in building operation and involve considerable financial and material expenses.

Depending on the purpose of the building, its space and planning solution, the site hydrogeological conditions various protection methods (structural, electrochemical, chemical, heat engineering, etc.) are used.

Recently the heat engineering methods have found wide application since they provide for a heat shield in the conjugation plane of a building with foundation soils, using different heat sources built in the floor structure.

Rational design of the heating system and reliability of its operation put forward the task to determine the regularity of the shield heating capacity distribution with the expenses reduced to minimum, taking into account: unsteady external climatic conditions considerably affecting formation of the soil temperature conditions in the peripheral zone of the building; presence of multitudes of heat-conducting insertions in terms of column and wall footings in the heat-insulating floor structure; moisture migration to the freezing front and phase transformations in humid soils; moistening of the floor heat insulation; frequent temperature changes in cooling chambers, etc.

The present study contains the analysis of various factors forming the thermal field in the heated floors and foundation soils, and the heat engineering design method of heating systems is considered here as well.

An element of the building conjugation with foundation soils is the floor structure. Therefore, for the purpose of analysis the process of heat and moisture transfer for the system "floor structures-foundation soils" should be considered. The general scheme of the building with indication of principal factors forming the thermal field is shown in Fig.1.

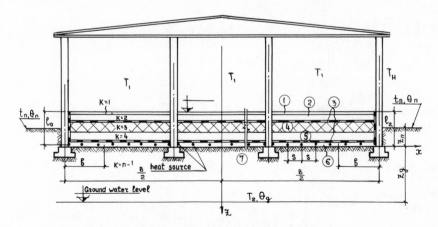

Fig.1. Structural diagram of the heated floor.
1 - floor finish; 2 - reinforced concrete slab; 3 - waterproofing course; 4 - heat insulating course; 5 - sand filling; 6 - heating plate with heat sources; 7 - foundation soil.

The materials of the floor structural elements are capillary-porous bodies. They are insulated from the soil and external walls by vapour barrier and water proofing course.

The temperature and dampness field of the system "floor structures-foundation soils" according to [1] is described by the heat and mass transfer equations.

$$C_{q_K} \gamma_{o_K} \frac{\partial t_K}{\partial \tau} = \text{div}(\lambda_{q_K} \cdot \nabla t_K) + \varepsilon_K \zeta \gamma_{o_K} C_{m_K} \frac{\partial \theta_K}{\partial \tau} + f(x,y,z,\tau), \tag{1}$$

$$C_{m_K} \gamma_{o_K} \frac{\partial \theta_K}{\partial \tau} = \text{div}(\lambda_{m_K} \cdot \nabla \theta_K + \lambda_{m_K} \delta_K^{ll} \cdot \nabla t_K), \tag{2}$$

$K = 1,2,\ldots,n$, $n + 1$ — numbers of the floor structure layers.

Initial conditions

$$t_K(x,y,z,0) = \varphi_1(x,y,z); \quad \theta_K(x,y,z,0) = \varphi_2(x,y,z). \tag{3}$$

The boundary conditions are established with the following considerations: 1) on the soil surface outside the building profile temperature and dampness potential are given as a function of time; 2) heat and mass tranfer takes place between the floor and air in the enclosed space according to Newton's law; 3) at the ground water level, its temperature and corresponding dampness potential are given; 4) along the building perimeter and up the vertical the floor structure is considered absolutely heat-insulated and waterproofed; 5) at the interface of the different floor structural layers the contact of capillary-porous bodies is observed; 6) the heat of phase transformations during transfer on the floor surface and the layer interface is negligible as compared to the heat flow.

The boundary conditions are assumed as follows:

$$\left. \begin{array}{l} [-\lambda_{q_1} \frac{\partial t_1}{\partial z} + \alpha_6 (t_1 - T_1)]|_{z=-\ell_2} = 0, \\[2mm] -\lambda_{m_1} (\frac{\partial \theta_1}{\partial z} + \delta_1^{ll} \frac{\partial t_1}{\partial z})|_{z=-\ell_2} = q_{m_0}, \end{array} \right\} \quad |x| < \frac{B}{2}, |y| < \frac{L}{2}; \tag{4} \tag{5}$$

$$\frac{\partial t_K}{\partial x}\Big|_{\substack{|x|=B/2 \\ |y|<L/2}} = \frac{\partial t_K}{\partial y}\Big|_{\substack{|y|=L/2 \\ |x|<B/2}} = 0, \quad K = 1,2,\ldots,n; \tag{6}$$

$$\frac{\partial \theta_K}{\partial x}\Big|_{\substack{|x|=B/2 \\ |y|<L/2}} = \frac{\partial \theta_K}{\partial y}\Big|_{\substack{|y|=L/2 \\ |x|<B/2}} = 0, \quad K = 1,2,\ldots,n; \tag{7}$$

$$\left. \begin{array}{l} t_{n+1}(x,y,z_n,\tau) = \psi_1(\tau), \\[2mm] \theta_{n+1}(x,y,z_n,\tau) = \psi_2(\tau), \end{array} \right\} \quad |x| > \frac{B}{2}, \text{ or } |y| > \frac{L}{2}; \tag{8} \tag{9}$$

$$t_{n+1}(x,y,z_g,\tau) = T_2 ; \tag{10}$$

$$\theta_{n+1}(x,y,z_g,\tau) = \theta_g ; \tag{11}$$

where
$$q_{m_0}(x,y,-\ell_2,\tau) = \begin{cases} const, & \theta_1(-\ell_2) > 100°M, \\ \beta_6(\theta_6-\theta_1)\big|_{z=-\ell_2}, & \theta_1(-\ell_2) < 100°M . \end{cases} \tag{12}$$

During heat and mass transfer contact conditions of capillary-porous bodies exist at the layer interface.

For the floor structural elements the heat transfer by vapour and water flows both in the thawed and frozen zones is very small due to several courses of water and vapour proofing. In the foundation soil as shown in [4,5] heat transfer by vapour or moisture can be neglected (with special measures to prevent moisture filtration). Therefore, heat transfer in the thawed zone occurs owing to conductive properties of materials without consideration of phase transformations. But in the frozen zone the effect of ice formation should be taken into account.

For the case under consideration Luikov's criterion $Lu = a_{m\ell}/a_{q\ell}$ is small almost for all materials of the floor structure (for concrete $Lu \approx 0.01$ [6], for mineral wool boards $Lu < 0.05$). An exception is the filling material ($Lu \approx 0.8 \div 1$) which, however, is waterproofed. Thus the process of heat transfer at any time may be considered to be occurring at constant dampness of materials.

The above mentioned allows to assume that in the process of the temperature field development the phase transition should be taken into account only when zero isotherm passes through the layer of material, and the temperature field develops at constant dampness of materials. Therefore, the set of equations (1), (2) and the corresponding boundary conditions are divided into two independent tasks: heat transfer (I) and moisture transfer (II).

$$\text{I.} \quad \frac{\partial t_\kappa}{\partial \tau} = a_{q\kappa} \nabla^2 t_\kappa + f'_\kappa(x,y,z,\tau), \quad \kappa = 1,2,\dots,n+1; \tag{13}$$

$$\left[\lambda^M_{q\kappa}\frac{\partial t^M_\kappa}{\partial n} - \lambda^T_{q\kappa}\frac{\partial t^T_\kappa}{\partial n}\right]\bigg|_{\xi=\xi(x,y,z)} = \varepsilon_{\kappa\Lambda}z_\Lambda u_\kappa \gamma_0 \frac{\partial \xi}{\partial \tau}, \quad t=0, \tag{14}$$
$$\kappa = 1,2,\dots,n;$$

with initial condition (3) and boundary conditions (4), (6), (8), (10);

$$\text{II.} \quad \frac{\partial \theta_\kappa}{\partial \tau} = a_{m\kappa}\cdot\nabla^2\theta_\kappa + a_{m\kappa}\delta'_\kappa\cdot\nabla^2 t_\kappa, \quad \kappa = 1,2,\dots,n+1 \tag{15}$$

with initial conditions (3) and boundary conditions (5), (7), (9), (11).

Here

$$f_\kappa'(x,y,z,\tau) = \frac{1}{C_{q\kappa}\gamma_{o\kappa}} \cdot f(x,y,z,\tau) \qquad (16)$$

$\gamma(x,y,z)$ – thawed zone – frozen zone interface,
n – normal to this interface,
M,T   – indices for frozen and thawed zones.

By solving the two problems we obtain the **regularity** of the heating system capacity distribution necessary to protect the foundation soil against freezing.

We are interested in steady temperature conditions of the building performance when initial conditions, in fact, insignificantly affect the thermal field.

It is shown in paper [4] that when the building length is much more than its width $L \geqslant 3B$, the thermal field can be regarded as two-dimensional. In actual conditions $z_n \approx 0$, therefore when investigating the effects of external temperature influence on the formation of the thermal field and the rate of the heat flow in this zone, we consider the thermal field of the semi-confined body (Fig.2) with the following conditions on its boundaries.

$$t(x,0,\tau) = \begin{cases} T_0, & |x| < B/2, \\ t_n + t_A \cdot \cos(\omega\tau + \varepsilon), & |x| > B/2; \end{cases}$$

$$\frac{\partial t}{\partial z}\Big|_{z \to \infty} = 0. \qquad (17)$$

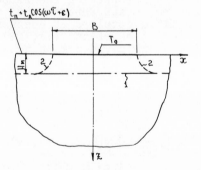

Fig.2. Design diagram: 1 – soil season freezing boundary in natural conditions; 2 – soil season freezing boundary under the building.

.   Temperature distribution down the soil depth in natural conditions discussed in [4] shows that yearly average temperature t_n of the soil surface does not coincide with the soil temperature on the level of annual zero amplitudes T_c due to phase transitions of the soil moisture in autumn/spring period.

By neglecting geothermal heat flow we may assume that below the level of soil season freezing ($z > H_M$) its yearly average temperature is constant and equals to T_c.

The soil temperature field below the level of season freezing can be obtained by means of conventional boundary conditions on the soil surface; these conditions form the same field without consideration of moisture freezing in soil. Using the condition of thermal wave propagation in a semi-infinite rod [3] the second expression in (17) may be presented as

$$t(x,0,\tau) = T_c + T_a \cdot \cos(\omega\tau + \varepsilon''), \ |x| > B/2 ,$$

where T_a — conditional amplitude of annual variation of temperature on the soil surface defined by formula

$$T_a = T_c \cdot \exp\left(H_M \cdot \sqrt{\frac{\omega}{2a}}\right); \ a = a_{q_{n+1}}; \ \lambda = \lambda_{q_{n+1}}; \qquad (18)$$

$$\varepsilon'' = \varepsilon' - H_M \cdot \sqrt{\frac{\omega}{2a}} \quad \text{— variation phase displacement, providing, it equals to } \varepsilon' \text{ at the depth } H_M.$$

The depth of soil season freezing H_M is defined by Lukyanov's formula [3].

Then problem (13), (14) with boundary conditions (4), (6), (8), (10) for $K = n + 1$ layer and $\tau \to \infty$ is reduced to the solution of equation

$$\frac{\partial t}{\partial \tau} = a\left(\frac{\partial^2 t}{\partial x^2} + \frac{\partial^2 t}{\partial y^2}\right) \qquad (19)$$

with boundary conditions

$$t(x,0,\tau) = \begin{cases} T_0, & 0 < x < B/2, \\ T_c + T_a \cdot \cos(\omega\tau + \varepsilon''), & B/2 < x < \infty, \end{cases} \qquad (20)$$

$$\left.\frac{\partial t}{\partial x}\right|_{x=0} = 0. \qquad (21)$$

The problem solution coincides with the solution of the initial problem (13), (14) in the soil thawed zone, and is somewhat distorted in the area of soil season freezing.

The solution of problem (19)-(21) with more general boundary conditions in terms of a triple integral is given in [3]. For the boundary conditions under consideration of (20) type the expression for the flow rate on the floor surface $q(x,0,\tau)$ is:

$$q_0 = q(x,0,\tau) = \frac{2}{B}\left\{\frac{2\lambda(T_0-T_c)\cdot B^2}{\pi(B^2-4x^2)} - \lambda T_a N \cdot \cos(\omega\tau + \varepsilon'' - \phi)\right\}, \quad (22)$$

$$N = \sqrt{N_1^2 + N_2^2}, \qquad (23)$$

$$N_1 = \sqrt{\frac{Pd}{2}}\left(1 + \xi\sqrt{\frac{Pd}{2}}\right)\cdot\exp\left(-\xi\sqrt{\frac{Pd}{2}}\right) - \frac{2}{\pi}\int_0^\infty \frac{\sin\mu\cdot\cos\mu x}{\mu}\cdot\exp\left(-\xi\sqrt{\frac{z+\mu^2}{2}}\right)\times$$

$$\times\left(\sqrt{\frac{z+\mu^2}{2}} + \xi\frac{z-\mu^2}{2}\right)d\mu \quad (\text{at } \xi \to 0), \qquad (24)$$

$$N_2 = \sqrt{\frac{Pd}{2}}\left(1 - \xi\sqrt{\frac{Pd}{2}}\right)\cdot \exp\left(-\xi\sqrt{\frac{Pd}{2}}\right) - \frac{2}{\pi}\int_0^\infty \frac{\sin\mu\cdot\cos\mu x}{\mu}\cdot\exp\left(-\xi\sqrt{\frac{z+\mu^2}{2}}\right)\times$$
$$\times\left(\sqrt{\frac{z-\mu^2}{2}} - \xi\frac{Pd}{2}\right)d\mu \qquad (at\ \xi\to 0), \tag{25}$$

$$z = \sqrt{Pd^2 + \mu^4}. \tag{26}$$

Fig.3 shows the dependence of flow amplitude $N\sqrt{2/Pd}$ on x and Pd ($0.6 \le x \le 1$, $Pd = 200 \div 1000$) and Fig.4 shows the dependence of the heat flow phase displacement on x and $\sqrt{Pd/2}$ at $\xi = 0(Z = 0)$.

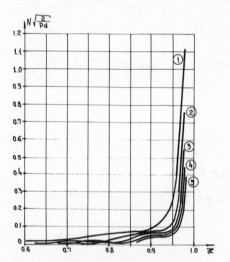

Fig.3. Dependence of the flow non-steady component amplitude of external periodic effect on Pd and x . 1 - Pd =200; 2 - Pd =400; 3 - Pd =600; 4 - Pd =800; 5 - Pd =1000.

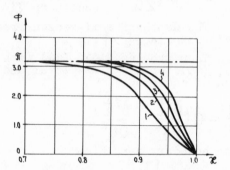

Fig.4. Dependence of phase displacement Φ on Pd and x .

. In expression (22) the first addend is the steady component of the heat flow and the second addend — non-steady component.

The data in Figs 3 and 4 indicate that the value of non-steady component at x differing from the unity more than by half wave length ($\Phi = \pi$) can be neglected. The influence of the non-steady component of the heat flow extends to the zone adjacent to the building external wall having the width.

$$\beta \approx 0.29 \sqrt{\frac{200}{Pd}}\cdot\frac{B}{2} = 2.9\sqrt{\frac{2a}{\omega}}. \tag{27}$$

With coefficient of temperature conductivity of soil $a = 2\cdot10^{-3} m^2/h$ the value of $\beta \approx 7$ m. This is the way of obtaining the heat flow rate between the heating plate and foundation soil.

The temperature in the plane of contact between the heating plate and the foundation soil has a number of local minimums due to discretion of heat sources and presence of heat-conductive insertions in terms of column and wall footings passing through the floor heat insulation (Fig.1). Let these points be assumed as critical ones. The heating system must maintain specified temperature (usually +2°C) in these points with minimum energy consumption.

In paper [7] we have formulated a variation problem to determine the heating system capacity.

Define

$$Q = \min_{f(x,y,\tau)} \int_F f(x,y,\tau)\, dF \;,\tag{28}$$

where $f(x,y,\tau)$ meets the problem solution of heat transfer through the floor structure with boundary conditions (4), (6)

$$-\lambda(x,y,z)\frac{\partial t}{\partial z}\Big|_{z=\ell_2} = q_0(x,y,\tau)\tag{29}$$

and restriction

$$\min_{(x,y)\in F} t(x,y,\ell_2,\tau) = T_0 \;.\tag{30}$$

Right there it has been shown that the solution of this problem can be presented as a population of similar problem solutions for individual floor structural assemblies.

The solutions of problems for the floor portions with and without heat-conducting insertions in terms of wall footings located in the central zone, are given in our papers [7,9] .

In the present study the solution is given for the problem of heating source capacity determination in the portion of the central zone with the heat-conducting insertion in terms of a column whose structural diagram is presented in Fig.5. It is evident that the temperature minimum at depth ℓ_2 (critical point) is on the column axis.

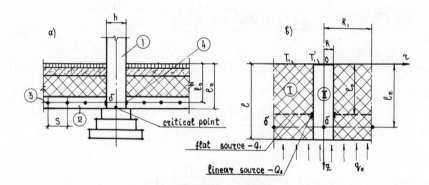

Fig.5. Structural (a) and design (b) diagram of floor heating in the vicinity of a column. 1 – column; 2 – heating plate; 3 – floor heating source; 4 – column heating source.

We assume that the column temperature at floor level is constant and equal to $T_1' = T_1 + \Delta T$. When sufficiently remoted from the column, the horizontal component of the heat flow is zero. The column has a linear heating source enveloping its perimeter. Square section of the column is replaced by the round cross-section and its radius is selected on the basis of its area invariability $R = h/\sqrt{\pi}$. ΔT - possible temperature deviation in the column section (Z=0) from the inside air temperature - T_1 .

Hence, the function of sources is:

$$f(x,y) = f(\tau) = Q_2 \cdot \delta(\tau - R) + q_1 \cdot \chi(\tau - R), \tag{31}$$

where Q_2 - linear capacity of the column heating source, w/m;
q_1^2 - average capacity of the floor heating, w/m².

Thus, the required capacity of the column heating Q_2 and floor heating q_1 is defined by the following equation:

$$\lambda_i \left(\frac{\partial^2 t_i}{\partial \tau^2} + \frac{1}{\tau} \frac{\partial t_i}{\partial \tau} + \frac{\partial^2 t_i}{\partial z^2} \right) + f \cdot \delta(z - l_0) = 0, \quad i = 1, 2, \tag{32}$$

with boundary conditions

$$t_1(\tau, 0) = T_1, \qquad R < \tau < R_1; \tag{33}$$

$$t_2(\tau, 0) = T_1 + \Delta T, \quad 0 < \tau < R; \tag{34}$$

$$-\lambda_i \frac{\partial t_i}{\partial z} \Big|_{z=l} = q_0, \quad i = 1, 2; \tag{35}$$

$$\frac{\partial t_1}{\partial \tau} \Big|_{\tau = R_1} = \frac{\partial t_2}{\partial \tau} \Big|_{\tau = 0} = 0; \tag{36}$$

$$t_1(R, z) = t_2(R, z), \quad \lambda_2 \frac{\partial t_2}{\partial \tau} \Big|_{\tau = R} = \lambda_1 \frac{\partial t_1}{\partial \tau} \Big|_{\tau = R} \tag{37}$$

and restrictions

$$t_2(0, l_2) = t_1(R_1, l_2) = T_0 . \tag{38}$$

The solution has been obtained by making us of the method of integral transforms with a special type nucleus.

$$Q_2 = \frac{\lambda_1(T_0 - T_1 - W_\sigma \cdot \Delta T) + q_1 R W_1 + q_0 R W_0}{W_2} \quad \text{w/m}, \tag{39}$$

$$q_1 \approx \frac{\lambda_1 \cdot (T_0 - T_1)}{l_0} + q_0 \quad \text{w/m}^2, \tag{40}$$

where

$$W_\sigma = 4 K_{\lambda_2} \cdot \int_0^\infty \frac{J_1(\mu)}{U^2(\mu)} \cdot \frac{ch \mu(L - L_2)}{ch \mu L} d\mu,$$

$$W_0 = L_2 - 4(K_{\lambda_2} - 1) \int_0^\infty \frac{J_1(\mu)}{U^2(\mu)} \cdot \frac{sh \mu L_2}{\mu \cdot ch \mu L} d\mu,$$

$$W_1 = -g(L_2) + 4 K_{\lambda_2} \int_0^\infty \frac{J_1(\mu)}{U^2(\mu)} \cdot \frac{G(L_2, L_0)}{ch \mu L} d\mu,$$

$$W_2 = 4 \int_0^\infty \frac{J_0(\mu)}{U^2(\mu)} \cdot \frac{G(L_2, L_0)}{ch \mu L} d\mu,$$

$$U^2(\mu)=[2K_{\lambda_2}+\pi\mu(K_{\lambda_2}-1)\cdot J_0(\mu)N_1(\mu)]^2+\pi^2\mu^2(K_{\lambda_2}-1)^2 J_1^2(\mu)\cdot J_0^2(\mu);$$

$$g(L_2)=\begin{Bmatrix} L_2, & L_2<L_0 \\ L_0, & L_2\geqslant L_0 \end{Bmatrix}; \quad G(L_2,L_0)=\begin{Bmatrix} ch\mu(L-L_0)\cdot sh\mu L_2, & L_2<L_0 \\ ch\mu(L-L_2)\cdot sh\mu L_0, & L_2\geqslant L_0 \end{Bmatrix};$$

$$K_{\lambda_2}=\frac{\lambda_2}{\lambda_1};$$

$$L_0=\frac{\ell_0}{R}, \quad L_2=\frac{\ell_2}{R}, \quad L=\frac{\ell}{R}.$$

Expression (40) is reduced to

$$Q_2=\gamma\left[\frac{\lambda_2}{\ell_0}(T_0-T_1-\Delta T)+q_0\right] \quad \text{w/m}, \tag{41}$$

where γ = coefficient accounting for two-dimensional character of the thermal field in the vicinity of the column, and depending on ℓ_2/R, λ_2/λ_1.

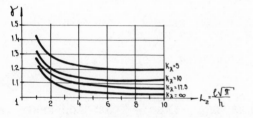

Fig.6. Dependence of parameter γ on $K_{\lambda_2}=\lambda_2/\lambda_1$ and $L_2=\ell_2\sqrt{\pi}/h$.

The results obtained enable the improvement of the methods of heat engineering design and the structural solution of heated floors, thus improving reliability and efficiency of the system of foundation soil protection in buildings with negative temperature of internal atmosphere against freezing and frost heaving.

Conclusions

1. The study of the heat and mass transfer process model in heated floors of buildings with negative temperature of internal media resulted in determining the possibility to consider the process of heat transfer with a fixed dampness field of the floor structure.

2. The width of the building peripheral zone with essentially non-steady thermal field due to external climatic effects, has been determined.

3. The analytical expression has been derived to determine the rate of heat flow between the heating plate and foundation soil both in the central (steady) and peripheral (non-steady) zones. For the peripheral zone the regularities of amplitude changes and the phase of heat flow annual variation have been found.

4. The expressions have been derived to define the required heating capacities for all portions of the floor structure with or without heat-conducting insertions in terms of column or wall footings.

5. The results obtained enable ⁚⁚ the improvement of the methods of design and the structural solution of heated floors in buildings with negative temperature of internal media thus improving reliability of the heating system operation and facilitating proper registration of heat transfer into enclosed spaces from foundation soils.

REFERENCES.

1. Лыков, А.В. Теоретические основы строительной теплофизики. Минск, изд-во АН БССР, 1961.

2. Лыков А.В. Явления переноса в капиллярно - пористых телах. М., ГИТТЛ, 1954.

3. Carslow H.S. Introduction to the mathematical theory of the conduction of heat in solids. New-York, 1945.

4. Порхаев Г.В. Тепловое взаимодействие зданий и сооружений с вечномёрзлыми грунтами. М., изд-во "Наука", 1970.

5. Чудновский А.Ф. Теплообмен в дисперсных средах. М., ГИТТЛ, 1954.

6. Цимерманис Л.Б. Термодинамические и переносные свойства капиллярно - пористых тел. Челябинск, Южно - Уральское книжное издательство, 1971.

7. Гиндоян А.Г., Грушко В.Я., Дуранов Е.Ф. Некоторые вопросы теплотехнического расчёта систем обогрева грунтов оснований зданий холодильников. Доклад на ХIУ Международном конгрессе по холоду. Москва, 1975 (шифр Д.1.16).

8. Лукьянов В.С., Головко М.Д. Расчёт глубины промерзания грунтов. Трансжелдориздат, М., 1957.

9. Гиндоян А.Г., Грушко В.Я., Пак М.А. Определение температурного поля оснований холодильников с системой электрообогрева. "Расчёты конструкций промышленных зданий". Труды ЦНИИпромзданий, вып.28, М., 1972.

ANALYTICAL AND EXPERIMENTAL DETERMINATION OF THE TEMPERATURE DISTRIBUTION IN STRATIFIED HOT WATER STORES

H. J. LEYERS, F. SCHOLZ, AND A. THOLEN

Kernforschungsanlage Jülich GmbH
Jülich, Federal Republic of Germany

ABSTRACT

Stratified hot water stores offer the advantage that most of their stored heat can be extracted close to the loading temperature. In stores of this kind during load changing processes a transition layer is built up which acts as a transient insulation barrier between hot and cold water. The development of this transition layer can be calculated assuming heat transfer by thermal conductivity only. But in reality the heat transfer through this transition layer will be also influenced by flow conditions of the load changing devices and natural convection. Therefore experiments have been performed to investigate the real behaviour of such stores. An existing well insulated square container (4.4 m x 4.7 m x 2 m high) was used. Comparisons between analytical and experimental results - so far available - indicate a good consistency only if a higher than the molecular thermal conductivity of water is assumed in the calculations. This effect seems to be due to increased heat transfer by natural convection.

INTRODUCTION

As a possibility to save energy it has been proposed by Schöll /1/ to construct unpressurized large volume water reservoirs in the form of lakes with floating surface insulation and waterproof sealing against the ground. They could absorb the heat being produced in summer time by a continuously operating power plant (heat with low exergy) and make it available for peak heat supply to buildings during winter time. As it is advantageous for thermodynamic reasons to withdraw the stored heat as near as possible to the heat input temperature, mixing of hot and cold water or other forms of heat transfer in the storage container should be restricted to a minimum. These requirements lead to build up either several separated water containers or to design a stratified container which could be a lake. The second solution is regarded as more economical. In a stratified water reservoir the density decrease due to temperature is utilized. The hot water is fed to the uppermost plane of the store and cold water is taken at the same time from the lowest plane.

Between hot and cold water a transition layer builds up increasing gradually in time and thus reducing the unwanted heat transfer from hot to cold water.

On the other hand, a thick transition layer is not desired because the volume of this zone, i.e. the heat contained in it, due to its low temperature can only be utilized to a certain degree. Therefore, the transition layer reduces the volume utilization of the water reservoir. This proves particularly disadvantageous for water reservoirs with a large horizontal surface in relation to height compared with present stores being designed as slender cylinders.

The growing of this layer considered in one dimension and taken account of limited reservoir height and thermal losses at top and bottom of the reservoir can be calculated during changes of loading process and for whole operating cycles. Various calculations have been carried out. Leyers /2/ has in his calculations - similar as Bloß and Grigull /3/ - reduced the cooling of the upper layers, which arises from thermal losses at the top of the store, by energy exchange with lower layers to an extent that negative temperature gradients are avoided. However, calculations taking account of free convection due to thermal loss at the sides can be performed at the present time and realistically only under great efforts /4/. Even more difficult will be a theoretical answer to the problem how the charging process devices should be constructed and arranged for to secure that at the beginning of the feeding process the hot water is distributed equally over the total surface at minimum mixing with the cold water in the container. According to existent studies, former experiences with spherical stores reveal that it is to be doubted if such a "layer formation" in containers with large horizontal surfaces is possible at all and can be kept stable later-on.

DESCRIPTION OF THE EXPERIMENTS

As a contribution to claryfy some of the mentioned problems experiments are performed with a square container (dimensions: 4.4 m x 4.7 m x 2 m height) constructed by Rheinisch Westfälische Elektrizitätswerke.

The container (Fig. 1) has been built like a welded basement oil tank mantled with a polyurethane foam layer (average thickness 50 cm). The basement walls, floor, and ceiling are in close contact with the foam layer. Therefore, visual inspections inside of the container or direct measurements at the outer insulating layer are impossible in our case.

By means of a loading loop constructed for the container the water can be drawn off at the bottom, heated up to a determined temperature, and fed back at the top.

The following load changing devices were installed:
-    One disk-shaped distribution cup and one exhaust cup
     in the center of the horizontal plane, 175 cm or 10 cm
     resp. above the bottom.
-    One similarly constructed quarter cup each in a corner
     of the container at same heights.
-    Three tube sections each at the face of the container with
     one horizontal bore are positioned also at the stated heights.

One of the three tube sections is located at the center of the front surface, the two others are positioned one third of container width to the left and to the right. The central bore and the two others have separated feed pipes. In all devices the discharge velocity is directed mainly horizontally.

As experiments were made mainly for the purpose of studying the loading processes, in the upper half of the container in 4 planes and near the bottom

in 2 further planes up to 20 thermocouples in each plane were installed in order
to measure the temperature distribution in the water. At some spots in vertical lines additional thermocouples were installed.

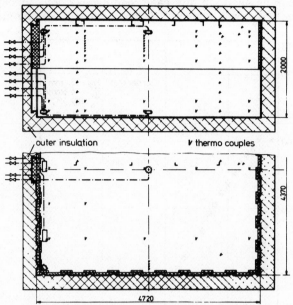

Figure 1
RWE-container with positions of thermocouples and various load-changing
devices

In order to minimize the heat conduction through the vertical steel walls,
which influence the temperature distribution in the water, 5 cm thick insulating plates of closed cell polystyrol foam were fastened to the inside of the
walls.

Prior to the loading tests the water in the store is brought to an equal
temperature of approx. 30 °C and a certain water level is set. Afterwards,
a constant volume flow of water is withdrawn from bottom of the container,
heated up by a control device to approx. 70 °C and fed back to the top by one
of the loading devices. After chosen time intervals the measured values (approx.
150 temperature readings and the water flow) are automatically recorded.

Because of time limitations the loading tests are stopped after a certain
time period as it can be assumed that beyond a certain thickness of the hot
water zone the mentioned parameters have no more further important influence
on the loading process.

According to the long load changing periods of seasonal heat stores the
vertical migration velocity is varied between approx. 0.5 cm and 3 cm per
hour.

In some cases, after the charging is stopped, discharging is not startet
at once but the equalization of temperature is observed for adequate time
periods. On these occasions the assumptions for the heat losses of the system
could be checked, too.

EXPERIMENTAL RESULTS

In Fig. 2 the measured vertical temperature distribution at the end of the charging process can be seen. In this case loading was done through the corner cup. The water level (H$_w$) was approx. 15 cm above the feeding cup, the mean

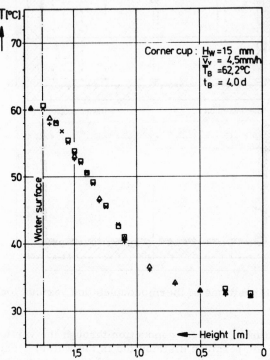

Figure 2
Vertical temperature profile in the water

vertical migration velocity ($\overline{V}_V$) approx. 0.45 cm per hour, and loading temperature (T$_B$) approx. 66 °C. The loading time (t$_B$) lasted about 4 days. A strong rise of temperature in the transition zone up to the water surface can be observed. Even at the surface the loading temperature is not yet reached. As the different symbols apply to different positions in horizontal planes it can be concluded already from this figure that the temperature in each horizontal plane is almost equal.

This is impressively confirmed by Fig. 3. It shows the horizontal temperature distribution over the length of the container with different symbols for breadth positions - right, middle and left - in different heights. As can be recognized, deviations from the mean value are very small in each height.

At start of the loading process the temperature differences in the horizontal planes are higher, as can be seen in Fig. 4 for loading times of approx. 0.5 and 1.1 days.

In front, i.e. in the left halves of the figures, the mean values for the upper zones are higher and the scattering is greater than at the back. The

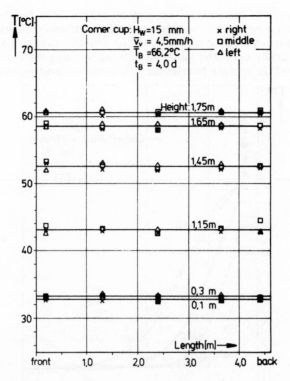

Figure 3
Horizontal temperature profiles

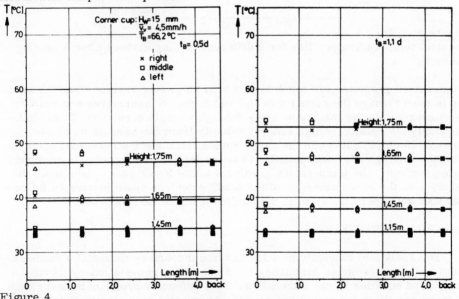

Figure 4
Horizontal temperature profiles after small loading times t_B

asymmetric position of the corner cup may be responsible for this result.

Fig. 5 gives a comparison between the effects of two different loading devices for 4 different loading times. The continuous curves apply to the loading through the corner cup, and the broken lines through the central bore. It can be seen - and this applies particularly to smaller loading times - that the

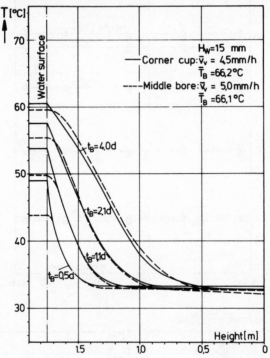

Figure 5
Vertical temperature profiles for 2 different loading devices after 4 loading times t_B

temperature gradients are much higher near the water surface when the loading is done through the corner cup and that higher temperatures are reached in comparison to the loading process through a single bore at the front wall. Obviously the relatively high exhaust velocity from the bore (in this case approx. 55 cm/s) causes in the upper zone a circulation stream that leads to a higher mixing and, consequently, to lower temperature gradients near the water surface. The temperature gradients in the lower zones are almost the same, i.e. the characteristics of the loading devices seem to have an influence only upon the upper zones and at the beginning of the loading process.

ANALYTICAL INVESTIGATIONS

The analytical calculations were carried through by means of a finite element computer program permitting solutions of the non-stationary thermal conduction equation in two dimensions. This computer program is described in more details in the mentioned report /2/ and was only slightly extended for our purposes here by incorporating additional terms for heat sinks ($\dot{q}_V$) which compensate for the lateral heat losses by the outer container insulation, i.e. the following applies:

$$\dot{q}_V \cdot F = \dot{q}_F \cdot U = -k \frac{T - T_o}{d} \cdot U \qquad (1)$$

I.e.:

$\dot{q}_V$	Heat sink	(W/cm^3)
$\dot{q}_F$	Heat-flux density through the outer lateral insulation of the container	(W/cm^2)
F	Cross-section of the container	(cm^2)
U	Periphery of the container	(cm)
k	Coefficient of thermal conductivity of outer insulation	(W/cm K)
d	Thickness of outer insulation	(cm)
T	Temperature at the inside of insulation (water temperature)	(^oC)
T_o	Temperature at the outside of insulation	(^oC)

The lateral thermal losses of the container above the water surface were added to the upward thermal losses, which means that calculations were made with an effective thermal loss of the container cover of:

$$\dot{q}_{F \, cover}^{eff} = \dot{q}_{F \, cover} \cdot (1 + \frac{U \cdot s}{F}) \qquad (2)$$

(s = Distance between water surface and container cover).

It has been furthermore assumed that no horizontal temperature gradient exists.

A measure for the applicability of equation (1) and (2) is the energy-balance factor (f_E) which is defined as follows:

$$f_E(t) = 1 + \Delta f_E(t) = \frac{\int_{water \, height} [T(x,o) - T(x,t)^{theory}] \, dx}{\int_{water \, height} [T(x,o) - T(x,t)^{experiment}] dx}$$

(t = time, $\Delta f_E(t)$ = Deviation from the ideal case).

COMPARISON BETWEEN EXPERIMENT AND THEORY

A comparison between calculation and experiment is given in Fig. 6 for an equalization process. The upper curve on the left side shows the measured vertical temperature readings at the beginning of an equalization process (t_A = 0 d). Based on time intervals of approx. 3.9, 6.8, 13.5 and 17.9 days the different symbols show the measured temperatures, and the continuous curves show the calculated temperatures.

In Fig. 6 the thermal conductivity of water was used, i. e. heat conduction through vertical walls and effects of natural convection were not considered. It has to be recognized that the consistency between theory and experiment is poor, which had to be expected.

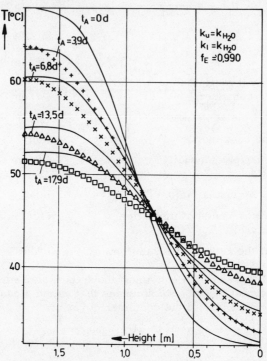

Figure 6

Temperature equalization in the container after 4 times t_A; comparison between calculations (continuous curves) and experiments (different symbols)

If the thermal conductivity in the steel walls had been considered by averaging their conductivity equally over the total cross-section of the container, a mean thermal conductivity of 1.42 times to that of the water would have been obtained. But such a great influence by the walls cannot be expected. Therefore, it was assumed - see Fig. 7, left - that in the upper half where the walls have a thin insulation against the water, a conductivity factor of 1.15 and in the lower half (no inner insulation) a conductivity factor of 1.3 should be used. In this case the consistency between theory and experiment is much better.

Still the temperature gradients of the experimental curves are lower which must be attributed to a higher effective thermal conductivity. The highest consistency was found when the thermal conductivity at the top was increased to the factor 1.42 and at the bottom to the factor 1.52 - see Fig. 7, right. Now, particularly in the lower part of the container, the measured values agree better with the calculated values. Deviations at the top can be ascribed to the simplified assumption of the mentioned energy equalization process which avoids negative temperature gradients. The temperature gradients in the transition zone agree now rather accurately.

By this high effective thermal conductivity the additional heat transfer by natural convection is taken account of.

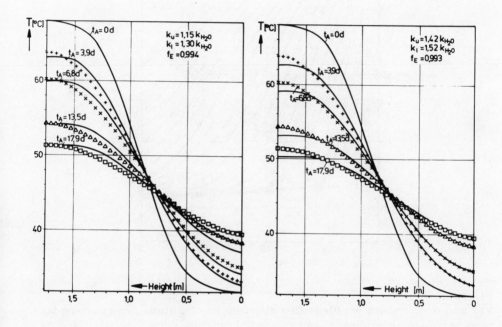

Figure 7
Temperature equalization in the container as Fig. 6 but using different thermal conductivities for the water in the upper half k_u and in the lower half k_l

In Fig. 8 a charging process is shown based on the same assumption regarding thermal conductivity as in Fig. 7 right. The experimental curves shown in broken lines harmonize very well with the calculated curves of comparable loading times. The slope of the experimental curves - especially at the beginning - is less than the slope of the calculated curves which can be ascribed to still higher effective conductivity probably due to flow processes. The mean vertical velocity of flow in this case was approx. 0.89 cm per hour. The temperature drop of the feeding flow between outside supply and central loading cup in the container was estimated to 1 K, i.e. a loading temperature of 67.7 instead of 68.7 °C was assumed. In the case of longer loading times the consistency between theory and experiment is very good.

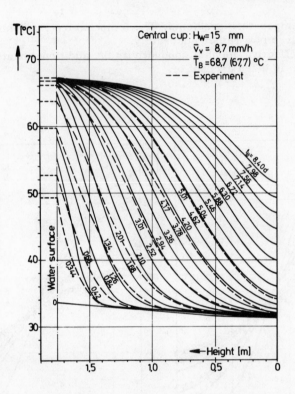

Figure 8
Vertical temperature profiles after different loading times; comparison between calculated (continuous) and measured (broken) curves

In summary it can be concluded from the experiments carried out so far,
that calculations of whole operating cycles with change-overs between load-
ings, rest periods, and dischargings can be made with satisfying accuracy
using the cited computer program.

ACKNOWLEDGEMENTS

The authors thank the Rheinisch Westfälische Elektrizitätswerke (RWE),
in particular Messrs. Dr. Dietrich and Kimpenhaus for making possible
these experiments and for their cooperation in connection with these investigations. We would like to extend our thanks also to Prof. Steimle, University Essen, and his assistants, Messrs. Gräff, Lammert and Nguyem for
their cooperation at the planning and performance of the experiments. We
would also like to mention representatively from KFA Messrs. Bauer,
Birmanns, Gräbner, Sauer, Schuster and Dohmen since without their con-
scientious cooperation the experiments would not have been successful.

REFERENCES

1. Schöll, G. 1974. Warmwasser-Großwärmespeicher. VDI report No. 223, 33.-38.

2. Leyers, H.J. 1976. Über das Temperaturverhalten von Jahreswärme-speicher-Seen. Jül-Report-1339, 1-38.

3. Bloß, S., Grigull, U. 1975. Temperaturverteilung in Seen. Wasser- und Abwasserforschung. 121-127.

4. Grigull, U. et al. 1976. Theoretische und experimentelle Untersuchun-gen über den Einfluß der Konvektion in Jahreswärmespeichern. Report of Lehrstuhl A for Thermodynamik, TU München, 1-28.

SOLAR RADIATION AND ITS USE FOR COOLING AND HEATING

THERMAL PROCESSES
OF SOLAR HOUSES

KEN-ICHI KIMURA

Department of Architecture
Waseda University
Shinjuku, Tokyo 160, Japan

ABSTRACT

An overview of the various thermal processes of solar house is described on
the basis of the author's experience in solar house design and building heat
transfer studies. The relationship between solar radiation transmitted through
windows and room air temperature variation is discussed with regard to the heat
storage effect of building structure. Examples are presented on the performance
evaluation of passive solar system. Practical problems on the system of space
heating and hot water supply integrated each other are discussed including the
problems of direct heating system versus heat pump assisted heating system.

INTRODUCTION

Construction of solar houses is expected to grow quite rapidly almost everywhere in the world as many people have become to recognize that solar heating of
building spaces and domestic hot water supplyis regarded very effective in saving
fossil fuels to conserve energy for our future generations. The reason for the
high effectiveness of solar heating is that the solar heat can be utilized at
lower temperature than for other means of solar applications. Though the total
amount of solar energy falling on the earth is enormous, the net amount of solar
energy permissible to be utilized for various modes of human activities of our
modern society is rather limited, taking account of the situation of coexistence
 with plants, animals and other living creatures on this planet. It may be considered reasonable to think about making use of the surface area already disrupted
by human beings such as roofs and walls of existing buildings and secondly those
areas to be inevitably disrupted such as those surfaces of buildings to be built
in the future, preserving meadows, fields, forests and water surfaces essential
for our daily lives as well as for the maintenance of the global energy balance.
Heat and mass transfer occurs in the thermal processes of solar houses from
collection of solar energy to the end use of energy as in the form of space
heating and hot water supply. It is often said that there are two basic types
of solar houses: passive solar house and active solar house. Passive solar house
is a type of solar house where heating or cooling the spaces can be achieved
imperfectly but satisfactorily without relying on mechanical means, whereas in
the active solar houses different mechanical components are combined to form a
solar heating, cooling and hot water supply system to be installed in the house.
Thermal characteristics of building structure very much effect on the performance of the solar heating and coolingboth in the passive and active solar
houses, but solar hot water can only be supplied through mechanical means. It
would be interesting, hterefore, to deal with this important problems with the

aid of existing theory and technology of heat and mass transfer in collabolation among architects, scientists and engineers in various fields.

BASIC SYSTEM OF SOLAR HOUSE

It is well known that collector and storage of solar heat are necessary to make effective use of solar energy for energy saving in buildings, regardless of the type of the system, active or passive. Fig. 1 shows the schematic diagram of energy flow pattern of a solar house. The heat storage is required to store the collected solar energy for a certain period of time because the period of collection and the period of energy demand are usually not coincided. In order to maintain comfort it would be necessary to provide auxiliary energy source to supplement as required when collection of solar energy could not meet the energy demand.

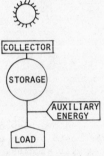

Fig. 1 Energy Flow of Solar House

Strictly speaking, there is no clear distinction between passive and active systems. Someone says that the house without moving parts may be defind as passive solar house, but this definition would fail to recognize movable panels as the components of the passive system. Since the natural environmental conditions are different between night and day and between summer and winter, no stationary device could be designed as optimal from the viewpoint of efficient use of solar energy.

GLASS WINDOW AS COLLECTOR

In the northern hemisphere the amount of solar radiation transmitted through glass windows on the south side is enormous on clear winter days and very much contributes to heating of the interior spaces. People think it a kind of natural blessing, but the south window itself must be recognized as one of the mostexcellent device of collecting solar energy as well as the means of daylighting and enjoyment of views.

From the thermal point of view, taking the south window as collector, one can easily reach the following conclusion the net heat gain from the window is transmitted solar radiation minus heat loss across the glass pane in proportion to the temperature difference between inside and outside as expressed by the following:

$$q(t) = \tau I(t) - K_G [\, \theta_r(t) - \theta_a(t)\,] \tag{1}$$

where $q(t)$: net heat gain from glass window unit area at time t [W/m^2]

τ : transmissivity of glass pane as a function of incident angle of solar radiation

$I(t)$: solar radiation incident on the window surface at time t [W/m^2]

K_G : overall heat transfer coefficient of glass window [W/m^{2}K]

$\theta_r(t)$: room air temperature at time t [^{0}C]

$\theta_a(t)$: outside air temperature at time t [^{0}C]

In a little more strict sense, the following equation must supercede eq.(1):

$$q(t)A_G = \tau I(t)A_{GC} - K_G A_G [\, \theta_r(t) - \theta_a(t)]$$ (2)

where A_G : glass window are including framing [m^2]

A_{GC} : unshaded glass area [m^2]

As $q(t)A_G$ Is instantaneous heat gain from windows, it often happens that
no heating is required during the sunny daytime in winter in the milder climatic
regions when solar radiation through windows exceeds the heat loss from opaque
walls and roofs. It is possible, however, that the daily cumulative value of
$q(t)$ becomes minus, because the first term of eq.(2) is zero during the night
and the second term may be even larger than the daytime. In this case the glass
window cannot be considered as solar collector. It is wise , therefore,to
install insulation panels inside or outside of the window to minimize heat loss
during the night,unless the owner fells that the operation of moving the panels
twice a day is troublesome, as shown in Fig. 2.

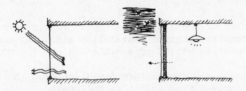

Fig. 2 Window Insulated
       during Night

Next question is when the windows should be covered by the insulation. If
the artificial lighting must be lit as soon as the window s are covered, use of
electrical energy for lighting must lead reconsideration of the time of covering
the windows. It then follows the windows may be kept open a little longer depending on the balance of heating and lighting requirements, disregarding the
fact that one would like to see the sunset.

Furthermore we must take account of the effect of room air temperature on
the second term of eq.(2). As a result of solar radiation transmitted through
glass windows, the room air temperature may get higher than the required level
of thermal comfort, when the excess amount of heat, which is a part of collected
solar energy,must be lost to the outside across the glass pane. Use of multiple
glazing reduces such losses but cannot make it zero.

ROOM AIR TEMPERATURE VARIATION DUE TO SOLAR RADIATION

To calculate the room air temperature variation resulted from the solar
radiation transmitted through glass windows, the thermal storage effect of building structure must be taken into consideration. It is convenient to use the
weighting factor technique by converting all of the time-dependent continuous
variables into the time series variables. This approach is currently widely used
in the computerized calculation of heating and cooling loads of a space as
represented be the procedure of the American Society of Heating, Refrigerating
and Air Conditioning Engineers [1] and that of the Society of Heating, Air Conditioning and Sanitary Engineers of Japan[2] and possibly the procedures of other
societies.

The basic equation of the room air temperature variation is expressed by the
following form of convolution, taking all of the heat flow per unit floor area
of the space concerned:

$$H_G(n) = H_L(n) + \sum_{j=0}^{\infty} W_z(j) \, T_r(n-j) + K_v(n) \, T_r(n) \tag{3}$$

where $H_G(n)$: heat gain from glass windows and plus internal heat generation by lighting and home appliances per unit floor area at time n $[W/m^2]$. The sampling time interval is usually taken as one hour as standard. Time n is an integer representing every hour; $t = n\Delta t$.

$H_L(n)$: heat loss from inside to outside through walls and roofs and by ventilation based on the constant reference room air temperature per unit floor area at time n $[W/m^2]$.

$T_r(n)$: room air temperature deviation from the reference room air temperature at time n [deg]. The reference room air temperature is usually taken as the comfortable temperature for occupants.

$W_z(j)$:weighting factors relationg the room air temperature deviation to the heat loss per unit floor area $[W/m^2 deg]$.

$K_v(n)$: variable coefficient of heat loss by ventilation depending on the variable rate of ventilation per unit floor area at time n $[W/m^2 deg]$, namely,

$$K_v(n) = C_p \gamma \, V(n) \tag{4}$$

where $V(n)$: rate of ventilation per unit floor area at time n $[m^3/m^2 s]$.
C_p : specific heat of ventilation air [J/kg deg]
γ : specific weight of ventilation air $[kg/m^2]$.

Then the room air temperature deviation from the reference temperature at time n can be derived from eq.(3) as follows:

$$T_r(n) = \frac{-1}{W_z(0) + K_v(n)} [\, H_L(n) - H_G(n) + \sum_{j=1}^{\infty} W_z(j) \, T_r(n-j)\,] \tag{5}$$

To simlify the calculation procedure, the weighting factors $W_z(j)$ must be approximated so as to be represented by the first three terms and the common ratio,viz.,

$$W = W_z(0) + W_z(1) + W_z(2) + cW_z(2) + c^2W_z(2) + \ldots$$

$$= W_z(0) + W_z(1) + \frac{1}{1-c} W_z(2) \tag{6}$$

where W : sum of the product of the overall heat transfer coefficient and the area of the room enclosure components per unit floor area $[W/m^2 deg]$.

$W_z(0)$, $W_z(1)$, $W_z(2)$: the first three terms of $W_z(j)$ $[W/m^2 deg]$.

c : common ratio of weighting factors $W_z(j)$ as defined by eq.(6).

The second term of the right hand side of eq.(3) represents the increase in heat loss due to the room air temperature deviation from the reference temperature on which the calculation of $H_G(n)$ and $H_L(n)$ are based. Third term represents the increase in the heat loss by ventilation due to the room air temperature deviation at time n.

Then the eq.(5) can be rewritten by a simpler form as in the following:

$$T_r(n) = \frac{1}{W_z(0) + K_v(n)} \{[H_G(n) - H_L(n)] - c[H_G(n-1) - H_L(n-1)]$$

$$+ [W_z(1) - cW_z(0)]T_r(n-1) + [W_z(2) - cW_z(1)]T_r(n-2)\} \tag{7}$$

The author tried to give appropriate values of $W_z(0)$, $W_z(1)$, $W_z(2)$ and cfor different types of the room for easy, approximate calculations and is proposing a temporary procedure as follows. Fig. 3 shows the chart to obtain$W_z(0)$ per unit floor area basis as a function of the ratio of the total interior surface area $S_s[m^2]$ to the floor area $S_F[m^2]$ of the space concerned. It may be determined that

$$W_z(1) = -7 \text{ W/m}^2\text{deg}$$

$$c = 0.9$$

for concrete structure. Calculating the values of $W_z(2)$ can be obtained from eq.(6).

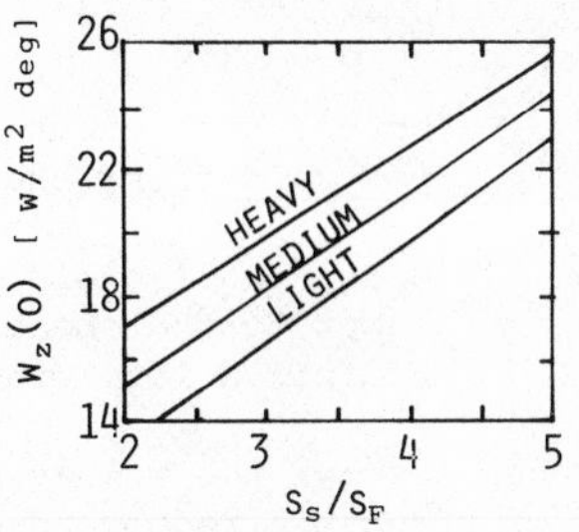

For example, when a small house has the following exterior enclosure:

Fig. 3 Chart to obtain $W_z(0)$ [3]

Floor area: $S_F = 8 \times 12 = 96 \text{ m}^2$

Room height: 2.5 m

Total interior surface area: $S_s = 2.5 \times (8 + 12) \times 2 + 96 \times 2 = 292 \text{ m}^2$

Total glass window area: $A_G = 20 \text{ m}^2$

Total exterior wall area: $A_W = 80 \text{ m}^2$

Total roof area: $A_R = 96 \text{ m}^2$

Overall heat transfer coefficient of glass window: $K_G = 2.5 \text{ W/m}^2\text{deg}$

Overall heat transfer coefficient of exterior wall: $K_W = 0.8 \text{ W/m}^2\text{deg}$

Overall heat transfer coefficient of roof: $K_R = 0.6 \text{ W/m}^2\text{deg}$

the values of the weighting factors $W_z(j)$ and c can be obtained by the following procedure:

$S_S/S_F = 292/96 = 3.04$

$W = (2.5 \times 20 + 0.8 \times 80 + 0.6 \times 96)/96 = 171.6/96 = 1.79$

$W_z(0) = 16.5$ from Fig. 3 for light weight construction

$W_z(1) = -7$ and $c = 0.9$ by assumption as in the above.

Substituting the above values into eq.(6),

$$1.79 = 16.5 - 7 + \frac{1}{1-0.9} \times W_z(2) \text{ and}$$

$W_z(2) = -0.77$ can be obtained.

It can be concluded that as long as $T_r(n)$ is positive the glass window is acting as solar collector from the theory described in the above.

HEAT STORAGE OF SOLAR RADIATION IN THE BUILDING STRUCTURE

Solar radiation transmitted through glass windows is partially stored in the building structure and comes back into the occupied space later on by convection and radiation from the floor and wall surfaces directly irradiated by solar radiation. In this sense $H_G(n)$ in the previous equations must be substituted by the following convolution form:

$$H_{GG}(n) = \sum_{j=0}^{\infty} W_G(j) \, H_G(n-j) \qquad (8)$$

where $H_{GG}(n)$:actual heat input to the space due to solar radiation from windows
per unit floor area at time n $[W/m^2]$

$W_G(j)$:weighting factors relating the solar heat gain to the actual heat input to the space at reference room air temperature due to solar radiation from windows [dimentionless]

Table 1 shows the values of $W_G(0)$, $W_G(1)$ and common ratio c_G for different weight of structure recommended for the purpose of approximate calculation[3].

Table 1   Weighting Factors $W_G(j)$ recommended by SHASE[3]

	STRUCTURE	$W_G(0)$	$W_G(1)$	c_G
WITHOUT VENETIAN BLINDS	HEAVY	0.3774	0.0551	0.9086
	MEDIUM	0.3990	0.0629	0.8593
	LIGHT	0.3728	0.0991	0.8420
WITH VENETIAN BLINDS	HEAVY	0.4828	0.0473	0.9086
	MEDIUM	o.4849	0.0539	0.8953
	LIGHT	0.4765	0.0827	0.8420

Using the values in Table 1, eq.(8) can be rewritten as follows :

$$H_{GG}(n) = c_G H_{GG}(n-1) + W_G(0) \, H_G(n) + [W_G(1) - c_G W_G(0)] \, H_G(n-1) \quad (9)$$

It is certain that the thermal capacity of the building structure helps to spread the peak heat gain by solar radiation from glass windows over a longer period of time, thus making the heat loss due to the temperature difference between inside and outside reduced. In most cases, however, the room air temperature tends to get excessively higher than required for many hours during the daytime in winter, causing an excessive heat loss through windows accordingly. It means that the thermal capacity of the floor slab and principal building structures is less than desirable.

Taking this effect into consideration, several investigators have attempted to install a very massive body intercepting and absorbing the excess amount of solar radiation to store it for a period of several hours and to use it for heating later in the evening.

For example, M.I.T. Solar House II [4] had a water tank inside of the glass window with a drape hung between the glass panes and the tank during the night as shown in Fig. 4(a). Odeillo Solar House [5] as shown in Fig. 4(b) has a 60 cm thick wall inside of the vertical glass pane with a provision of holes in the upper and lower parts for natural circulation of warm air. The stored heat is naturally dissipated to the interior space toward the evening without any mechanical means. It has been proved, however, that this arrangement would lose so much of the heat from the outer surface of the wall to the outside from the sunset to the nighttime. Another disadvantage of this system may be that a larger portion of the south facade is occupied by the opaque wall, though this could be overcome by a proper design of a house. In order to avoid these two drawbacks, a solar house proposed in Adelaide, Australia [6] provides a storage to receive directly the solar radiation transmitted through glass windows in winter as shown in Fig. 4(c). Theoretically this will make an excellent thermal effect, but practically there would be such problems that the storage might not be able to absorb the solar radiation of different incident angles or furniture might partially intercept it.

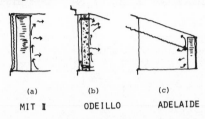

Fig. 4 Typical Passive
Solar Houses

(a)   (b)   (c)

MIT II   ODEILLO   ADELAIDE

All of these solar houses might be called as passive solar houses, as only natural means are employed to harness the solar energy; the water tank of M.I.T. Solar House II, the thick concrete wall of Odeillo Solar House and the interior tank of the proposed Adelaide House are all act as collector, storage and room heater by themselves. It is important to note, however, that provision of thermal insulation between the window and the storage is desirable for efficient use of solar energy.

The Pheonix Solar House [7] designed by Hay and Yellott is known as a passive solar house and it has a water pond on the roof. The thermal insulation panels are made to cover the roof pond during the winter night and summer daytime and be removed during the winter daytime and summer night as shown in Fig. 5. The idea was developed later by Hay and realized as Atascadero Solar House [8],where the moving operation of insulation panels are made by motor drive; someone says that this is an active system. Neverthe-less it is reported that the natural air conditioning has been successful.

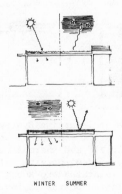

WINTER   SUMMER

Fig. 5 Pheonix Solar House

THERMAL EVALUATION OF PASSIVE SOLAR HOUSE

It is quite difficult to assess the effectiveness of passive solar houses based on a common criterion, because the thermal systems are all different and in many cases non-linear. The finite diffence may be a straight forward approach to the solution of the problem. The thermal system involves the components among which heat conduction, radiation and convection take place. Thermal diffusion

and heat transfer by natural convection also take place in the water tank and it
brings complexity if rigorous solution is attempted.

Making the thermal network for the system consisting of all of the building
components among which heat transfer and heat accumulation occur, a set of simultaneous equations can be established. Fig. 6 shows an example of the thermal
network of a passive system as illustrated in Fig. 7 [9]. This might better be
called a semi-passive system.

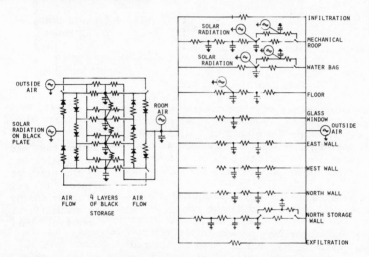

Fig. 6 Thermal Network of a Semi-Passive Solar House

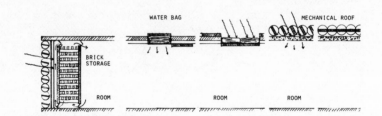

Fig. 7 A Semi-Passive System Proposed for Example Analysis

Here the brick storage is treated as four masses each of which has thermal
capacity in itself and thermal resistance with adjacent masses. On the roof there
are two kinds of semi-passive solar systems: the water bags with movable insulation as used in Atascadero House and the mechanical roof consisting of an array of
pivoted insulation boards to invite solar energy between them to be directly
stored in the concrete roof slab. There are ordinary glass windows, too. The
results of a simulation under the actual weather conditions of Tokyo for several
consecutive days are depicted in Fig. 8, where discharge of stored heat into the
occupied space is started at 2 p.m. It is found that the room air temperature
can be kept considerably higher (8-10 degrees) than the outside air temperature
without auxiliary energy. Alternative disigns can be tested by modifying the
thermal network.

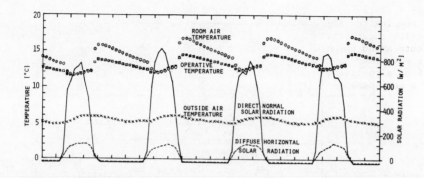

Fig. 8   Resuls of Simulation with the Semi-Passive System

Another approach may be attempted by experiments. The author have made a
special type of a curtain wall acting as collector, storage and heater as illust-
ated in Fig. 9. Solar radiation through glass is absorbed by the black painted
glass tubes in which Glauber's salt ($Na_2SO_4.10H_2O$) is contained. The crysterized Glauber's salt gradually changes to liquid as it absorbes solar energy while
the temperature stays the same around 32°C and the temperature gradually rise
after the entire substance is dissolved until equiliblium is reached. During
this period of sunshine hours the
insulation panel is placed inside of
the curtain wall to store as much as
solar radiation as possible. Before
sunset the insulation panel should
be placed between the panel and the
glass pane so as to prevent the
stored heat from being lost to the
ambient air. In the evening heat is
dissipated into the room space from
the warm tubes and thus solar heating
is made and the Glauber's salt then
changes back to crysterization.
Thus the solar heating is achieved
and the cycle resumes. The use of
heat of fusion type of salt such as
Glauber's salt very much helps to
reduce the volume of heat storage,
which is very important and nice
for architectural design of solar
houses.

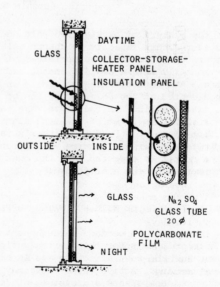

Fig. 10 shows an example of the
experimental results [10] from which
the twice phase changes as described
above can be recognized. It was not
successful, however, in the sense
that the salt was separated after
a number of cycles. It was seen that
the dissolved salt was not entirely
crysterized and the temperature went
down below the phase change temperature

Fig. 9 Collector-Storage-Heater
          Integrated Curtain Wall

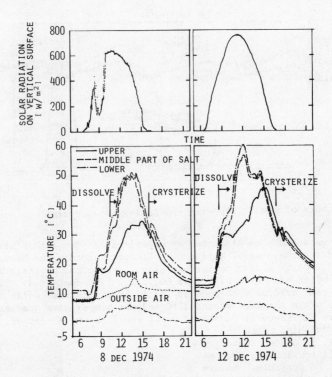

Fig. 10 Experimental Results with Collector-Storage-Heater
Integrated Curtain Wall

with the mixture of crysterized salt, dehydrated salt and water. This consequent
reduced the heat capacity to absorb solar radiation on the following day. It has
been recommended that the container of the heat of fusion type of salt be placed
horizontally with a small diameter of tubing so as to increase the heat transfer
surface and to avoid non-uniformity in temperature within the container, as being
experimented at the Solar House One of the University of Delaware, U.S.A.[11]
The test described in the above were made along with this recommendation, but it
seems that even 20 mm of the tube diameter still is not small enough to achieve
perfect dissolution and the Gluber's salt eventually may remain in the state of
separation. Various studies are currently being conducted by a number of scientists and it is expected that the most appropriate substance and method will be
recommended in the near future.

ACTIVE SOLAR HOUSE WITH HOT WATER SUPPLY SYSTEM

Hot water is needed throughout the year and heating the space is necessary
only on cold days for the limited period of the year in many countries. This
means that the cost effectiveness is higher in solar hot water supply than solar
space heating. Active solar house may be defined as the house equipped with
mechanical components and systems of solar energy collection and storage together
with auxiliary heating system and in the sense described in the above the system
is incorporated with solar hot water supply system. Solar cooling is not referr

to in this article and the discussion is concentrated to the solar space heating
and hot water supply.

There are many different systems of solar heating and hot water supply
proposed and actually built. None of the active systems built so far seem economically feasible, if the future oil price is taken into account, in terms of the
pay back period of the first cost of solar equipments versus the expected annual
energy savings within the expected life of solar equipments. The energy conserving design of buildings would make the pay back period of solar equipments
even longer. It should be stressed here that what is more important is to use
conventional fuels as little as possible than to live on economics in order to
maintain our future generations in the clean and safe environment. In this sense
it is worth while discussing the performance of economically unjustifiable solar
heating system as it does save energy when it is installed.

There are two basic kinds of active solar heated house with regard to the
combination of heating the space and hot water: separate system and integrated
system. The separate system is easy to design, to install and to operate, as
everything can be handled separately, but it is considered wasteful that the
collectors for space heating cannot be used for hot water supply during the non-
heating season when auxiliary heat may be called for on cloudy days. The integrated system, on the other hand, can be designed so that maximum use of solar
energy can be achieved for space heating and hot water supply, but integration of
both systems requires heat exchanger and related control devices which tends to
bring complication of the system and to be resulted in unexpectedly higher cost
than initially predicted.

The space heating system can be classified into two from the viewpoint of
thermal processes: direct heating system and heat pump system. The direct heating
sytem is simple and has been used in many solar houses such as M.I.T. Solar House
IV [12] and Löf House in Denver [13]. It is necessary for this system to use the
heating media at as low temperature as possible in avoiding the use of auxiliary
energy. Panel heating is particularly suitable for this purpose as the liquid
temperature can be as low as 35 - 40°C.

Fig. 11 shows the diagram of space heating system at the KEP Solar House of
Japan Housing Corporation, where auxiliary heater is installed to heat only one
of the three water tanks connected in series and the floor panel heating system

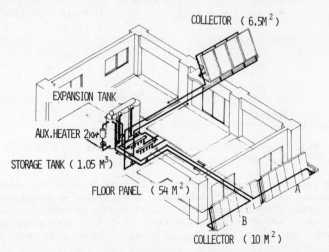

Fig. 11   Direct Solar Heating System - KEP Solar Experimental House

is used. Fig. 12 shows the results of the experiment at the KEP Solar House.
It was foud that about 50 - 60 % of the total heating energy during this period
of one week in February 1976 was shared by solar energy with a collecor of 16.5m^2
for floor area of 89m^2 under no occupancy conditions. The operation time of the
circulating pump turned out much longer than expected because of the heat deliver
from the storage to the heating panels at such low temperature.[14]

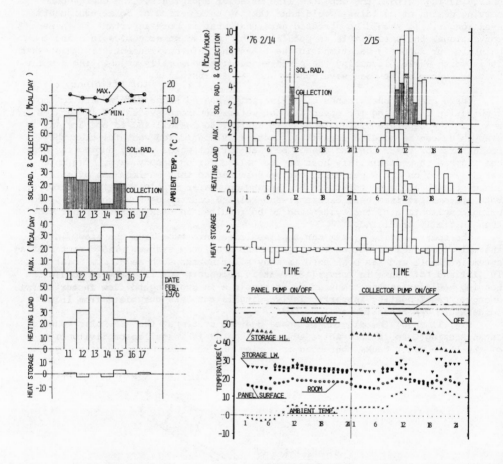

(a) Daily Variation   (b) Hourly Variation of Temperature
    of Energy   and Heat Flow

Fig. 12   Experimental Results of Solar Space Heating at
           KEP Solar House

Heat exchanger must be provided somewhere to have the solar heat exchanged
to hot water from the solar tank directly connected to heating system or to the
heating circuit from the hot water tank directly connected to the hot water
sevice system. Integration of auxiliary system is also quite difficult as the
temperature required by hot water is usually higher than that of space heating.

HEAT PUMP IN SOLAR HEATING

In the case of heat assisted solar heating system the collector temperature can be lowered than the case of direct heating system. This makes the collector efficiency higher and the utilizability of solar radiation on the long term basis becomes higher as the lower intensity of solar radiation can be utilized on which extensive studies were made earlier by Liu and Jordan [15]. Fig. 13 shows the approximated linear relationship between collector efficiency and the temperature difference between collector inlet and ambient air over the intensity of solar radiation on the collector surface based on the so-called Hottel-Bliss-Whilier equation that states:

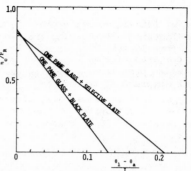

Fig. 13 Collector Efficiency of Flat-Plate Collector

$$\eta_c = \frac{q_c}{I} = \frac{F_R}{I}[\; \tau\alpha - K\frac{\theta_1 - \theta_a}{I}\;] \tag{10}$$

where η_c : collector efficiency

q_c : collection per unit collector area [W/m^2]

I : solar radiation on collector surface of unit area [W/m^2]

F_R : heat removal efficiency associated with fin efficiency of collector plate, bond conductance between tube and plate and flow factor taking account of temperature distribution of collector plate from inlet to outlet [16]

$\tau\alpha$: product of transmissivity of glass covering and absorptivity of collector plate

K : overall heat transfer coefficient from collector plate to ambient air [W/m^2deg]

θ_1 : collector inlet temperature [°C]

θ_a : ambient air temperature [°C]

As the heat pump is a tool of energy transfer from lower temperature side to higher temperature side, the advantage of the heat pump taking heat out of the solar heated water over the outside air source heat pump cannot be evaluated only by the increase in the value of the coefficient of performance (COP) if heating energy were delivered through the heat pump all the time. It must be considered of course that the air source heat pump does not work when the outside temperature gets down too low, though a type of heat pump working at the air source of -7°C is currently available. It is advantageous, therefore, to design the system so that direct heating can be made whenever the tank temperature is high enough to be suited for direct heating.

Experiences with the author's solar house [17] show that the switch-over to heat pump operation from direct heating operation is not so simple as conceived from the standpoint of manual operation. Moreover, direct heating may fail to

follow efficient thermal process because the heat in the tank supplied by auxiliary electric heater during the night might possibly be brought to the collector
in the next morning. On account of a lack in collector area (24 m^2 for floor
area of 150 m^2) the heat exchange from the solar heated water was very scarce
during the winter as the tank temperature remained as low as 0 - 20°C in the
evening.Fig.14 shows the result of power consumption of Kimura Solar House on
monthly average basis of daily electric consumption for space heating [18].
The house was calibrated without collector in the first winter and it has been
proved that about 60 % of energy for space heating was shared by solar energy.

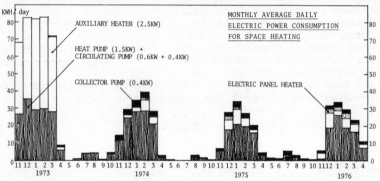

Fig. 14 Monthly Average Daily Electric Power Consumption of Kimura
        Solar House

     The obvious advantage of solar heat pump system has been realized recently
by two Japanese Manufacturers: One is the dual purpose heat pump which can pick
up heat either from solar water or from outside air depending on the conditions
by automatic switch-over device and the other type of the system uses the air
source heat pump as auxiliary to pick up heat from outside air and supply it to
the rock bed storage. These are reported to be functioning quite well[19],[20].
There are other schemes of solar houses using heat pump in Japan on the ground
that it also can serve as cooler in the summertime.
     It must be stressed here that the advantage of heat pump should not be
justified by the dual effect of working as heater in winter and cooler in summer
but the heat pump is simply a heating apparatus for making effective use of low
temperature energy, because the use of electricity for cooling in the summer
daytime must be avoided as much as possible. It can be envisaged that heat pump
willmore effectively and widely be used in combination with solar collection
sytem of various schemes utilizing low temperature energy in the colder regions
for heating purposes.

LONG-TERM HEAT STORAGE

     As long as the collector is installed at a certain tilt angle suited for
the most effective use of solar energy for yearround hot water supply and winter
space heating, it naturally follows that there should be an excess amount of
solar energy unused during the period from summer to autumn. In this respect
the long term heat storage is considered worth while to be attempted.
     In the past the annual heating load has been much larger than the annual
hot water load in most of the Japanese single family houses and the long-term
storage has never been justified. In the recent years, however, owing to various
energy conservation incentives are making space heating load less and less,
whereas use of hot water tends to increase in these days as central hot water
supply system is becoming popular in many households.

It is roughly estimated that the annual space heating load of a single family house in Japan could be decreased down to 3 MWh in Tokyo area if proper insulation were applied and the annual hot water heating load of average family in Japan is about 3.5 MWh which might increase in future. In this order of magnitude in the heating load variation pattern, storage of the summer heat in the underground may be regarded possible.

The water storage tank of 30 - 40 m^3 is considered capable of keeping the summer heat for winter heating in northern Europe as being experimented at Zero Energy House [21] in Denmark and Aachen Solar House [22] in West Germany. Mathew Solar House [23] in Oregon,U.S.A. has a water storage tank in the underground below which an air void layer is provided so that the heat can be stored in the soil beneath the tank.

The use of the summer heat for winter heating would of course reduce not only the annual total heating load by auxiliary energy source but also reduce the collector area. The underground heat storage is being tested by Nakajima at his house [24] and at the experimental house of the Building Research Institute [25] in Japan. The success of the long term storage would certainly contribute to the improvement of the economical feasibility of solar space heating.

Fig. 15 shows the annual variation pattern of the space heating and hot water heating loads of a single family house versus solar radiation on the collector surface and collection under the weather conditions near Tokyo. It can be seen that the excess amount of collection in summer is comparable with the shortage of energy requirements in winter both expressed in hatched area, allowing for some loss from storage tank. These are predicted features.

Fig. 16 shows the schematic diagram proposed for the solar house. It is planned that the excess heat may be stored in the underground and the stored heat can be taken out of it if the house calls for heat. The hot water is supplied through a pair of cylindrical small tank immersed in the solar tank functioning as heat exchanger. The unsteady state thermal diffusion process in the underground of semi-infinite solid is an interesting problem and this may be approximated by a solution of partial differential equation of heat conduction with cylindrical ordinates.

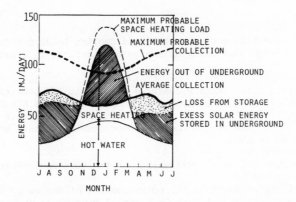

Fig.15 Annual Variation of Heating Energy Requirements
and Collection for a Proposed House with Long-
Term Storage

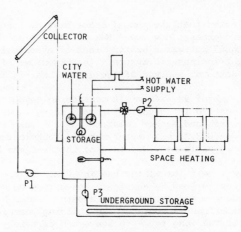

Fig. 16   Schematic Diagram of Solar Space Heating and Hot
          Water Supply System Proposed

SUMMARY

     The thermal processes of solar houses are much varied with cases by cases
as seen in some examples presented in this paper. It is important to note that
the thermal processes of solar houses must be understood not only with the behaviour of the mechnical sub-systems but also with the nature of heat transfer in
building structures. The optimality in solar house designs depends on the logical estimation of the variation of temperature and heat flow at various points
so that the environment of the house can be comfortably maintained and the hot
water be properly supplied with minimal consumption of auxiliary energy owing to
the sound harnessing of solar energy.

REFERENCES

[1] Procedure for Determining Heating and Cooling Loads for Computerized Energy
Calculations, ASHRAE Task Group on Energy Requirements for Heating and Cooling,
1971.

[2] Computerized Calculation Procedures of Dynamic Air Conditioning Load Developed by SHASE of Japan, The Second Sub-committee of the Air Conditioning
Standards Committee of SHASE of Japan represented by H. Saito & K. Kimura,
Second Symposium on the Use of Computers for Environmental Engineering Related
to Buildings, Paris 1974.

[3] SHASE Handbook of Air-Conditioning and Sanitary Engineering, 1975, p. II-50.

[4] Dietz, A.G.H. & Czapek, E.L., Solar Heating of Houses by Vertical South Wall
Storage Panels,Paper Presented to 56th Annual Meeting of the American Society of
Heating and Ventilating Engineers, Dallas, Texas, Jan. 1950.

[5] Trombe, F., Heating by Solar Radiation, CNRS Report B-1-73-100, 1973.

[6] Morse, R.N., CSIRO, Australia, personal communication.

[7] Hay, H.R. & Yellott,J.I., Natural Air Conditioning with Roof Ponds and Movable Insulation, ASHRAE Transactions, Vol.75 Part I, 1968.

[8] Niles, P.W.B., Thermal Evaluation of a House Using a Movable-Insulation Heating and Cooling System, paper presented to ISES Congress, Los Angeles, July, 1975.

[9] Kimura,K. & Okuyama, H., Study on Natural Air Conditioning with Numerical Solution Method of thermal Network, Transactions of Architectural Institute of Japan, Oct. 1976, p.333

[10] Kimura,K. Udagawa, M. et al, Experimental Study on Collection-Storage-Heater Type of Curtain Wall using Latent Heat Storage, paper presented to the Technical Meeting of the Society of Heating, Air-Conditioning and Sanitary Engineers of Japan, Oct. 1975, p.41.

[11] Böer,K.W., A Combined Solar Thermal and Electrical House System, International Congress-The Sun in the Service of Mankind, Paris,July 1973.

[12] Engebretson, C.D.,The Use of Solar Energy for Space Heating M.I.T. Solar House IV, U.N. Conference on New Sources of Energy, E/CONF.35/S/67, Apr.1961.

[13] Löf, G.O.G.,El-Wakil,M.M. & Chiou,J.P., Residential Heating with Solar Heated Air - Colorado Solar House -, ASHRAE Journal, Oct. 1963.

[14] Kimura,K. Udagawa,M. et al, Operation Results of Space Heating and Hot Water Supply Systems of Solar Experimental Multi-family Housing, Transactions of A.I.J., Oct. 1976.

[15] Liu,B.Y.H. and Jordan,R.C., A Rational Procedure for Predicting the Long-Term Average Performance of Flat-Plate Solar Energy Collectors, Solar Energy, 7[2], 1963.

[16] Whillier,A., Design Factors Influencing Solar Collector Performance," Low Temperature Engineering Application of Solar Energy" edited by Jordan,R.C., Technical Committee on Solar Energy Utilization of ASHRAE, 1967.

[17] Kimura,K., Design and Yearround Performance of Kimura Solar House, ISES Los Angeles Meeting, 1975, Extended Abstract 42-4.

[18] Kimura,K., Investigation on the Annual Energy Consumption of All Electric Solar House, Transactions of A.I.J., Oct. 1976.

[19] Sakai,J. et al, Solar Space Heating and Cooling with Bi-Heat Source Heat Pump and Hot Water Supply System,Extended Abstract 45-1, ISES, Los Angeles, 1975.

[20] Koizumi,H., Kawada,Z. et al, Space Heating Performance of Toshiba Solar House I, Second Technical Meeting, Japan Solar Energy Society, Dec. 1976.

[21] Esbensen,T.V. & Korsgaard,V., Dimentioning of the Heat Balance and the Solar Heating System in the Zero Energy House in Denmark, Report Meddelelse NR, 31A, Thermal Insulation Laboratory, Technical University of Denmark, DK-2800, Denmark.

[22] The Experimental House, brochure issued by Philips, Aachen, West Germany.

[23] Reynolds, J.S. et al, The Atypical Mathew Solar House at Coos Bay, Oregon, ISES Los Angeles Meeting,1975.

[24] Nakajima, Y. & Ohashi, K., Onthe Underground Heat Storage with Solar Energy, Second Technical Meeting of Japan Solar Energy Society, Tokyo, Dec. 1976.

[25] Tsuchiya, T. & Seto, H., A Model Experiment on the Utilization of Summer Solar Energy for Heating with the Underground Heat Storage, Paper Presented to Architectural Institute of Japan Meeting, Nagoya, Oct. 1976.

ECONOMIC AND SOCIAL IMPLICATIONS OF SOLAR ENERGY*

RUBIN RABINOVITZ AND FRANK KREITH

Department of Chemical Engineering
University of Colorado
Boulder, Colorado, USA 80309

The development of solar-energy systems is often singled out as an example of a benevolent technology. Solar energy promises solutions to problems that are the legacy of other energy technologies: depletion of non-renewable resources, economic dependency on a few countries with reserves of fossil fuels, increases in air pollution, environmental deterioration, and the safety hazards associated with nuclear power. Although solar energy is as yet only marginally competitive with conventional energy sources, the future promise of solar-energy research seems great: sunlight is readily available and costs nothing. {1}

It is often assumed that, in the United States, it will be some time before solar-energy systems are widely competitive with fossil-fuel systems. However, such an assumption is persuasive only when energy production is considered exclusively in terms of dollar costs. If fossil-fuel technology is viewed in terms of the energy lost in the process of energy production, a somewhat different picture emerges. Fossil-fuel energy production is wasteful because of many intermediate stages--mining, fuel transportation, steam production, and so forth--which themselves consume energy. A solar installation, on the other hand, can eliminate many of these intermediate steps. This means that the total energy consumption of a nation could decline once a changeover to solar energy systems in suitable applications, like space and water heating, had been completed.

Conventional energy systems are wasteful in still another way. When electricity is used for space or service water heating, the process involves a thermodynamic cycle in which heat is converted into electricity and then back into heat again. Burning coal for heating or cooking may be less convenient than using electricity, but it is far more efficient in terms of utilization of available energy. It has been estimated that in the United States we consume about 58% of our energy as heat; of this, about half is used in heating applications below the boiling point of water. {2} This is a range eminently suited to solar-energy systems. Moreover, the same point is underlined in an extensive study of first- and second-law efficiencies, conducted by members of the American Physical Society, showing how energy policies must take into account the type of work to be performed. A significant finding of this study is that the second-law

---

*The authors wish to express their appreciation to the National Science Foundation for assistance in completing this study; the work was done as a part of "Ethical and Human Values Assessment Procedure: A Pilot Project," N.S.F. Grant Number OSS76-16410.

efficiencies of space heating, water heating, refrigeration, and air conditioning with conventional equipment and fossil fuels are all less than six per cent. {3} On the other hand, it has been estimated that a solar tower plant with six hours of thermal storage and a capacity factor of 0.41 will deliver ten times as much energy during its life as the capital energy required to construct the plant. {4} A simple solar hot water heater installed at 40° latitude will, on the average, return its energy investment in usefully delivered heat in about one and a half years of operation. {5}

A far-sighted energy policy, then, would match energy sources and the tasks to be performed. In terms of energy costs rather than dollar costs solar technology is already very promising for some applications; and, with rising fossil-fuel prices, the likelihood of at least a partial changeover to solar-energy technology increases even more.

Such considerations make solar energy attractive as a partial solution to our imminent problems of energy shortages; and, in the climate of benevolent tolerance created by its promise, solar-energy technology has been granted considerable latitude to develop freely. Another factor that has contributed to this freedom is the commonly offered argument that a free-market economy--which permits supply and demand to regulate energy prices--is the best solution for energy-shortage problems. Given these factors, it becomes apparent why there has been relatively little scrutiny of the social effects that might accompany important developments in solar-energy technology.

Generally, those areas of science and technology which are identified as potentially harmful receive public attention whereas those which seem innocuous (radio astronomy, for example) are usually scrutinized exclusively in terms of research costs versus potential intellectual gains. Solar-energy technology is an example of a field where a comprehensive scrutiny of humanistic, ethical, and other social effects is usually considered superfluous.

It might be remembered that not too long ago nuclear-energy technology was also considered a benevolent technology; it was heralded as the benign alternative to nuclear-weapons technology. This exemplifies a common occurrence: when a technology is in the initial stages of application, few of the problems that accompany large-scale application are apparent. This was true of nuclear-energy technology, and it may also be true of solar-energy technology. Scrutiny of possible side-effects is therefore necessary, no matter how benevolent a technology seems, once large-scale application of the technology becomes an imminent possibility.

Solar-energy technology touches on a number of general issues which are to some degree applicable to other fields where scrutiny of social implications may in the past have seemed unimportant. {6} Foremost among these is a criterion of widespread effects: it would certainly seem reasonable to examine the possible side-effects of any technology which touches the lives of a large number of people. Solar-energy technology meets this criterion because of two factors: the universality of energy consumption in our society, and the diluted and intermittent form in which solar energy is naturally available. The first point seems self-evident: a new source of energy potentially affects every member of an energy-hungry society. The second point is based on the likelihood that a changeover to solar energy will require large amounts of land for the location of solar collectors, as well as great investments of capital and raw materials

in order to build the new installations. Every acre of cultivated land in a
sense is part of an extant collection system for the utilization of solar
energy; this is one indicator that large collection areas and tremendous amounts
of material are necessary to build future solar-energy systems. {7} It seems
clear, then, that the effects of a changeover to solar-energy production will
affect a large segment of our society.

These are very general considerations. More detailed estimates of the
impact of a changeover to solar-energy technology vary widely. For example, one
prediction of the cost of capitalization of such a changeover has been made by
Charles A. Berg, who investigated the cost of a collector suitable for domestic
hot-water heating. According to his estimate, the capital cost of a device
collecting 3.5 kwh per day of solar energy in the form of low-temperature heat
is between $15 and $18. Berg concludes that the capital expense for providing
one per cent of the American national energy requirement by using such a system
would be approximately $3 billion; to produce the same amount of energy by
increasing the supply of natural gas would cost $10 billion; by adding conven-
tional electrical-power systems, $16 billion. {8} A more recent estimate,
which deals with a solar-energy system with wider applications and includes
installation expenses and the cost of a backup system, predicts an outlay of
between $25,000 and $250,000 for a solar-energy system which will provide the
energy equivalent of a barrel of oil per day (about 67 kilowatt of heat). {9}
Such costs are greater than the full expense of increasing oil productivity
($3,000 - $10,000 for the same barrel-of-oil-per-day increment), but may be
cheaper than increments in coal-electric systems ($150,000 per barrel per day)
or in nuclear-energy systems ($200,000). {10}

Such figures, even if they prove to be accurate, can be compared only
when all of them pertain to increments in energy productivity; changeover costs
indicate only the expense of maintaining the status quo in energy production.
Moreover, a changeover may involve additional economic factors, such as outlays
for disposing of obsolete equipment, unemployment in the industry being super-
seded, and the raw-materials costs of the readjustment. {11}

Another kind of problem may arise as traditional patterns of capitalization in the energy industry are transformed after a changeover to solar-energy
technology.   The financing of conventional energy equipment has been the province
of large corporations, but individual homeowners are often the ones who borrow
the money to install solar energy equipment. It is possible, therefore, to
anticipate a shift in energy-financing patterns involving increased risk and
paperwork for lending institutions, the cost of which will be passed on to the
consumer in the form of higher interest rates. {12}

A homeowner who wishes to participate in a changeover to solar energy
may have the responsibility, not only for the financing of solar equipment, but
also for making important decisions regarding its utilization. With fossil fuel
systems it is possible for a contractor to specify the desired amount of energy
to be supplied from a given device, such as a gas heater or oil burner, and to
demand that the manufacturer guarantee that his equipment meets the specifications. This approach simplifies design requirements in the building industry
and makes it possible for a homeowner to assume that heating equipment, once
properly installed, will always operate in a predictable fashion.

With solar-energy systems the situation is quite different. A solar
device installed in a region where solar energy is plentiful can deliver

considerably more useful energy than the same device in an area of poor insolation; hence energy-output specifications are not uniform, but vary appreciably according to location. The amount and the cost of the useful energy supplied by a solar device depends not only on its location but also on its utilization. For example, a solar device for heating service water is used for the entire year; the same device, as part of a home heating system, will be used little in the warmer half of the year. Consequently, though the cost of the solar device may be the same in both cases, the amount of energy delivered per dollar invested will be greater in the first case.

These examples can serve to illustrate that accurate projections for the amortization of the expense of solar installations can become extremely complicated. Such projections also depend on several other factors which are difficult to specify or predict: future increases in the price of fuel, long-range weather forecasts, maintenance costs for relatively untested systems, and complicated engineering calculations for the heat losses in the building where the device is to be installed. This sort of analysis--requiring the services of a highly-trained consultant--may prove to be too expensive for many homeowners or contractors. Yet to go ahead without such calculations is to risk installing the equipment improperly and cause financial losses.

In such a situation, the consumer may have no choice but to accept the predictions made by the manufacturer of the device he is about to buy. Since even a careful engineering analysis, made in good faith, must to some degree be hypothetical, a manufacturer can easily present exaggerated claims in a context of scientific reliability. Some manufacturers of solar-energy devices have already indulged in such exaggerations; the end result, of course, may be to damage the credibility of the industry so that a changeover to solar energy becomes more difficult. {13}

Another economic problem is one which occurs in other developing technologies and grows out of the need to choose among alternative research approaches. Such approaches generally fall into two categories: basic research, programs geared towards discoveries which will supersede the existing knowledge in a field; and applied research programs which utilize the existing knowledge in a field and attempt to exploit it efficiently.

In thermal-conversion technology (where solar radiation is converted to heat) the possibility of important innovative discoveries seems remote and effort is concentrated on applied research. It is therefore sometimes assumed that basic research is superfluous in the field as a whole because the probability of developing successful devices based on other principles seems so small as to make basic research in areas like photochemical conversion purely academic. However, most of the food and fuel we consume are directly or indirectly the products of a photochemical process, photosynthesis. The efficiency of energy conversion in photosynthesis is only about one percent; this compares unfavorably with fossil-fuel conversion (thirty-five to sixty percent efficiency) and solar-thermal conversion (five to twenty percent efficiency). {14} The discovery of an inexpensive photochemical process which utilizes solar radiation more efficiently than photosynthesis could change the economics of applied solar-energy technology radically. Clearly, a potential exists for innovations based on photochemical conversion, and such innovations could easily render other forms of collection obsolete.

Most of the emphasis in current solar-energy technology is on applied research and centers on problems related to perfecting the efficiency of thermal-conversion systems or to lowering the cost of photovoltaic systems which convert solar radiation directly into electricity, since the greatest promise of economically competitive systems seems to lie in these areas. In thermal-conversion systems solar radiation falls on either a flat-plate or focusing collector; and the radiant energy is used to heat a working fluid which then delivers useful energy. Research in this area is devoted to developments which improve one of the links in the chain which involves the collection, transmission, storage, and utilization of solar energy. Typically, researchers deal with problems like the following: measurements of the quality of solar radiation; optimum locations for collectors; absorptivity in collection materials; heat loss of materials used in solar energy systems; improving the durability of collector materials or lowering their cost; and innovations in the design of collection, transmission, conversion, and storage units.

The main problem with photovoltaic systems is economic: until the 1970's the cost of energy from a scaled-up version of the photovoltaic systems used in space probes was about a thousand times that of energy from a conventional fossil-fuel system. {15} The price now is lower, but still not competitive with other systems; and hence much of the research in this area centers on attempts to lower the costs of photovoltaic devices.

Given the conflicting claims of those who follow these and other approaches in the various areas of solar energy research, it often is difficult to predict with accuracy which of the many alternatives will prove to be superior. It may be that an economically competitive innovation of the sort promised by those who anticipate major discoveries will never be found. Possibly, conventional methods of solar-radiation collection will have been operative for many years--long enough to make it worthwhile having installed them--before a revolutionary discovery occurs. Or, a sudden breakthrough could make extant installations obsolete, and the wisdom of hindsight will be used to demonstrate that the effort expended on applied research was wasted. Questions of this sort come to the attention of those with the responsibility for funding such research. But they are often beyond the capability of national energy-policy planners, who usually consider solar-energy technology entirely in terms of its best-known area, thermal-conversion systems. This follows a general pattern in the funding of energy technology in this country: those areas with the greatest short-term success prospects usually receive a disproportionate amount of support. {16}

Decisions of this sort clearly illustrate the dangers of a narrow approach to questions raised by a developing technology, no matter how benevolent. The problem is compounded when researchers themselves--even in the same field--underestimate the promise of a rival approach. Researchers are naturally tempted to make exaggerated claims when they seek financial support for their work, and members of funding agencies sometimes lack the expertise to choose among the claims of rival approaches. Conversely, those who are knowledgeable may be prejudiced because they know one area in a discipline better than others.

The inflated claims of researchers can eclipse the development of viable approaches; and they often raise expectations which, if they are not fulfilled, weaken the credibility of all researchers in a field. In solar energy technology, such claims can raise unrealistic expectations which defeat a conservative energy policy, for they lead to the inference that the depletion of fossil fuels will coincide with the perfection of solar-energy systems. Predictions of this sort also often fail to take into account the costs--in energy and in capital investment--of the process of conversion from fossil-fuel systems to solar-energy systems.

Similar problems, of course, may arise in any developing technology, but the reputation of a benevolent technology may provide an umbrella under which simplistic policies or unscrupulous behavior can flourish more easily. Condemning individual instances of such conduct will do little to alleviate the overall problem; instead, far-reaching solutions, like tangible incentives for reliable estimates of solar-energy system specifications, or a reexamination of research-funding practices, are necessary.

Usually a cluster of social problems arises with the changes introduced by a new technology. Technological changes often have a great economic impact, and moral questions which are related to financial gains and losses in certain sectors quickly arise. Moreover, transitions involving large segments of a population are in many instances accompanied by a blurring of the ethical issues that accompany them, for it is notoriously difficult to determine responsibility for moral decisions that come as a result of a far-reaching corporate activity. The effects of such possibilities can often be tempered or controlled if the scope of the technological changes and the concommitant ethical problems are to some degree anticipated.

Some of these moral issues have already begun to emerge. For example, it seems likely that fossil-fuel and nuclear systems will persist until--and even after--solar systems become economically attractive. {17} Given a free economy, either solar systems which can compete actively with fossil-fuel systems will be developed, or the rising price of fossil fuels will make extant solar systems competitive. Such solar-energy systems, even if expensive, have obvious advantages: they are clean; once built, they do not deplete natural resources; they are often relatively easy to recycle; and they are less hazardous than nuclear-energy systems. This raises important questions: should an energy policy persist which, by making fossil fuels available at relatively low prices, delays a transition to solar-energy systems? Should conventional fuels be taxed in order to pay the changeover costs for solar-energy systems? {18} Such questions are often ignored because economic plans which call for the reduction of the avail-ability of fossil fuels are politically unpopular. But an energy policy which ignores these questions and yields easily to the demands of political expediency is clearly short-sighted.

It would not be overstating the case to say that energy policy in the United States is often restricted to short-term considerations. Insufficient attention is being given to the long-range question any new technology raises: what changes--cultural and social as well as economic--can be expected if the new technology becomes dominant? An example of the kind of change which can be expected in the wake of a changeover to solar-energy technology is a possible shift in size, quality, and location of population centers.

Historically, towns and cities have been located near waterways. Water-ways in the past have been useful in a number of ways: for defense, transporta-tion, sanitation, agriculture, or as an energy source. Today, in a city with a conventional energy system, the availability of water is also important for providing the heat sink in the condensation step of a Rankine-cycle power plant.

The availability of cooling water is only one of several geographical factors important in locating solar energy systems. Two other considerations are local insolation conditions and the cost of the land on which the collectors will be installed. The local conditions which must be evaluated include the number of cloudy days per year, latitude, and the amount of haze. Average humidity can be an important factor: installations in equatorial zones, with their high humidity, are less efficient than those in arid regions of both hemispheres with latitudes between twelve and thirty-five degrees. This indicates that some deserts will be favorable areas for solar-energy installations: insolation is high and the land is cheap. Because it is inefficient and expensive to transport energy over long distances, most large communities are located fairly close to their major sources of power. A transition to solar-energy technology therefore raises the possibility of population relocations.

If new population centers are largely dependent on solar installations for their power, they may be smaller than cities which use conventional energy systems. Fossil-fuel and nuclear-power stations are usually most efficient when their power output is between 1000 and 2000 megawatts; in America, stations of this size might be expected to supply the electrical-energy needs of between 500,000 and one million people. {19} The optimum size for solar-power stations, according to a recent study, is between 20 and 300 megawatts; such installations would support proportionately smaller populations. {20} The amount of space needed for an array of collectors for larger installations--one estimate is fifteen to twenty square miles for a 1000-megawatt installation--makes it un-likely that they will be build near large cities. {21} A changeover to solar-energy technology may therefore encourage the development of smaller population centers. In addition, the production of solar-energy units for home use may permit some of those who have been dependent on large power stations to settle in remote and sparsely populated areas. A change in population density, a movement away from coastal areas, a settlement of previously uninhabited areas--these are among the possible effects of a transition to solar-energy systems. And, of course, if widespread redistributions of population do occur, these inevitably will have profound social effects.

Sudden economic gains and losses may be anticipated as solar-energy development makes once-worthless land valuable. The development of previously unsettled areas may threaten fragile ecological systems. Great arrays of collec-tors may prove to be very unappealing aesthetically. {22} The shapes of houses built to incorporated fixed flat-plate collectors may be drably uniform: the optimum angle at which such collectors are mounted depends on the latitude, and all of them will face the same direction. Legal disputes may arise when a new building's shadow diminishes the sunlight available on adjacent property. New growth may be encouraged by the promise of solar-energy technology to such a degree that the new systems cannot keep pace with growing energy demands.

It becomes evident, then, that even in the case of a benevolent technology there can be many unanticipated side-effects. Innovative developments in many areas of technology must be tested extensively--often with a comprehensive systems-analysis approach--before they are given wide application. But at the moment, while there are great economic incentives for the development of various areas of renewable energy technology, there has been little funding for research on the results of successful developments in this technology. This almost guarantees that the only intensive study of the anticipated changes will be made by those who thereby hope to realize immediate profits. When the main force shaping technological development is economic, the results may lead to haphazard growth, pollution, sporadic unemployment, wasted raw materials, aesthetic losses, and environmental damage. Long-term losses for the majority have too often been the price of shortsighted policies. This, in the past, has been represented as the inevitable price of technological development. But it is only inevitable when we repeat the old error of assuming that benevolent technologies flourish best in an atmosphere of benign neglect.

NOTES

1. Frank Kreith and Jan F. Kreider, "Preliminary Design and Economic Analysis
    of Solar Energy Systems for Heating and Cooling of Buildings," Energy, 1
    (1976), 63-76. Jan F. Kreider and Frank Kreith, Solar Heating and Cooling
    (New York: McGraw-Hill, 1975).

2. Amny B. Lovins, "Energy Strategy: The Road Not Taken?" Foreign Affairs,
    October 1976, p. 78. According to Lovins, in Western Europe "the low-
    temperature heat alone is often half of all end-use energy."

3. William D. Metz, quoting a report of the American Physical Society completed
    in the summer of 1974 in Science, 188 (May 23, 1975), 820. Commenting on
    the wastefulness of the six-percent figure, Metz says: "In other words, each
    of these processes {space heating, water heating, etc.} requires almost 20
    times as much as the minimum energy to do the task consistent with the laws
    of physics."

4. Alvin F. Hildebrandt and Lorin L. Vant-Hull, Solar Energy Laboratory, Univer-
    sity of Houston, Houston, Texas 77004, personal communication. The data for
    the nuclear power payback were taken from ERDA-76-1, Appendix B.

5. Testimony by Frank Kreith before the National Research Council Energy Commit-
    tee, Feb. 5, 1976 (to be published by the National Academy of Sciences). See
    also "Report to the Oregon Energy Council," Office of Energy Research and
    Planning, Office of the Governor, Salem, Oregon, Jan. 1, 1975. At 40 de-
    grees latitude a solar collector at a 45 degree angle will receive on the
    average 2,000 Btu/day-ft^2. At 30% efficiency 1,000 ft^2 will deliver about
    250 MBtu/year. The energy required to build and install such a collector,
    assuming that the energy input is equally distributed between labor and
    material (see J. Krenz, "Energy per Dollar Value of Consumer Foods and
    Services," IEEE Trans. on Systems, Man and Cybernetics, v. SMC-4, July 1976,
    pp. 386-88) was estimated to be 385 MBtu.

6. Recently there has been an attempt within the context of the 1969 Environ-
    mental Protection Act to assess ecological implications of alternative
    energy sources, including solar-energy sources; S. Beall, et al., An
    Assessment of the Environmental Impact of Alternative Energy Sources (Oak
    Ridge, Tennessee: Oak Ridge National Laboratory, 1974). This survey, how-
    ever, only begins to touch on some of the issues that concern us here.

7. It should be noted, however, that thermal solar-energy systems have attained
    a higher rate of efficiency in energy utilization than that of photosynthesis
    and hence will require a commensurately smaller collection area unless there
    is a changeover to processes that involve a photosynthetic intermediate step.
    See Beall, et al., pp. 36-84.

8. Charles A. Berg, "Energy Conservation through Effective Utilization,"
    Science, 181 (July 13, 1973), 135-36.

9. Bruce Anderson, "A Strategy: Capital Feasibility," Solar Age, February 1977,
    p. 2.

10. Lovins, p. 69; Anderson, p. 2.

11. One estimate of the raw-materials requirements of a changeover to solar energy technology (and possible consequent shortages in raw materials) has been given by Allan L. Frank, "From Washington," Solar Age, February 1977, p. 4. Quoting a recent study by the U. S. Geological Survey, Frank lists the following materials as being required to build solar collectors sufficient to replace 0.6 quadrillion Btu's by 1990: 5,200,000 tons of glass; 9,910,000 tons of iron; 80,000 tons of manganese.

12. A discussion of the attitudes of lending institutions towards solar-energy projects can be found in Melicher, Ronald, "Lender Attitudes Toward Solar Heating and Cooling of Buildings," Journal of Business Research, May 1976, pp. 187-195. See also Scott, Jerome, Ronald Melicher, and Donald Sciglimpaglia, "Demand Analysis, Solar Heating and Cooling of Buildings," Phase I Report, NSF Grant # GI-42508, December, 1974.

13. Donald E. Carr discusses the possibility that shoddy manufacturing techniques in the solar-energy industry could give it "a bad reputation it could never live down." Energy and the Earth Machine (New York: W. W. Norton and Company, 1976), p. 149.

14. S. S. Penner and L. Icerman, Energy, I (Reading, Mass.: Addison-Wesley Publishing Co., 1974), 12, 13. Recombinant DNA research, however, may produce substantial increases in crop productivity per unit area and quantity of insolation while at the same time decreasing reliance on synthetic commercial fertilizers. See Jean L. Marx, "Nitrogen Fixation Prospects for Genetic Manipulation," Science 196 (May 6, 1977), 638-41.

15. John Holdren and Philip Herrera, Energy (San Francisco: Sierra Club, 1971), p. 112.

16. See, for example, Lovins, pp. 65-66.

17. See Frank Kreith, "Time Scale of Fossil Fuels," Colorado Engineer, 73 (January 1977), 22-25.

18. Tax incentives for solar installations have been adopted by a number of state governments; see David Morris, "Local Report," Solar Age, February 1977, p. 5. But the measure we have suggested would go far beyond these modest programs--even if similar programs were adopted nationally.

19. According to Jerold H. Krenz, "on per capita basis the average U.S. electrical energy consumption rate is approximately one kilowatt--approximately six times the world's average: an installed capacity of twice the average power is necessary to provide for . . . fluctuations in demand." (Energy: Conversion and Utilization {Boston: Allyn and Bacon, 1976}, p. 86).

20. Colorado State University and Westinghouse Electric Corporation, Solar Thermal Electric Power Systems, I (Washington: National Science Foundation, 1974), 36-37.

21. Beall, et al., p. 45.

22. Donald E. Carr (p. 122) describes an incident where a plan to build large windmills in the White Mountains of New Hampshire raised objections (from the local Audubon Societies) about the "visual pollution" that would occur.

SOLAR HEATING AND COOLING EXPERIMENTS AT THE JOINT RESEARCH CENTRE-ISPRA OF THE EUROPEAN COMMUNITIES

E. ARANOVITCH, M. LEDET, C. ROUMENGOUS, AND D. van ASSELT

Commission of the European Communities
Ispra Establishment, Italy

1. INTRODUCTION

At the Joint Research Centre of the Commission of the European Communities, situated at Ispra, Italy, the EC is carrying out a four-year programme on Solar Energy. An important part of this programme is devoted to the applications of solar energy for habitat, mainly concerning the production of hot domestic water, space heating and space cooling.

For these purposes a laboratory for the study of solar systems has been constructed and became operational in December 1976 (Fig. 1). This laboratory, characterized by its flexibility, has been conceived for:

- the testing of components and subsystems,
- the evaluation of performances of complete solar systems,
- the demonstration and teaching activities.

The aims of the programme are for one part to support industry with specific testing of solar components, for another part to set up methodologies to evaluate the performances of solar systems, which then, through adequate mathematical models, are to be transposed to various climatic conditions in Europe.

This report is intended to give an overall view of activities with some illustrative examples without entering into specific details.

2. MEASUREMENTS AND ANALYSIS OF SOLAR DATA

One of the essential bases for the evaluation of a solar system lies in the availability of proper solar data. The Meteorological Station of the JRC, which has a ten-year period collection of data of global solar radiation on horizontal plane, is upgrading its installations with instrumentation for the measurements of direct solar radiation, diffuse solar radiation and global radiation at inclinations of various angles.

The analysis of solar data is being performed with the following aims:

- calculation of global radiation at a given inclination from measured data

725

obtained on a horizontal plane,
- evaluation of percentage of diffuse light,
- determination of "typical" sunny days, representative of monthly averages.

In this respect, some LIU and JORDAN types of correlations are being tested.
For instance, the following models are being investigated:

$$H_i = \gamma H_o + (\beta - \gamma) H_o \left(\frac{H_o}{H_{ox}}\right) \quad \text{or} \tag{1}$$

$$H_i = \gamma H_o + (\beta - \gamma) H_o \sin^2 \frac{\pi}{2} \frac{H_o}{H_{ox}} \tag{2}$$

$$\gamma \simeq (1 - \frac{i}{\pi}) \tag{3}$$

From these formulas the global radiation H_i on an inclined plane defined by
the angle i, can be derived from the measured global radiation H_o on a horizontal plane and the calculated value of the extra-terrestrial global radiation
H_{ox} on a horizontal plane. β is a coefficient which relates trigonometrically
the direct radiation on a given inclined plane to the direct radiation on a horizontal plane

$$H_{ir} = \beta \cdot H_{or} \, .$$

The validation of such correlations is still in the preliminary phase but their
use will be important for the design calculations of solar systems. As the
correlations are not linear, the determination of the optimum time constant
over which the global radiations are measured (hour, day, week or month)
has to be determined.

3. HIGH EFFICIENCY COLLECTORS

If the performance of a black paint single-glazed collector is taken as reference, substantial gain on efficiency can be obtained by using what is commonly called high efficiency collectors.

Mostly in collaboration with some industrial and research institutes, the JRC
has made studies and performed tests on:

- honey-comb cells,
- selective surfaces,
- Vee-corrugated surfaces.

3.1 Honey-comb Cells

Honey-comb cells, also called Francia structures, placed between the absorber and the glass cover, reduce considerably heat losses due to natural
convection and radiation, but the overall transmission coefficient, specially
for incidence of light beam above 30°, is also reduced. Systematic testing
has been carried out with polycarbonate cells for various ratios of the height

over diameter of the cells, with an optimum around 4 or 5. Baehr[1] investigated the repartition of heat losses by natural convection and radiation for various inclinations of the cells.

One of the problems with such cells, at this time, lies in the cost. They must be able to withstand the equilibrium temperature in case of no-coolant flow conditions, which exclude most of the cheap plastic materials.

3.2 Selective Surfaces

The development of selective surfaces has been progressing rapidly in the last two years and it can be expected that they will soon be produced on an industrial basis. Still, more information is needed on the long term behaviour of such surfaces. At the JRC-Ispra, performance tests have been carried out on selective surfaces developed by the Centre d'Etudes Nucléaires de Grenoble (black chromium on steel) and Alumetal (black nickel on aluminium) (Fig. 2).

3.3 Vee-corrugated Surfaces

Tabor[2] and Hollands[3] suggested the interesting possibilities of Vee-corrugations, which, by multiplying the number of incident reflections will increase the effective absorptivity of a surface (Fig. 3). Such geometries are particularly interesting when combined with a selective surface, where it is difficult to maintain at the same time a very low emissivity in the infrared spectrum with a uniform high absorptivity in the light spectrum (Fig. 4).

The JRC, in collaboration with Alumetal, has developed a Vee-corrugated collector which has been operative for more than a year and which is now being manufactured[4]. It is made of modular extruded aluminium profiles which have undergone a black nickel surface treatment (Fig. 5). The modular structure gives a great flexibility as far as dimensions are concerned.

The development of these high efficiency collectors is needed for solar cooling systems where cooling machines require working temperatures of the order of $85^{\circ}C$, but they are also interesting for the production of hot domestic water and space heating in moderately insolated regions as it is the case for most of Europe, with a relative gain over an ordinary collector, which can be of the order of 50%.[5]

4. SYSTEM STUDIES

The testing and analysis of performance of complete solar systems are essential before any reliable conclusions on the technology and the economics of solar energy in habitat can be achieved.

Many aspects, such as maintenance, reliability of components, corrosion, interface problems between subsystems, cannot be predicted theoretically.

Moreover, it is important that mathematical models, necessary either for transposition from one solar system to another, or one climatic region to another, be validated by systematic testing.

As said already, the laboratory for system studies was conceived as a flex-
ible facility where components and subsystems can easily be replaced. The
description summarized here should not be considered as fixed in time.

4.1 Systems Description

4.1.1 The Building

The laboratory consists of a one-storey building of 160 m^2 of ground surface
(20 x 8 m), divided into three zones:

- an experimental zone where the indoor components and circuits are concentrated,
- a zone for the analysis and treatment of data (data acquisition system),
- a zone for comfort studies used as a conference room.

The wall facing south is inclined at 60°. The walls are made of a sandwich
structure with two brick layers and 12 cm of insulating material. 27 m^2 of
windows are double-glazed.

4.1.2 Solar Collectors

64 m^2 of solar collectors can be accommodated on the wall facing south of
the laboratory. This wall is inclined at 60°. 80 m^2 of collectors can be put
on a flat roof with variable inclinations.

At this time 40.5 m^2 of collectors have been mounted on the inclined wall,
consisting of:

- 14 m^2 of black paint, stainless steel, one-glass cover collectors,
- 6.5 m^2 of honey-comb combined with aluminium roll bond collectors,
- 6 m^2 of Vee-corrugated, selective aluminium collectors,
- 14 m^2 of Vee-corrugated, black paint and coloured collectors.

In preparation is the mounting on the roof of:

- 1 array of 20 m^2 of black chromium selective one-glass cover collectors,
- 1 array of 20 m^2 of Vee-corrugated black nickel selective collectors.

One of the two arrays will be combined with a reflective parabolic mirror.
These two arrays will be mounted for a solar cooling experiment, assisted
with an Arkla absorption machine.

4.1.3 Thermal Storage

The storage system consists of:

- an insulated tank of 2000 litres, for short term storage,
- a well-insulated basement with a capacity of 50,000 litres, for long term
  storage.

Heat exchangers made of copper tubes, are mounted in the two storage devices for transfer of heat from the collector circuits to the space heating systems.

4.1.4 Auxiliary Heating and Heat Pump

- Electrical Heating:

A 7 kW electrical resistance placed in the storage tank, switches on when the temperature in the tank becomes inferior to prefixed values.

- A Greg heat pump of 8 kW can be used as an auxiliary electrical system, either for heating or for cooling experiments.

4.1.5 Absorption Air Conditioner

An Arkla WF-501 absorption air conditioner, based on a lithium-bromide cycle operates with an inlet temperature of hot water of the order of 90°C with an expected capacity of 3 tons of cooling effect.

4.1.6 Regulation Systems

A Danfoss regulation system offering various possibilities, is used on the primary solar collector circuit. Essentially it cuts off the primary circuit from the storage tank when the solar collectors are not in a position to deliver calories.

A Honeywell regulation system is used on the secondary circuit for the space heating system. It consists essentially of a three-way regulated mixing valve which keeps the inside ambient temperature at desired value.

4.1.7 Space Heating Systems

Three different space heating systems will be experimented:

- a floor heating system which functions at low temperature; it consists of copper pins embedded in the floor with a spacing of 25 cm. This spacing can be made to vary by shutting off certain pins;
- a thermal convector system using either hot or cold water;
- a thermo-ventilation system with a heat recuperator.

At this time, only the first system has been tested.

4.2 Measurement System

The basis of the measurement system is to provide the needed data for energy balance models. The instrumentation will vary from one experiment to another.

Environmental

- total insolation on horizontal, 45° and 60° planes,
- temperature, outside ambient,

- relative humidity,
- wind velocity and direction.

Collector System

- temperature total array inlet, and temperature difference between outlet and inlet,
- mass flow rate total array.

Storage System

- tank temperatures at three different levels.

Auxiliary Heating Systems

- electrical consumption and energy balance for heat pump.

Space Heating Systems

- temperature levels,
- mass flow rates.

Building

- local temperatures by thermocouples,
- zone temperatures by thermography.

Data Acquisition System

Most data will be treated on a 3050 B Hewlett-Packard Data Acquisition System.

4.3 System Thermal Performance

A global system can be defined by:

- the collector array,
- the storage system (large, small, water, salt, etc.),
- heat pump (yes/no),
- space heating system (floor, thermo-convector, thermo-ventilation).

As can be seen from the preceeding paragraph a great number of possible combinations can be investigated. For the year 1977 two systems have been considered:

- a low temperature system with floor heating (winter 1977),
- a cooling experiment with the lithium-bromide absorption machine which will become operative during the summer of 1977.

4.3.1 Solar System Combined with Floor Heating

This choice was based principally on the idea of working at low temperatures in order to achieve a high efficiency for the solar collectors. From this point of view the combination of a solar system with a floor heating system looks attractive. Another advantage of this combination is that the floor heating system in itself can constitute a storage device of a relevant heat capacity. Another aim of the experiment was to evaluate the comfort of such a solution.

During the past years in Europe there has been a certain prevention against
floor heating, mainly because of poorly insulated houses, which required
higher working temperatures leading to discomfort for feet.

It is not within the scope of this report to give a detailed analysis of this experiment. The main results are presented here.

The system essentially consisted of the 40.5 m^2 of collectors mentioned previously, and of the 2000 l water tank. The 7 kW heating auxiliary ensured
that the average temperature in the tank never went below 27°C. The secondary circuit leading to the floor heating system was regulated by the mixing
valve to maintain an inside ambient temperature around 21°C.

Under normal operating conditions the energy balances concerning the primary circuit (collectors) and the secondary circuit (floor heating system) were
represented by the equations:

$$\eta_o \cdot I.A_1 = U_1 A_1 (T_1 - T_a) + (mc)_1 \frac{dT_1}{dt} + U_{12} A_{12} (T_1 - T_t) \qquad (4)$$

$$U_{12} A_{12} (T_1 - T_t) = (mc)_t \frac{dT_t}{dt} + U_T A_T (T_T - T_{ia}) + L \qquad (5)$$

The first equation states that the absorbed energy is equal to the sum of energy lost to the outside ambiance plus the transient energy plus the energy
transferred to the storage tank. The second equation states that the energy
transferred to the storage tank is equal to the transient energy in the tank
plus the tank heat losses plus the house load.

Performance monitoring of this system started the 50th day of the year 1977
and lasted 8 weeks.

Preliminary performance tests on subsystems were carried out to determine
the average value of the coefficients in relations (4) and (5). For instance,
the water tank disconnected from the other circuits was heated up and allowed to cool naturally in order to determine its heat capacity and heat loss coefficients. In similar manner the array characteristics were determined by
sub-system testing.

The average heat transfer coefficient K_b between the inside ambiance and
the outside ambiance was found to be, on the average, equal to 0.36 kcal/m^2.
h.°C with people living in the building (opening doors and windows). Without
people it was found to be equal to 0.28 kcal/m^2.h.°C.
Some global results are presented in Figs. 6 and 7.

Fig. 6 shows the dayly incident global radiation on an inclined plane at 60°.
Fig. 7 gives the repartition between solar and electrical auxiliary heating
on a weekly basis. These results can be summarized in the following way:

Total energy balance for 7 weeks (it was felt the 8th week should not be included because of warm weather):

total load:	2917 kWh	100%
active solar:	1431 kWh	49%
passive solar:	471 kWh	16%
auxiliary heating:	1015 kWh	35%

Global incident radiation
on collector array   5184 kWh

Average efficiency of active solar system: 28%.

These results, obtained for a period of moderate sunshine, confirmed the
advantage of operating at low temperatures. Subsequent analysis showed that
the overall performance could have been slightly better with a larger storage
tank, excluding at this time, cost optimization.

ACKNOWLEDGEMENTS

The authors wish to thank Mr. Gandino for his contribution to the meteorological data measurements.

NOMENCLATURE

A_1 total collector array surface (m^2)

H_i integrated global solar radiation on inclined plane defined by i

H_{ir} integrated direct solar radiation on inclined plane (kWh/m^2)

H_o integrated global solar radiation on horizontal plane (kWh/m^2)

H_{or} integrated direct radiation on horizontal plane (kWh/m^2)

H_{ox} integrated extra-terrestrial so-lar radiation on hor. plane (kWh/m^2)

i angle of inclination (^{o}C)

I instantaneous global radiation (kW/m^2)

L load (kW)

$(mc)_1$ heat capacity of primary circuit $(kWh/^{o}C)$

$(mc)_t$ heat capacity of storage tank $(kWh/^{o}C)$

T_a outside ambient temperature (^{o}C)

T_{ia} inside ambient temperature (^{o}C)

T_1 temperature of primary cir-cuit (^{o}C)

T_t temperature of storage tank (^{o}C)

$U_{12}A_{12}$ heat transfer coeff. of heat exchanger in storage tank $(kWh/^{o}C)$

U_tA_t heat transfer coeff. between storage tank and inside am-bient temperature

U_1 heat transfer coeff. per unit surface between collector and outside ambient temperature

γ coefficient

β coefficient

η_o average transmission-absorp-tion coeff. for collector array.

Fig. 1 - General View of Solar Laboratory

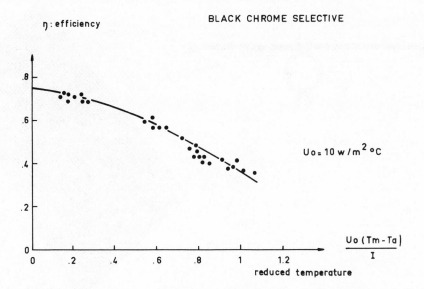

Fig. 2 - Instantaneous Efficiency Curve of Black Chromium Selective
Collector

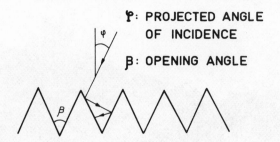

φ: PROJECTED ANGLE OF INCIDENCE

β: OPENING ANGLE

Fig. 3 - Vee-Corrugated Surface

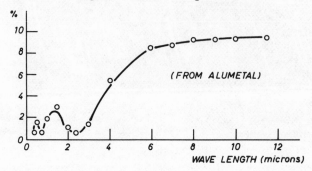

Fig. 4 - Reflectivity of Black Nickel Absorber

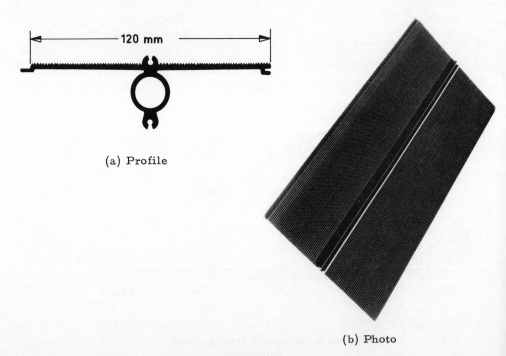

(a) Profile

(b) Photo

Fig. 5 - Modular Extruded Aluminium Absorber

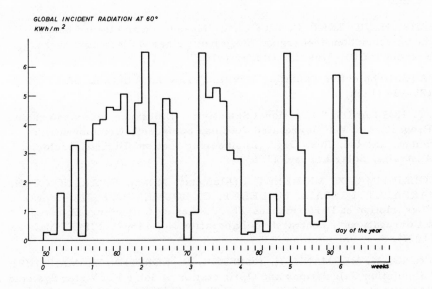

Fig. 6 - Daily Incident Global Radiation on Collector Array

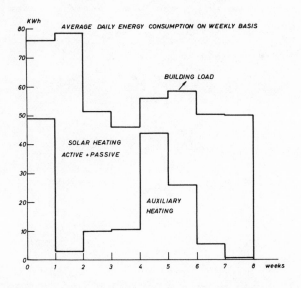

Fig. 7 - System Performance and Building Load

REFERENCES

(1) A. BÄHR, H. PIWECKI, L. RIGOLINI; "Messungen an einem Ebenen thermischen Sonnenenergieauffänger mit einer Zellstruktur nach Francia", EUR-5760-d, October 1976

(2) H. TABOR; "Selective Radiation", Bull. Res. Counc. Israel, 5A(2): 129-134 (1956)

(3) K. G. T. HOLLANDS; "Directional Selectivity, Emittance, and Absorptance Properties of Vee Corrugated Specular Surfaces", Commonwealth Scient. and Ind. Res. Org., Engineering Section, Highett, Victoria, Australia, Solar Energy, (1963)

(4) P. COLAIEMMA, X. MICHELETTI (Alumetal, Milan), E. ARANOVITCH, F. FARFALETTI CASALI, M. LEDET, C. ROUMENGOUS (JRC-Ispra); "Description et Performances Thermiques de Collecteurs Solaires à Corrugations", contrat de collaboration No. 145-75-PIPGI, Mai (1977)

(5) R. BRUNO, W. HERMANN, H. HÖRSTER, R. KERSTEN, F. MAHDJURI; "Simulation Calculations and Optimization of Solar Hot Water Systems", "Arbeitsgemeinschaft Solarenergie E. V. (ASE)", Febr. 4th, 1977.

SOME ASPECTS OF THE ROLE OF SOLAR RADIATION IN DETERMINING DWELLING HEATING AND COOLING ENERGY REQUIREMENTS

P. J. WALSH

CSIRO Division of Building Research
Highett 3190, Victoria, Australia

ABSTRACT

Every building acts as a solar collector. This paper describes and quantifies the so-called 'passive' role played by incident solar energy in determining heating and cooling requirements for some typical Australian dwellings. Quantification is achieved through computer simulation of building thermal behaviour. The requirements for such a model, including climatic data input, are described in some detail. Use of the sol-air temperature and shading coefficient concepts enables the influences of solar radiation to be simulated. Since the necessary radiation measurements are often unavailable for a given location, various theoretical estimates have been investigated. In addition, the relative contributions of conduction and transmission solar heat gains have been examined for certain types of construction.

NOMENCLATURE

t_{sa} sol-air temperature for a surface ($^{\circ}$C)

t_{oa} outdoor air temperature ($^{\circ}$C)

I total (global) solar irradiance incident on a surface of given inclination (W/m^2)

D diffuse solar irradiance (W/m^2)

H subscript denoting horizontal surface

C subscript denoting clear sky conditions

α solar absorptance of a surface

ε emittance of a surface to long-wave radiation

σ Stefan-Boltzmann constant

h combined (convective and radiative) surface film heat transfer coefficient (W/m^2K)

v wind speed (m/s)

m sky cloudiness number or total cloud cover (oktas)

K cloud cover factor

r horizontal sunbreak rating for a vertical surface (%)

ϕ function of σ, ε, m, r and t_{oa} (eqn. 2)

L_0 calculated load over a specified period with no solar radiation (GJ)

L_m calculated load over a specified period with measured solar radiation (GJ)

L_c calculated load over a specified period with shading coefficients zero (GJ)

L_f calculated load over a specified period with no solar radiation or occupant activities (GJ)

F   fractional contribution from total solar heat gain

F_c fractional contribution from conduction solar heat gain

F_t fractional contribution from transmission solar heat gain

INTRODUCTION

     Solar energy influences the thermal environment of all buildings.   It does this in both direct and indirect ways.   Its indirect influence comes about because the sun is the ultimate climatic determinant, modifying outdoor temperatures, pressures and cloud patterns.   Solar energy plays a more direct role however through the effects of radiation incident on a building, i.e. every building acts as a solar collector.   There are two influences at work here, viz., solar heat gain by conduction through building elements and solar heat gain by transmission of radiation through transparent elements.   In general, these influences are beneficial when heating is required and detrimental when cooling is required.   Note that the effects of conduction solar heat gain are additional to those due to air-to-air conduction heat transfer.

     In recent years successive 'energy crises' have highlighted the necessity for conservation of existing non-renewable energy resources.   Buildings in Australia consume about 25% of total primary energy usage, about two-thirds of this amount being used in dwellings, mainly for space heating and cooling and water heating.   Ballantyne [1] has highlighted the considerable scope for energy conservation in this area and indeed the resultant national economic gains would be considerable.   It is estimated [2] that by the turn of the century, a continuance of present energy trends would result in an annual national energy import bill of $A8 x 10^9.

     This paper focusses on the role played by incident solar radiation in meeting the energy requirements for space heating and cooling of a typical dwelling in Melbourne, Australia.   The availability of solar radiation for both clear sky and actual conditions for this location has been previously discussed [3].   A computer-based mathematical model of the thermal performance of a dwelling is used in the present investigation.

     Though the computer is now a readily available tool to assess the effects of passive design measures on the heating and cooling energy requirements of buildings, care needs to be exercised in its use.   Errors may arise from assumptions used in modelling heat flows within a building, from imprecise values of thermal resistances and capacitances and from other sources.   One must be satisfied that the accumulation of such errors is acceptable. Furthermore, Ayres [4], suggests that accurate calculation of building energy requirements necessitates "hour-by-hour calculations of all building loads over a full year of 'typical' weather data".   This in turn requires that the calculation technique be efficient as well as accurate, since computer costs might otherwise be prohibitive.

     The mathematical model used here has been previously outlined [5].   It has certain advantages of efficiency which enable hourly temperatures or loads over one year to be computed in about 6 seconds CPU time on a CDC Cyber 76 machine.   As input data it requires details of building layout and construction, solar absorptances and emittances of outdoor surfaces, shading devices, any internal loads generated by appliances and occupants, and coincident hourly values of dry bulb temperature, wind speed, cloud cover, direct and diffuse components of solar irradiance on a horizontal plane. A demonstration has been conducted [6] in which temperatures measured inside a model building were compared with those calculated with a related technique. Errors were shown to be satisfactorily small suggesting that the method could reasonably be used in thermal calculations for buildings.

Compilation of the required climatic data for purposes of calculation of energy requirements presents considerable difficulties, particularly with regard to solar radiation.

In Australia the Bureau of Meteorology measures surface climatic parameters at three hourly intervals and solar radiation continuously, with integration at half hourly intervals. The surface data is measured at clock time for the given location whilst the solar data is measured at mean solar time. This time difference means that coincident surface and solar data must be 'created'. This problem is compounded by the introduction by some Australian States of daylight saving during the summer months. During those periods the standard measurement times of surface climatic data are altered. Furthermore, surface data must be interpolated to produce hourly values.

The Bureau network of stations measuring surface climatic data over the Australian continent is very large in size but the same is not true for solar data. At present half-hourly global radiant exposure on a horizontal plane is being measured at nineteen stations and diffuse data is being measured at about half of these. The solar network does not include three of the six State capital cities and neither does it include the Federal capital, Canberra. For some of those places, data is available from Universities and other institutes. Expansion of the Bureau's solar network is unlikely in the near future.

MODELLING THE SOLAR CONTRIBUTION

(i) Sol-Air Temperature

As already stated, solar radiation influences the thermal behaviour of a building through solar heat gain by conduction through building elements and by transmission through transparent elements. The first of these effects is modelled by use of the sol-air temperature (or equivalent outdoor temperature) concept, in which dry bulb temperature, short-wave and long-wave radiation exchange are incorporated into a single equivalent boundary condition. It is defined [7] as the equivalent outdoor temperature which will cause the same rate of heat flow at the surface and the same temperature distribution through the material, as results from the outdoor air temperature and the net radiation exchange between the surface and its environment.

The concept is of immense value in heat transfer calculations since the sol-air temperature is independent of the bulk properties of the given material and under suitable assumptions, it may be calculated directly from external climatic data. Rao and Ballantyne [7] concluded that "the effect of errors in estimating sol-air temperatures would be considerably reduced by building components, and for most purposes sufficiently accurate predictions of indoor air temperatures are possible".

The exact form of the relationship used here [8] is:

$$t_{sa} = t_{oa} + [\alpha I - \phi(t_{oa} + 273.15)^3]/h \tag{1}$$

$$\text{where } \phi = \begin{cases} 4\sigma\varepsilon(14-0.2t_{oa})\ (1-m/9) & \text{for horizontal surfaces} \\ 8\sigma\varepsilon(5-0.1t_{oa})\arctan\ [100/r](1-m/9)/\pi & \text{for vertical surfaces} \end{cases} \tag{2}$$

$$\text{and } h = 11.3 + 3.31v. \tag{3}$$

The method used for the calculation of radiation on vertical surfaces with horizontal sunbreaks is outlined by Ballantyne and Spencer [9]. The direct and diffuse irradiance on vertical surfaces must be calculated separately by algebraic means from the values on a horizontal surface, given a knowledge of the bearing of the surface, the solar altitude and azimuth, and the ground reflectance. The horizontal sunbreak rating, r, for a given vertical surface is defined as (100 x horizontal projection/vertical drop of the surface).

(ii) Shading Coefficient

The second solar effect, viz., transmission through transparent building elements such as windows, is modelled by means of the shading coefficient concept, ([10], p.478) in which the solar heat gain is described as a fraction of that through unshaded 3 mm thick flat glass. The amount of solar radiation either transmitted, or absorbed and then radiated to indoors, by the reference glazing material, called the solar heat gain factor, depends on the angle of incidence (and hence solar position) and is calculable according to known functions ([10], p.398). By definition 3 mm glass with no internal shading device has a shading coefficient of 1. Similar glass with white internal holland blinds drawn would allow a solar heat gain of about 30% of that through the unshaded glass and thus has a shading coefficient of 0.3. Spencer [11] tabulates solar heat gain factors for clear skies as part of a general tabulation of solar position and radiation for various Australian capital cities. In the present work it is assumed that all radiation passing to the interior of the building is absorbed by the floor.

(iii) Incomplete solar data

How are we to proceed when calculating building thermal performance for a location for which complete solar radiation measurements are not available? Two alternatives are to use either no radiation at all or theoretical clear sky radiation. For purposes of calculating heating loads these two estimates should provide upper and lower limits respectively whilst for cooling loads, they should provide lower and upper limits respectively.

Several mathematical models of clear sky radiation are currently available. The model of Spencer [11] is based on the work of Rao and Seshadri [12]. This model relates clear sky radiation to solar position and atmospheric attenuation by dust particles, ozone and water vapour. Clear sky irradiance may be calculated for any location for which estimates of particulate dust content and mean monthly precipitable water vapour figures are available.

For non-clear sky conditions the situation is vastly more complex. This is principally because radiation can no longer be simply related to solar position and atmospheric attenuation. Clouds may drastically alter radiation patterns, generally increasing the diffuse and decreasing the direct component.

For situations in which no measurements at all are available, the only possibility seems to be to set up a radiation model based upon measured cloud amounts and types together with knowledge of reflective, absorptive and trans-missive properties. The success of such modelling depends very much upon the time scale over which radiation is required. Norris [13] reports an examin-ation of several methods of cloud classification in none of which was a high correlation with measured solar radiation found. He believed this to be due largely to the fact that cloud measurements are somewhat unreliable, relying heavily upon subjective assessment. He concluded that if cloud measurements were to be used to predict values of solar radiation, only monthly or longer averages could be obtained with reasonable accuracy.

Nevertheless, ASHRAE [14] recommends such a method for the calculation of hourly heating or cooling loads for buildings. A cloud cover factor

K defined as I_H/I_{CH}, has been calculated as a function of various cloud types and may be used to calculate I_H from a knowledge of I_{CH}. The fundamental problems with such cloud models is their dependence on cloud data, the reliability of which for Australian localities is unknown.

(iv)   The Bugler Method

For locations for which only global radiation measurements on the horizontal plane are available, it is necessary to split this measurement into direct and diffuse components. A promising method is that of Bugler [15] in which, from an extensive correlation study of radiation data for Melbourne, relationships have been established which determine the component parts of global radiation measured on the horizontal plane as a function of the cloud cover factor K, as:

(a)   for K<0.4, $D_H = 0.94I_H$

(b)   for 0.4<K< 1.0, $D_H = \dfrac{[1.29-1.19K]I_H}{[1.00-0.35K]}$

(c)   for K>1.0, $D_H = 0.15I_H$ (4)

Then the direct component is simply $I_H - D_H$.

Bugler's method differs from the work of Liu and Jordan [16] in that it gives actual, not average, values. Furthermore, that work was applicable to daily rather than hourly data, making it less suitable for calculation of building energy requirements. Bugler indicates that his method represents a universally applicable empirical correlation of global and diffuse irradiance.

We now present the results of some numerical experiments conducted to ascertain whether or not the Bugler method would be satisfactory for our purposes of calculating theoretical heating or cooling loads of a typical dwelling over a complete heating or cooling season. This we could do by calculating hour-by-hour loads for a given construction in Melbourne (latitude 38.0°S) using firstly measured direct and diffuse data, then using radiation as computed from (4) with global data. As well, it was felt useful to calculate supposed limiting loads resulting from use of clear sky radiation [11] or use of no radiation at all.

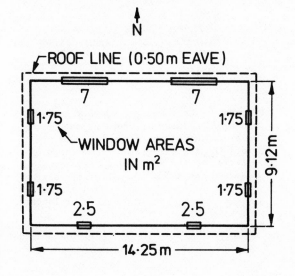

Figure 1.   Plan of the standard house used in the thermal calculations. Plan area is 130 m².

All calculations are referred back to a reference or standard house
(Figure 1). Based on outdoor dry bulb temperatures, the Melbourne heating
season is considered to extend from April to November inclusive and the
cooling season from December to March inclusive. The heating season chosen
is that for 1968, the cooling season is for 1967/68. For calculation of
heating loads, all shading coefficients are set to 1 (no blinds) whilst for
calculation of cooling loads they are set to 0.3 (blinds drawn). All opaque
building surfaces have α = 0.7 and ε = 0.9, whilst ground reflectance is set
to 0.2. Calculations are of <u>whole-house</u> loads, with the thermostat setting
at 20°C and 23°C for heating and cooling load calculations respectively. In
each case a 1°C thermostat differential is applied.

The standard house has brick veneer walls with reflective foil laminate
in the wall cavity; suspended timber floor with 75% carpet, 25% vinyl tiles
as covering; tiled roof with 50 mm rockwool batts above the ceiling. Total
plan area is 130 m^2. The ventilation rate chosen is 1 air change per hour
and the house is assumed to be occupied by a family of four with typical usage
pattern of household appliances. The heat arising from this usage pattern
is incorporated into the calculations as a direct heat flow to indoor air.

Figure 2 indicates the variation of monthly load total (GJ) over the
specified heating season for continuous (i.e. plant switched on for 24 hours
each day) heating of the standard dwelling for each of these four cases. The
seasonal load total is closely proportional to the area under the relevant curve
and in each case is expressed as a fraction of that resulting from use of
measured radiation. Also shown are such variations for intermittent (i.e. plant
switched on between 1600 and 2200 each day) heating, and for continuous cooling.

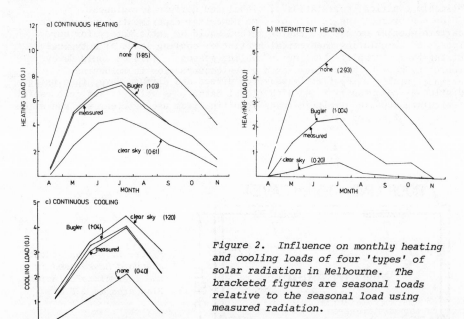

Figure 2. Influence on monthly heating
and cooling loads of four 'types' of
solar radiation in Melbourne. The
bracketed figures are seasonal loads
relative to the seasonal load using
measured radiation.

    The case where zero radiation is assumed clearly results in gross errors
in load totals for all forms of heating and cooling, as does use of theoretical
clear sky radiation for heating.   Clear sky radiation gives a better
approximation to cooling load totals, a result which is hardly surprising and
which confirms previous findings [5].
    Most significantly, however, use of the method of Bugler [15] to predict
direct and diffuse radiation components gives remarkably close agreement to all
load totals calculated using measured radiation.   Since the relationships (4)
were derived using Melbourne radiation data, it could be argued that this
agreement should be expected.   It is required to check the relationships using
data from another location.   This has been done for Port Moresby (latitude 9.5°S)
in Papua New Guinea, a tropical region with different radiation patterns to
those of the generally temperate Melbourne region.   Continuous hour-by-hour
cooling loads have been calculated for the standard dwelling over the year
November 1971 to October 1972 under the four previous radiation conditions.
In this case the thermostat setting is 25°C, again with a 1°C thermostat
differential.   Variation of monthly load totals (GJ) is shown in Figure 3.
Again the excellent agreement provided by the equations (4) is evident.

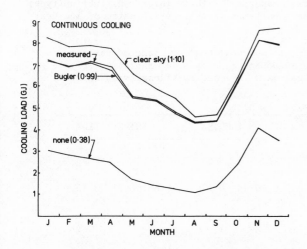

*Figure 3. Influence on
monthly cooling loads of
four 'types' of solar
radiation in Port Moresby.*

    It is concluded that the method of Bugler [15] may be used with confidence
in the assessment of heating and cooling energy requirements for those locations
for which only global radiation measurements are available.

SOLAR CONTRIBUTION TOWARDS HEATING REQUIREMENTS

    Various authors [eg. 17,18,19] have examined the effects of solar radiation
on heating and cooling energy requirements for buildings.   Where thermal
calculations are used they often do not account in a dynamic fashion for all
the relevant climatic parameters.   Here an attempt is made to realistically
quantify the contribution of incident solar radiation towards reducing the
energy requirements for space heating of a typical Melbourne house.
    For a given time period, a number of different types of calculated heating
load provide useful information about this contribution.   These include (1)
the design load, L_m, calculated using measured direct and diffuse solar
irradiance;   (2) a load, L_0, calculated with solar irradiances set to zero as
in the previous section, thus eliminating the effects of solar radiation;

(3) a load, L_c, calculated with all shading coefficients set to zero,
i.e. allowing only conduction solar heat gain through building surfaces.
Clearly the aim of effective thermal design is to make L_m as small as possible,
ideally zero.

The contribution of incident solar radiation is towards reducing the load
L_0 and must be assessed with respect to that load.   The solar contribution is
expressed in absolute terms as (L_0-L_m) or as a fractional contribution, F, as
$(L_0-L_m)/L_0$.   Ideally we seek to make F equal to unity.   Furthermore, the
fractional contribution, F_c, from *conduction* solar heat gain is $(L_0-L_c)/L_0$
whilst that from *transmission* solar heat gain, F_t, is $(L_c-L_m)/L_0 = F-F_c$.
Thus, for a given building, it is possible to distinguish the relative
contributions of conduction and transmission solar heat gains.

The loads, L_m, L_0 and L_c have been calculated over the eight month
Melbourne heating season for (a) the standard house of Figure 1, (b) the
standard house with wall and roof insulation removed, (c) the standard house
with external walls of cavity brick, internal walls of single brick whilst
the floor is a concrete slab-on-ground with floor covering of 100% vinyl
tiles.   The envelope of house (a) is characterized by a high thermal resistance
to heat flow, that of (b) by a low thermal resistance whilst (c) is characterized
by walls and floor of high thermal capacity.   These three construction types
ought to show noticeably different patterns in their utilization of incident
solar radiation.

TABLE 1:   *Fractional contribution to seasonal heating loads of various*
construction types in Melbourne made by solar radiation (F),
by conduction solar heat gain (F_c) and by transmission solar
heat gain (F_t).

CONSTRUCTION	CONTINUOUS			INTERMITTENT		
	F	F_c	F_t	F	F_c	F_t
Case (a), high thermal resistance	0.46	0.14	0.32	0.66	0.21	0.45
Case (b), low thermal resistance	0.38	0.21	0.17	0.55	0.32	0.23
Case (c), high thermal capacitance	0.50	0.17	0.33	0.56	0.20	0.36

Table 1 shows seasonal values of F, F_c and F_t whilst Figure 4 indicates
monthly variation of the quantities F and F_c for continuous and intermittent
heating modes respectively for case (a),   and   Figure 5 shows seasonal values
of L_m, L_0 and L_c for all three cases.   Also shown in Figure 5 are seasonal
loads, L_f, calculated with all fortuitous heat gains (i.e. those due to effects
of solar radiation and to occupants and their use of household appliances)
neglected, thus allowing an assessment to be made of the total contribution
from such gains.

The following points are noted:
(1)   In all cases, solar radiation makes a greater relative contribution in
the intermittent mode than it does in the continuous mode, since heating occurs
at times when the indoor environment is still considerably influenced by the
day's input of solar radiation and does not occur during the early morning
hours when solar effects are minimal and heating requirements are greatest.

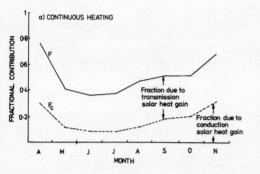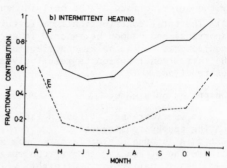

Figure 4. Monthly fractional contribution of total solar heat gains (F) and conduction solar heat gain (F_C) towards reducing the 'zero radiation' heating loads for the standard house in Melbourne.

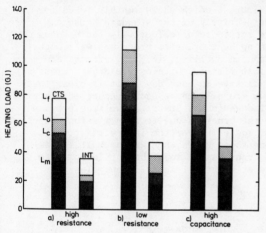

Figure 5. Seasonal values of the heating loads L_m, L_C, L_0 and L_f for continuous (CTS) and intermittent (INT) modes in Melbourne. The standard house (a) is examined and also variants in which (b) thermal resistance of the building envelope is low, and (c) thermal capacitance is high.

However, for case (c) the difference is quite small. The desirable rapid response of the heated space for the intermittent mode is retarded by a high level of thermal capacitance. In fact, whereas for the continuous mode L_m is lower for case (c) than for case (b), the reverse behaviour is observed for the intermittent mode. Note too that the ratio of the value of L_m for the intermittent mode to that for the continuous mode is about 0.25 (equal to the fraction of the day for which heating is required in the intermittent mode) for cases (a) and (b), but the ratio for case (c) is close to 0.5.

(2) For the intermittent mode the solar contribution fraction is greatest for case (a) whilst for the continuous mode it is greatest for case (c) in which incident solar radiation can best be stored for later use. However, the high-resistance construction is preferred for both modes since L_m is thereby minimized. Thus it is important, if stored solar energy is to be effectively utilized, that the building envelope be well insulated.

(3) The effects of transmission solar heat gain dominate the total solar effect in cases (a) and (c) but not in (b). Here the low level of thermal resistance in the building envelope enables conduction transfer of solar energy absorbed

by the outer surface of the building envelope to proceed more readily.

(4) The total contribution from fortuitous heat gains is of considerable importance and cannot be ignored in any calculation technique for heating energy requirements.   The solar contribution dominates over that from the occupants. The latter contribution is relatively more useful in a well-insulated construction.

THE EFFECT OF WINDOWS

It is clear that, taken in isolation, transmission solar heat gain reduces heating energy requirements.   Nevertheless, maximization of transmission solar heat gain through use of large window areas is not necessarily compatible with a lessening of heating energy requirements, since unless checked substantial heat losses, particularly at night, will occur through windows during the heating season.   Insulation is most commonly provided in Australian dwellings by close fitting curtains, preferably with a backing of low emittance, drawn across windows at night.   Such insulation is not provided in the standard house of Figure 1.

These points are illustrated in Figure 6 which shows monthly loads for continuous and intermittent heating respectively of the standard house and variants in which (i) window areas are set to zero (effectively the windows are replaced by material through which no heat transfer processes will occur) whilst wall areas remain as in the standard dwelling, ie., the effects of windows are entirely removed;   (ii) close fitting curtains are drawn across all windows between the hours 1700-0900 inclusive on each day.   It is clear that if heat losses at night through windows are allowed to occur unchecked, then for the continuous heating mode the windows in the standard dwelling provide a definite thermal disadvantage, ie., the benefits due to transmission solar heat gain are outweighed by these heat losses.   For the intermittent mode, however, heating is not required during those hours when window heat losses are greatest, and thus the effects of transmission solar heat gain dominate. For both modes, windows well-insulated at night will considerably lessen heating energy requirements.

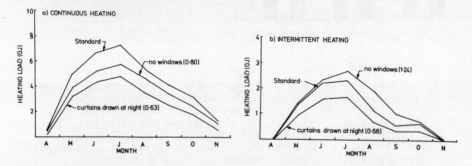

Figure 6.   Monthly heating loads in Melbourne for the standard house and variants in which all windows are removed or tight fitting curtains are drawn at night.   The bracketed figures are seasonal loads relative to that for the standard house.

CONCLUSIONS

It is possible to describe the hour-by-hour influences and effects of solar radiation incident on a building using a computer-based mathematical model.   Such passive effects represent the most basic use of solar energy within the built environment.

Lack of adequate solar radiation data presents theoretical difficulties in this respect. A method developed by Bugler [15] appears to be adequate for calculation of energy requirements in those situations in which only global radiation data is available, and appears to be applicable to most locations. On the other hand, use of either theoretical clear sky solar radiation or of zero solar radiation can result in large errors in load estimation.

To make the best use of incident solar radiation for heating purposes, the building envelope should be of high thermal resistance. This particularly applies to windows which should be insulated at night with tight fitting curtains or insulated shutters to reduce heat losses and thus conserve daytime solar heat gain.

REFERENCES

1. Ballantyne, E.R. 1975. Energy costs of dwellings. Thermal Insulation (Aust.), 1:7-14.

2. Read, W.R., and Wooldridge, M.J. 1976. Solar heating - domestic applications. Arch. Sci. Rev., 18: 31-34.

3. Ballantyne, E.R. 1976. Solar energy - the resource and its availability. CSIRO Div. Build. Res., Rep.47.

4. Ayres, J.M. 1974. Energy calculations. ASHRAE J., Dec., 71-72.

5. Scanes, P.S. 1974. Climatic design data for use in thermal calculations in buildings - estimated clear sky solar radiation versus measured solar radiation. Build. Sci., 9:219-225.

6. Muncey, R.W. and Holden, T.S. 1967. The calculation of internal temperatures - a demonstration experiment. Build. Sci., 2: 191-196.

7. Rao, K.R. and Ballantyne, E.R. 1970. Some investigations on the Sol-Air temperature concept. CSIRO Div. Build. Res., Tech. Paper 27.

8. Höglund, B.I., Mitalas, G.P. and Stephenson, D.G. 1967. Surface temperatures and heat fluxes for flat roofs. Build. Sci., 2:29-36.

9. Ballantyne, E.R. and Spencer, J.W. 1969. Solar radiation incident on building surfaces and solar heat gains through windows. Proc. Symp. environmental physics as applied to buildings in the tropics, Vol. II, Central Building Res. Inst., Roorkee, India.

10. American Society of Heating, Refrigeration and Air-Conditioning Engineers, Inc. 1972. Handbook of fundamentals.

11. Spencer, J.W. 1974. Melbourne solar tables - Tables of solar position and radiation for Melbourne (latitude 38°S) in S.I. units. CSIRO Div. Build. Res., Tech. Paper 7 (2nd Ser.).

12. Rao, K.R. and Seshadri, T.N. 1961. Solar insolation curves. Indian J. Met. Geophys., 12 : 267-272.

13. Norris, D.J. 1968. Correlations of solar radiation with clouds. Solar Energy, 12 : 107-112.

14. American Society of Heating, Refrigerating and Air-Conditioning Engineers, Inc., 1975. Subroutine algorithms for heating and cooling loads to determine building energy requirements. ASHRAE Task Group on energy requirements.

15. Bugler, J.W. 1975. The determination of hourly solar radiation incident upon an inclined plane from hourly measured global horizontal insolation. CSIRO Solar Energy Studies Unit, Rep. 75/4.

16. Liu, B.V. and Jordan, R.C. 1960. The interrelationship and characteristic distribution of direct, diffuse and total solar radiation. Solar Energy, 4 : 1-19.

17. Wooldridge, M.J. 1972. Solar radiation and its effects on air conditioning load. CSIRO Div. Mech. Eng., Report E.D.12.

18. Davies, M.G. 1976. The contribution of solar gain to space heating. Solar Energy, 18 : 361-367.

19. Wilberforce, R.R. 1976. The effect of solar radiation on window energy balance. Intl. CIB W67 Symp. on Energy Conservation in the Built Environment, UK.

HOUSES AS PASSIVE SOLAR COLLECTORS

J. B. SIVIOUR

Electricity Council Research Centre
Capenhurst, Chester, CH1 6ES, United Kingdom

ABSTRACT

Measurements show houses to be solar collectors with efficiencies of 6-7% for space heating. The useful contribution amounts to 2000-3000 kWh/yr, depending on the level of insulation.

Calculations show that concentrating glazing as far as is reasonable in south-facing walls would increase the useful contribution by about 300-400 kWh/year. But even when south-facing, double glazing in winter loses more heat than the solar gain through it. Insulating shutters used when dark and on dull days would improve the heat balance.

INTRODUCTION

The sun provides a useful contribution to space heating in houses, and by measuring internal and external temperatures with no other heating, it is possible to quantify how much. Such measurements have been made in the ECRC test houses, which are representative of British housing and cover a wide range of insulation levels. The tests were made over periods throughout the year.

Based on these measurements detailed calculations have been made to assess the solar heating which is available during the heating season and how much could be expected to be used in conventional houses. The results include the effects of orientation, and the use of double glazing and window shutters, and are compared with the effects of insulation.

The variability of insolation daily is illustrated for January and problems of utilization identified which such variability can cause.

THE ECRC TEST HOUSES

The six test houses are shown in figure 1. All are identical in layout internally except that they are alternately left- and right-handed. They are two-storey buildings, have three bedrooms, a floor area of 85m² and volume of 200m³. Details of their construction and instrumentation can be obtained from references 1 and 2. A summary of their characteristics relevant to this paper is given in Table 1.

The weather data recorded on site are wet and dry bulb screen temperatures, windspeed and direction, and insolation on the horizontal plane.

749

Figure 1.   Experimental houses at Capenhurst.   Their size and layout are typical of British housing.   These frontages face south-east.

Table 1.   House heat losses, component type and areas externally
Overall roof area in plan 48m^2

House number	Theoretical fabric heat loss	Glazing			Total wall areas (m^2) including windows and type (2) facing			
		Type (3)	Area (m^2) facing					
	kWh/K day		NW	SE	NE	NW	SE	SW
8	5.42	Single	4.1	5.5	-	30F	30F	40B
10	5.9	Single	5.3	6.7	40B	30B	30B	-
12	5.42	Single	5.3	6.7	-	30F	30F	40B
14	5.9	Single	5.3	6.7	40B	30B	30B	-
16 (i)	3.96	Double	5.3	6.7	-	30F	30F	40B
(ii)	2.76	Double	4.1	4.3	-	30F	30F	40B
(iii)	2.08 (1)	Double	4.1	4.3	-	30F	30F	40B
18	7.25	Single	5.3	6.7	40B	30B	30B	-

(1)   The value in the solar heating calculations is much lower (see Table 2) after allowing for temperatures in the basement and instrument cavity.

(2)   F timber framed construction, B brick construction

(3)   Solar transmittance factors taken from IHVE Guide, on a daily basis are:

(single glazing, plain 0.75, frosted 0.7, composite 0.74 for SE facade
(double glazing, plain 0.65, frosted 0.6, composite 0.64 for SE facade

Internal temperatures are usually recorded at eight positions in each house, shielded from direct solar radiation.   Tests are made with the houses unoccupied but fully furnished, which includes fitted carpets except in the kitchen, bathroom and WC which have plastic floor tiling.

EXPERIMENTAL FINDINGS

Measurements

Experiments were carried out at periods throughout the year in four of the test houses with only the sun providing heat.   Internal temperatures were measured continuously on chart recorders in the following eight positions: lounge, kitchen, dining area, hall, landing and the three bedrooms.   Externally the dry bulb temperature, insolation on the horizontal plane and windspeed were measured hourly.   Each experiment ran for a number of consecutive days and averages of the recorded data for the whole experiment used to minimize the effects of changing temperatures and the thermal capacity of the houses.   The results are summarized in Table 2 and can be obtained in more detail from reference 3.

During the winter tests, in house 16, temperatures were measured in the loft, basement and across the party wall.   This was necessary to assess accurately the small solar heating and modified the heat loss characteristics given in Table 1 to those given in Table 2.

Analysis

Insolation on the vertical surfaces (item 9, Table 2) is obtained from the average daily measurements on the horizontal plane (item 8) using Figure 2. Insolation on the whole exterior (item 10) is then calculated using the areas given in Table 1.

Solar energy transmitted through the glazing (item 11) is also obtained from item 9 using the glass areas and transmittance factors in Table 1.   For the glazing facing south-east the composite values are used to allow for the frosted glass in the front door.

The total solar heating is equal to the fabric and ventilation losses during tests (item 13).   Fabric losses are calculated from items 2 and 6. The ventilation rates (item 12) are based on measurements (reference 3) and the ventilation heating requirement, VEN, equals 1.6 VΔT kWh/day for these houses.

Interpretation

The total solar heating (item 13) is in all cases larger than the calculated heat transmission through the glass (item 11).   The ratio (item 14) varies considerably from 1.24 to 1.92, with a weighted average of 1.49.   This can be interpreted as on average 2/3 of the total solar heating is transmitted through the glazing and 1/3 through the structure.

The structural heating is given in item 15 and worked out as a percentage of the daily total insolation on the houses (16).   The weighted average is 2% with a variation from 1.19 to 3.41%.   It is interesting to note that the averages for each house are higher at 2.3% for the all brick construction than the values of 1.95 for house 8 and 1.8 for house 16 with double glazing, both having timber framed front and back walls.

On average therefore the houses are solar collectors with efficiency of 6%.   The average for the brick houses with single glazing is 6.9% and for the well insulated house with double glazing 5.4%.   The lower value with double glazing is a result of its lower solar transmittance.   The higher value with single glazing agrees better with an earlier analysis (reference 5) which did not include results from house 16.

Table 2. Summary of the experimental findings on solar heating in the ECRC test houses

		Test House 8				Test House 14			Test House 16					Test House 18				
1. Test House number		8				14			16					18				
2. Theoretical fabric heat loss	kWh/K day	5.42				5.9			3.96			1.06	1.29	7.25				
3. Month		Apl.	July	May	June	Apl	May	June	Mar.	Apl/May	July	Dec/Jan	Feb.	Mar.	May	June	June	July
4. Number of consecutive days		13	13	11	9	13	11	9	15	19	5	10	13	14	17	9	13	7
5. Median day number		106	196	146	156	146	156	196	81	128	198	363	38	85	143	156	175	185
6. Average internal external temperature difference	ΔT K	4.0	4.4	5.0	5.6	5.0	5.6	4.9	3.21	5.13	6.54	3.0	2.35	2.1	4.2	4.9	4.8	3.4
7. Average windspeed	W m/s	1.9	3.7	3.1	4.9	3.7	3.1	4.9	4.8	4.3	3.9	3.0	3.35	1.6	2.7	4.9	2.1	5.3
8. Average daily insolation on horizontal	SH kWh/m²day	3.32	4.94	4.87	4.98	4.94	4.87	4.98	1.93	4.12	4.95	0.5	0.72	2.3	4.68	4.98	4.73	4.4
9. Average daily insolation on vertical surfaces NE/NW	kWh/m²day	1.40	2.24	2.2	2.27	2.24	2.2	2.27	0.84	1.77	2.23	0.24	0.36	0.86	2.1	2.27	2.12	2.0
SE/SW	kWh/m²day	2.43	3.0	2.96	2.92	3.0	2.96	2.92	1.59	2.67	3.0	0.6	0.66	1.92	2.85	2.92	2.73	2.62
10. Total insolation on house exterior	kWh/day	371	514	477	486	514	477	486	230	438	515	73	92	228	458	486	458	430
11. Calculated heating through glazing NW	kWh/day	4.3	6.8	8.8	9.0	6.8	8.8	9.0	2.9	6.1	7.7	0.64	0.96	3.4	8.3	9.0	8.4	8.0
SE	kWh/day	9.9	12.2	14.5	14.3	12.2	14.5	14.3	6.8	11.4	12.9	1.65	1.82	9.3	13.9	14.3	13.4	12.8
Total	kWh/day	14.2	19.0	23.3	23.3	19.0	23.3	23.3	9.7	17.5	20.6	2.29	2.78	12.7	22.2	23.3	21.8	20.8
12. Average ventilation rate	V ac/hr	0.27	0.38	0.37	0.38	0.38	0.37	0.38	0.31	0.29	0.27	0.25	0.26	0.19	0.36	0.38	0.37	0.35
13. Total solar heating Ventilation VEN	kWh/day	1.73	2.7	2.96	3.4	2.7	2.96	3.4	1.6	2.4	2.8	1.2	0.98	0.6	2.4	3.0	2.8	1.9
Fabric FEN	kWh/day	21.68	23.8	29.5	33.0	23.8	29.5	33.0	12.7	20.3	25.9	3.18	3.03	15.2	30.4	35.5	34.8	24.7
Total	kWh/day	23.4	26.5	32.5	36.4	26.5	32.5	36.4	14.3	22.7	28.7	4.4	4.0	15.8	32.8	38.5	37.6	26.6
14. Total heating ÷ heating through glazing (13÷11)		1.65	1.39	1.39	1.56	1.39	1.39	1.56	1.47	1.30	1.39	1.92	1.44	1.24	1.48	1.65	1.72	1.28
15. Structural solar heating (13-11)	kWh/day	9.2	7.5	9.2	13.1	7.5	9.2	13.1	4.6	5.2	8.1	1.9	1.0	3.1	10.6	15.2	15.6	5.8
16. Solar heating ÷ total insolation (15÷10)	%	2.45	1.46	1.93	2.69	1.46	1.93	2.69	2.0	1.19	1.57	2.9	1.33	1.36	2.31	3.13	3.41	1.39
Average for each house	%	1.95				2.3			1.8					2.3				

Weighted averages: Total heating ÷ heating through glazing (item 14)   1.49

Solar heating ÷ total insolation (item 16)   2%

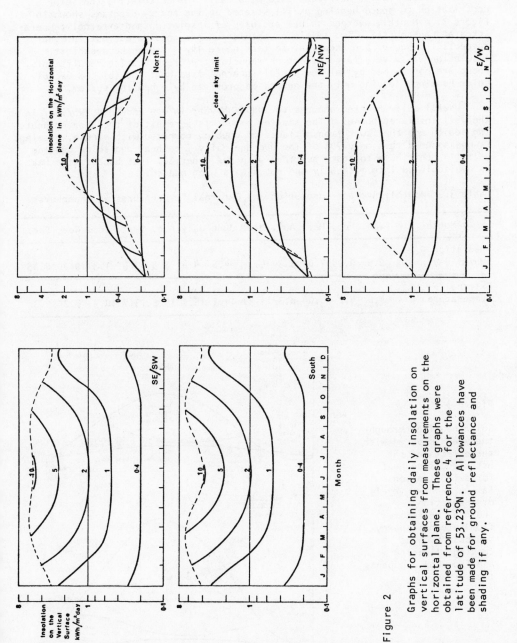

Figure 2

Graphs for obtaining daily insolation on vertical surfaces from measurements on the horizontal plane. These graphs were obtained from reference 4 for the latitude of 53.23°N. Allowances have been made for ground reflectance and shading if any.

SEASONAL SOLAR CONTRIBUTION TO SPACE HEATING

    The above results have been used as a basis for calculating the solar
contribution to space heating as illustrated in the energy diagrams shown in
Figure 3.   Monthly average values are used of insolation and external temperatures at Capenhurst.   Two levels of insulation are considered, but in other
respects the houses are identical to test house 14.   It can be seen that
insulation has a very large effect on reducing the gross heating requirements.
These are partly met by (a) the useful internal free heat and (b) the useful
solar heating, leaving only the net to be provided by a heating system.

    Useful solar heating amounts to 3000 kWh/year in one and 2000 kWh/year in
the well insulated house.   The smaller amount is a result of the shorter heating season and the lower transmittance of double, compared with single, glazing.
It was assumed that only 70% of the average daily free heat and solar heating
usefully contribute to space heating, with the other 30% lost because the heat
is available so intermittently and not always when needed.

Table 3.   Monthly Average Insolation and External Temperatures at Capenhurst

Month	Jan.	Feb.	Mar.	Apl.	May	June	July	Aug.	Sept.	Oct.	Nov.	Dec.
Insolation kWh/m^2 day	0.5	0.98	2.04	3.3	4.1	4.8	4.4	3.5	2.4	1.37	0.7	0.45
External temperature $^{\circ}$C	4.0	4.2	6.0	8.3	11.3	14.1	15.7	15.6	13.7	10.3	7.2	5.3

Figure 3

Solar heating through
the year compared with
the effects of the
estimated free heat
for a four person
family and two levels
of insulation.
The free heat comes
from internal energy
sources, such as
cooking, lighting
and from the occupants
themselves.

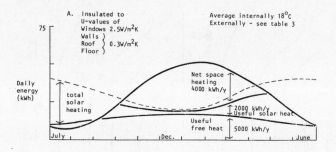

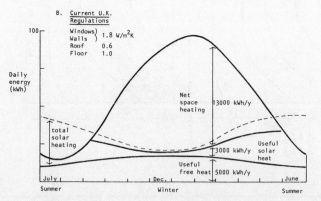

EFFECTS OF ORIENTATION ON SEASONAL SOLAR HEAT GAINS

The orientation of semi-detached houses will have some effect on both the structural and glazing solar gains.   Some calculated results are summarized in Figure 4 for the well insulated house above, so the heating season extends from October to April.

The structural gains have been taken to be 2% of the average monthly insol-ation on the whole exterior, calculated using Table 3 and Figure 2.   The gains through the double glazing have also been calculated from monthly average data for a total glass area of 12m^2 split in the ratio of 2:1 as shown.   This ratio covers reasonable limits for daylighting in houses.   The east/west split does not matter when calculating on a whole house basis.

The results are that the house with southerly windows has a seasonal gain of around 400-500 kWh more than the other three.   At 70% utilization this would be a useful contribution of 280-350 kWh, about 7-9% of the net need of 4000 kWh.   This is an improvement of roughly 14-18% of the useful contribution of 2000 kWh/year in Figure 3 which applies to an east/west orientation of windows and includes some solar heating in May, June and September.

April gains have been given separately because they amount to around 30% of the total at a time of the year when net heating requirements are small. Solar heating could be easily in excess of total need on bright days, yet be very small on dull days, resulting in a useful contribution of less than 70%.

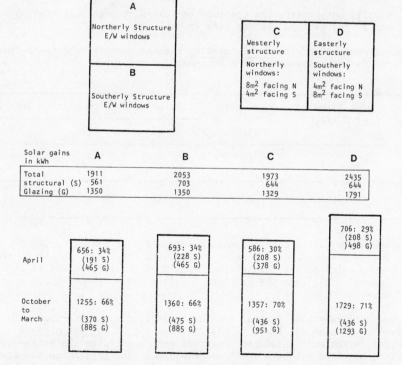

Figure 4.   Effects of house orientation on structural (S) and solar gains through double glazing (G) from October to April

ENERGY BALANCE THROUGH GLAZING

(a) Seasonal

Energy transfer through glazing is shown in Figure 5 for $1m^2$ facing south, calculated from average insolation and external day and night-time temperatures for Capenhurst.   Average losses exceed gains in winter except when shutters are used.   Day and night-time temperatures of 19°C and 16°C are used internally 5.5°C and externally 2.5°C.   Allowances are made for curtains and insulating shutters (U = 0.3 W/m^2K overall) being drawn when dark.

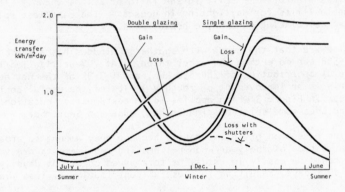

Figure 5.   Daily solar gains and transmitted losses through south-facing glazing based on the average weather data in Table 3

On a seasonal basis with <u>double glazing</u> the balance (in kWh/m^2) for north and east/west as well as south-facing becomes:

Period	Loss allowing for curtains	Loss with insulating shutters	Solar gains with glazing facing		
			S	E/W	N
October to March	123	65	135	73	50
October to April	141	77	186	112	71

Only southerly facing glazing shows a net gain over the season.   For northerly facing glazing there is deficit even with insulating shutters.   Based on these simple calculations then glazing should be omitted, unless it is southerly facing.   But then the problems of glare and overheating can arise on bright days.

The seasonal reduction in heat loss using insulating shutters works out to be 64 kWh/m^2 or 770 kWh for $12m^2$.   This is about 20% of the net need of 4000 kWh/year in the well insulated house and more than twice the extra useful solar heating obtained with the south orientation.   Shutters which may already be used in summer could possibly be modified for winter use.

(b) Daily

Insolation and temperatures vary considerably from day to day and during the day, and therefore the results above based on average data should not be taken as definitive.   To illustrate the effects of daily variability, insolation for a January has been examined from two sites, Bracknell and Capenhurst.

Measurements on the horizontal plane are shown in Figure 6a overleaf in both date order and increasing magnitude.   About an equal number of days receive more insolation than average and about an equal number less.   The dullest days receive about 30% of the average, and the brightest days about twice the average or more.

A similar distribution is found for a vertical surface facing north (figure 6b) but with an average insolation of only 183 W/m^2 day.   The insolation needed to balance the 9 hour daytime heat losses through glazing taking average January temperatures externally is 550 Wh/m^2 day on double glazing and 910 Wh/m^2 day on single.   Insolation is every day less than these values, so using monthly averages is as applicable as using daily measurements.

For the south-facing surface the measurements are very variable, figure 6c. Half the days receive less than half the average and one day exceeds four times the average, a result of the low solar altitude in January.   Insolation now exceeds the level required to match transmission losses through double glazing on nearly half the days, assuming average temperature differences.   On the dull days insulated shutters would be thermally advantageous even during the daytime.   This would eliminate daylighting, but on such dull days artificial lighting may well be needed anyway.   A compromise on such days could be to partially draw back the shutters to provide a view out and some background lighting when the room is not in constant use.

The high variability of insolation is shown in terms of accumulative energy and days in Figure 7.   The horizontal plane measurements for both sites are

Figure 7

Examples of the accumulative distribution of insolation in January from measurements on horizontal surfaces at two sites and two vertical surfaces.   The value of total insolation is for the whole month.

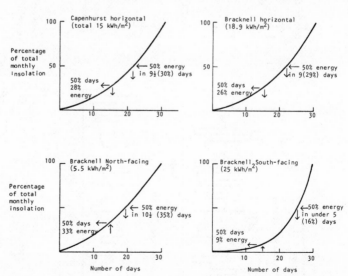

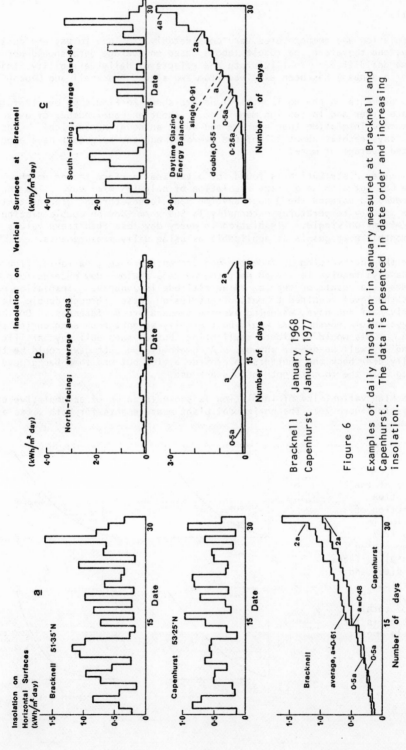

Figure 6

Examples of daily insolation in January measured at Bracknell and Capenhurst. The data is presented in date order and increasing insolation.

Bracknell   January 1968
Capenhurst   January 1977

similar, with half the energy incident over about 30% of the days, and half the days receiving just over 1/4 of the energy. The north-facing measurements at Bracknell show a similar distribution. The south-facing measurements however are much more skew. Half the energy is received in less than five days. Half the days share only 9% of the energy incident during the month.

CONCLUDING REMARKS

The calculated results of annual energies presented in this paper were made using monthly average insolation and external temperatures for Capenhurst. These are very close to the averages for the whole of Britain. The average internal temperature of 18°C allows for higher temperatures in living rooms when occupied and lower temperatures elsewhere.

Measurements in typical houses show that they are solar collectors with efficiencies of around 6% through the year. From these results it is estimated that the useful solar contribution to space heating is 2000 kWh/year in a well insulated house with double glazing and 3000 kWh/year in a house meeting the current Building Regulations in Britain.

Insulation itself has a far larger effect, reducing the net space heating requirement, after allowing for the solar contribution and internal free heat of a four person family from 13000 kWh/year to 4000 kWh/year. Generally there is a net loss of heat during the heating season even through double glazing. Insulating shutters drawn on all windows when dark would reduce heat loss through double glazing by a calculated 770 kWh over the heating season. Concentrating the glazed areas in walls facing south would increase theoretically the useful solar heating by 300-400 kWh/year above the values for other orientations.

However measurements show that insolation in south-facing vertical surfaces is very variable. In January daily values vary by more than 0.24-4 times the average so that on bright days rooms facing south in a well insulated house could overheat.

REFERENCES

1. Siviour, J.B. 1975. Construction instrumentation and heat loss data of the ECRC Experimental Houses. ECRC/N839.

2. Siviour, J.B. 1977. Measurements on the thermal performance of a house with different levels of insulation.

3. Siviour, J.B. 1977. Insolation and the performance of houses as passive solar collectors.

4. Basnett, P. 1975. Calculation of solar radiation on vertical surfaces from readings on a horizontal plane. ECRC/N846.

5. Siviour, J.B. 1976. The efficiency of houses in the use of solar energy for space heating. Sun at Work in Britain, No. 3, June.

EXPERIMENTAL STUDY OF PASSIVE AIR-COOLED FLAT-PLATE SOLAR COLLECTORS: CHARACTERISTICS AND WORKING BALANCE IN THE ODEILLO SOLAR HOUSES

J. F. ROBERT, J. L. PEUBE, AND F. TROMBE

Laboratoire de l'Energie Solaire
66120 Odeillo Font–Romeu
France

The natural convection of air in classical flat-plate solar collectors is generally considered only for the evaluation of the heat
losses of the collector.

This phenomena can also be used to restore to the load the solar
energy collected or to produce air mouvments into a volume.

The collectors of the ODEILLO solar houses and solar chimneys are
examples of application of natural convection to solar processes.

The collectors of the solar houses (fig.1 and 2), generally called
"TROMBE wall Collectors", (1) (2) (3) (4) are set in a vertical East/
West plane (Vertical south wall) because of the many advantages of that
arrangement. Energy is collected between a double selective covering
(double layer of glass, for example) and the absorbent surface of a
heavy masonry wall. One part of the energy is transmitted to the air
of the hot house that, in turn, is warmed, rises through natural convection, and enters the space to be warmed by an opening in the upper
part of the wall. Fresh air is pulled into the system naturally, by
an opening at the level of the floor, in the lower part of the wall.
Natural thermal convection circulation is thus established that is set
off the first moment the sun falls on the facade, and continues efficiently until two or three hours after sunset, depending on the quantity of energy received during the day and the climatic conditions
outside.

Another part of the energy is transmitted by conduction through
the storage material (the wall) and manifests itself, with a time
lapse that is a function of the nature and thickness of the wall, by
a rise in temperature on the wall's interior surface. That energy is

thereupon spread through the space to be heated by conduction/convec-
tion of air, and by radiation toward the other interior surfaces.

    In the solar chimney, convection is used to induce air mouvment
into the solar collector and the associated room and volume. It can be
used, for instance, to move the air of a solar dryer or to pull out
the air of a room. In the last case, it is possible to treat the air
entering the room to compensate the evacuated one by means of dessica-
tion   in humid climates, or evaporative cooling in arid areas.

    Experimental and industrial solar collectors are used for heating
six solar houses in ODEILLO (one 1962 experimental house, two 1967
prototype houses and three 1974-1975 privately owned houses). An experimental building is equiped with a solar chimney and a humidifying
system.

    A theoratical study of the convection in such solar devices was
initiated in 1976 and experimental studies were realised since 1974 in
the different solar collectors and with the help of a simulation prototype.

I - GENERAL STRUCTURE OF THE FLOW

    A Trombe wall collector may be considered as a vertical plane duct
inside which flow is driven by a thermosiphon. The actual flow, rather
intrcated to describe, may be divided in several regions (Fig.3a):

A) an entrance length in the duct, located at the bottom of the wall,
in whichtwo boudary layers are developping, after an initial separation
due to contraction of the flow in a pipe with sharp angles. It may be
noticed that this separation could be prevented by a suitable profile
of the entrance design, but it does not seem to be of interest to do
such a change.

B) a right-angle deviation of flow which induce a second separation
just at the bottom of the wall collector.

C) an entrance length for flow inside the vertical plane duct, where
air begins to be heated in some kind of thermal boundary layer growing
fromthe separation zone at the bottom of the wall.As Prandtl number
is of order of one, some kind of dynamical boundary layer is also
growing from the bottom of this channel, but on the two sides of this
one (wall and glass cover). The thickness of these boundary layers
are growing up to the width of the channel. The flow and heat transfer
are then being established.

D) a zone of fully-developed flow and heat transfer which we shall
discuss more thoroughly in next paragraph.

E) a second elbow with sharp angles

F) a duct flow very much pertubated by the elbow flow.

    For the active part of the flow, that is to say, those regions C
and D in which the necessary motrice pressure is induced by a phenomenon
of thermosyphon, we have to get a better knowledge of the fully-developed
flow and of the entrance zone. Unfortunately, this last one is very
stongly influenced by the elbow flow number B. Vrey little is known on
these regions and a lot of studies have to be performed.

The fully- developed zone of flow and heat transfer is probably much easier to study, but the main feature is that the motrice part of flow is entirely controlled by the vertical temperature repartition which is strongly dependant of mass flow rate in duct. We shall discuss this point in next paragraph.

The passive part of flow leading to the head losses is easier to study : zones A,B,E, and F nearly isothermal, have head losses which are nearly independant of heat transfer phenomena and which may be estimated with usual engineering methods. Of course precision may be not very high, but order of magnitudes may be obtained.

The head losses in parts C and D are much more difficult to know. Actually in these regions, a flow is developing in a quite completely different way from a head loss in an isothermal pipe or in an thermosiphon with a constant temperature (i.e. a chimney). The velocity profiles are thoroughly different from the previous flows cited because of heat transfer; hence, it is very uncertain to use the Moody diagram or some equivalent formulas.

We have also to take into account the nature of flow?For the regions A,B,D, and E, as far as head losses are singular, the value of some Reynolds number characteristic of the flow is not the major feature. On the contrary, in zones D the occurence of turbulence is very important, because it leads to a thorough change in heat transfer and friction mechanisms. Nearly nothing is known in this field.

II - THEORETICAL STUDY OF FULLY DEVELOPED LAMINAR FLOW.

II.1 Let us consider a plane vertical channel (with notations of figure 3b) in which a Boussinesq fluid is flowing. The equations of flow and heat transfer are, under the assumptions of a laminar flow with boundary layer approximations :

- mass conservation:

$$(1) \qquad \frac{\partial u}{\partial x} + \frac{\partial v}{\partial y} = 0$$

- Navier-Stokes equations:

$$(2) \qquad \rho_0 \left(u \frac{\partial u}{\partial x} + v \frac{\partial u}{\partial y} \right) = - \frac{d\pi_g}{dx} + \rho_0 \beta g (T - T_0) + \mu \Delta u$$

- energy equation:

$$(3) \qquad \rho_0 c_p \left(u \frac{\partial T}{\partial x} + v \frac{\partial T}{\partial y} \right) = \lambda \Delta T$$

with:

ρ_0 , T_0 : reference volumic mass and temperature
$\pi_g = \pi + \rho_0 g x$ motrice pressure
β : expansion coefficient for gas
g : gravity intensity
T : temperature
c_p : heat capacity at constant pressure
μ : dynamic viscosity
λ : thermal conductivity

The reference temperature T_0 and the corresponding volumic mass ρ_0 are chosen in such a way to give a physical sense to the motrice pressure. For example, in the case of air entering at temperature T_0 in the duct and flowing out of the duct in an atmosphere at this same temperature T_0, the interpretation of the motrice pressure p_g is obvious : it should remain constant in such a fluid at rest and at this constant temperature T_0. Thus boundary conditions on pressure p_g will be merely that difference is given :

(4) $$p_g (0) - p_g(\ell) = \Delta p_g$$

This pressure difference Δp_g may be due as well to a non uniform temperature in the atmosphere outside the plane collector , as to a flow in this atmosphere (wind, for example).

The other boundary conditions are :

- zero velocity on walls :

(5) $y = \pm e$ $u = v = 0$

- heat flux density is assumed to be constant and given on each wall :

(6) $$\begin{cases} \lambda \left(\dfrac{\partial T}{\partial y} \right)_{y=-e} = -q_1 \\[2mm] \lambda \left(\dfrac{\partial T}{\partial y} \right)_{y=e} = q_2 \end{cases}$$

- entrance temperature is equal to T_0.

II.2.   The equations and the boundary conditions (1) to (6) may be written under adimensional form with the following non dimensional variables, generally used in study of free convection :

(7)

$$x_+ = \frac{x}{\ell} \qquad y_+ = \frac{y}{e} \qquad u_+ = u\,\frac{e}{\nu} \qquad v_+ = v\,\frac{\ell}{\nu}$$

$$p_{g_+} = \frac{p_g}{\rho \left(\frac{\nu}{e}\right)^2} \qquad\qquad T_+ = \frac{\lambda e\,(T-T_0)}{q_1+q_2}$$

It is then easy to obtain the non dimensional equations :

(8)

$$\frac{\partial u_+}{\partial x_+} + \frac{\partial v_+}{\partial y_+} = 0$$

$$u_+ \frac{\partial u_+}{\partial x_+} + v_+ \frac{\partial u_+}{\partial y_+} = -\frac{\partial p_{g_+}}{\partial x_+} + \frac{\beta g e^4(q_1+q_2)}{\lambda \nu^2} T_+ + \Delta_+ u_+$$

$$u_+ \frac{\partial T_+}{\partial x_+} + v_+ \frac{\partial T_+}{\partial y_+} = \frac{\lambda}{\rho\, c_p} \Delta_+ T_+$$

with the following boundary conditions :

$$(9) \begin{cases} \left(\dfrac{\partial T_+}{\partial y_+}\right)_{y_+=-1} = \dfrac{q_1}{q_1+q_2} \qquad \left(\dfrac{\partial T_+}{\partial y_+}\right)_{y_+=1} = \dfrac{q_2}{q_1+q_2} \\[2mm] y_+ = \pm 1 \qquad u_+ = v_+ = 0 \\[2mm] x_+ = 0 \qquad T_+ = 0 \\[2mm] p_{g_+}(0) - p_{g_+}(1) = \dfrac{e^2 \, \Delta p_g}{\rho \, v^2} \end{cases}$$

These equations show that the problem is depending upon the five following non dimensional parameters ρ:
- the reduced length of duct : $\dfrac{\ell}{e}$

- the non dimensional pressure difference : $\dfrac{e^2 \, \Delta p_g}{\rho \, v^2}$

- Prandtl number $\quad Pr = \dfrac{\mu \, c_p}{\lambda}$

- Grashof number : $\quad Gr = \dfrac{g \, \beta \, e^4 (q_1 + q_2)}{\lambda \, v^2}$

- ratio of heat fluxes : $\dfrac{q_1}{q_2}$

II.3. With such a number of parameters, it seems that discussion of this problem is not easy. Fortunately the evolution of temperature along the x-axis may be easily written . If T_m is the mean temperature difference defined by the equation :

$$T_m = \frac{\rho_0 \int_{-e}^{+e} u T \, dy}{q_m}$$

with $q_m = \int_{-e}^{+e} \rho_0 u \, dy$ mass flow rate

it is easy to show that this temperature obeys the following law :

$$(10) \qquad T_m - T_0 = \frac{q_1 + q_2}{q_m \, c_p} \, x$$

But mass flow rate q_m is a quantity which can be known only if problem is completely solved. So, it follows from equation (10) that mean temperature T_m is a linear function of abscissa x , if heat flux densities are constant. This is a very different repartition from the case of an usual chimney or thermosyphon where temperature is nearly constant.

II.4. The linear dependance of mean temperature T_m with abscissa x

suggests to try the following solution for the problem of fully-developed flow and heat transfer :

$$(11) \begin{cases} u = u(y) \\[3mm] T = T_0 + Ax + \theta(y) \end{cases}$$

This solution is valid only far enough from the inlet of the duct, where the normal component of flow has vanished. Taking (11) into equations (2) and (3), we obtain, after separation of variables x and y :

(12) $-\dfrac{dp_g}{dx} + \rho_0 \beta g A x = B$

(13) $\rho_0 \beta g \theta + \mu \dfrac{d^2 u}{dy^2} = -B$

(14) $\rho_0 c_p A u = \lambda \dfrac{d^2 \theta}{dy^2}$

where B is an unknown constant.

Boundary conditions (5) and (6) are valid without any change.

Integration of equation (12) and condition (4) give the motrice pressure (15) along the x axis in this fully developed zone of flow :

(15) $p_g (x) = p_g(0) + A \rho_0 \beta g \dfrac{x^2}{2} - Bx$

For sake of simplicity, we shall assume that there is no entrance length for flow and heat transfer (i.e. pipe is long enough). A condition between unknown conctants A and B is then obtained from condition(4) and relation (15) :

(16) $\Delta p_g = - A \rho_0 \beta g \dfrac{\ell^2}{2} + B\ell$

From equations (13) and (14), it is easy to obtain a linear differential equation of order four with constant coefficients for anyone of the two unknown functions u or θ .

Four boundary conditions deduced from (5)and (6) have to be used .

We need two other conditions for the determination of the solution because constants A and B have to be calculated. Equation (6) gives one of them. The other one is obtained by writing that the mean temperature is given exactly by combination of relations (10) and (11), in order to fix the origin of temperature at the inlet of the duct :

(17) $\displaystyle\int_{-e}^{+e} u \, \theta \, dy = 0$

It may be noticed that this relation is independant of the existence of an entrance length for flow and heat transfer.

II.5. Analytical and numerical integration of the previous system of equations is possible, but rather tedious. Such a kind of solutions has already been studied by S. OSTRACH [6] , but with constants A and B assumed to be known ; this assumption leads to a problem much more easy to solve.

Two features of interest must be noticed :

- temperature is a linear function of x coordinate,

- motrice pressure is a quadratic function of x coordinate : this means that if we consider a duct twice longer than another one, with the same heat flux density (so heat flux is twice greater), the mass flow and the temperature profile will not be the same. Problem is not linear with the length 1 of the duct.

The non dimensional parameters governing the flow may be obtained as previously by changing the variables under non dimensional form with relations (7) . Developing the non dimensional solution leads to the important fact that only three non dimensional parameters (instead of five) are effectively present for the case of fully developed flow and heat tranfer :

- A modified Rayleigh number : $\quad Ra = \dfrac{\beta g e^5 (q_1 + q_2)}{\nu a l \lambda}$

(with a : thermal diffusivity)

- A pressure parameter : $\quad P = \dfrac{\lambda \, \Delta p_g}{P_0 \, \beta g \, e \, l (q_1 + q_2)}$

- A parameter for the ratio of heat fluxes on the two walls :

$$Q = \dfrac{q_1 - q_2}{q_1 + q_2}$$

The pressure parameter P is a characteristic ratio of the relative effects of external pressure difference and of buoyancy.

II.6. The complete numerical determination of previous solution would be of interest for plane wall collector, because the air flow is induced mainly in fully developed flow (zone D). Nevertheless, it is necessary to have a better knowledge of experimental conditions in which flow is laminar or turbulent, in order to use the correct relations in the various zones of the flow. So , a preliminary experimental study has been performed, under practical conditions for use of air collectors, but with geometrical characteristics simpler than those of Trombe collectors.

III - PRELIMINARY EXPERIMENTAL STUDY OF THE SOLAR CHIMNEY.

Because of the many problems related to the theoratical approach
of the natural convection processes in passive solar collectors, a
preliminary experimental study was realized to investigate the influence of the different working parameters of the solar collector
and solar chimney processes (fig.4).

A concrete slad 5cm thick, 1m wide and 3m high is installed vertically in supports at about 1m above the floor level. The back of the
wall is equiped with electrical resistances and insulation (glass-
wool and styrofoam). The heat input can be varied by variying the
electrical imput to the resistances. An initial length of 1m is not
heated and realizes an inlet zone. On the front face, an aluminum
or transparent polycarbonate cover can be slided in and out along rub-
ber gaskets to adjust the spacing between the plate and the cover. To
realize the working conditions of the TROMBE wall solar collectors,
horizontal inlet and outlet ducts can be adjusted to the simulation
prototype. For the study of the solar chimney the inlet and outlet
openings are the bottom and top surfaces between hot plate and cover.

Temperature measurements are realized with embeded thin (0,1mm)
thermocouples at the surface of the hot wall and the cover at different heights (fig.4) and air temperatures are measured with thin
(0,1mm) aluminum-schielded thermocouples.

The air velocity is given by a precalibrated low velocity hot wire
anemometer DISA 55 D 80/81. The measuring range is between - 30 and
+ 30cm/s in the LVA (Low Velocity anemometer) mode and it can be increased to - 60 to + 60cm/s with a 60° incidence angle between the
hot wire and the direction of the flow.

A preliminary set of experiments was realized in May and June 1977.
The main properties measured are the wall and cover temperatures at
different heights, the electric imput to the heating resistances, the
ambiant temperature in the room, and the velocity of air in the inlet
zone of the chimney (the average velocity $\bar{V}$ is equal to 0,81 V_{max},
according to the results of different former measurements with the
Log-Linear method). Air temperatures at different locations into the
air gap are also measured but, because of the impossibility to achieve simultaneous velocity and temperature measurements, they were not
taken it account for the results of this experimental study.

The electrical heat imputs used for the first set of experiments
were 200 W/m^2, 400 W/m^2 and 635 W/m^2 for a 2m^2 heating zone. For each
value of the heat imput, the distance between the plate and the cover
was varied from 5 to 20 cm (5,7.5,10,15,20cm). For each experimental
configuration, the different measurements were realized after a 24
hours stabilisation time to reach steady-state conditions, and the
heat losses through back and lateral insulation and through the cover
were computed to determine the amount of energy Q dissipated in the
chimney.

The mean outlet temperature is then

$$T_{out} = \frac{Q}{\dot{m}.C_p} + T_{in}$$

where $\dot{m}$ is the measured mass-flow rate, C_p the specific heat of air and T_{in} the mean inlet air temperature (room temperature). The average air temperature in the chimney is

$$\bar{T}_f = \frac{T_{in} + T_{out}}{2}$$

and the mean convection coefficient $\bar{h}$ is

$$\bar{h} = \frac{Q}{A(\bar{T}_p - \bar{T}_f)}$$

where A is the heating area and $\bar{T}_p$ the average hot plate temperature.

For different values of the spacing between plate and cover, are plotted in figure 5 the values of the mass flow-rate computed from the velocity measurements in the inlet zone of the chimney, in figure 6 the Reynolds number calculated for the average velocity $\bar{V}$ and the hydrolic diameter d_H of the channel and in figure 7 the values of $T_p - T_{room}$.

In figure 8, are plotted the values of the Nusselt number N_u computed fo $\bar{h}$ and the hydrolic diameter d_h, in terms of the Crashof number

$$G_r = \frac{g \; \beta \; d_h^5 \; Q_u}{\lambda \; \nu^2 \; H}$$

where g is the gravity acceleration, β the cubic expansion coefficient of air, Q_u the heat flux dissipated per unit of surface in the chimney and H the height of the heating part of the channel.

The characteristics of air were determined for the average temperature between hot plate and air

$$\frac{\bar{T}_p + \bar{T}_f}{2}$$

The values correlate as

$$Nu = 1,91 \; Gr^{0,18}$$

with a correlation coefficient $r^2 = 0,97$.

IV - EXPERIMENTAL STUDY OF THE SOLAR COLLECTORS FOR HOUSE HEATING.

Some performance characteristics of the solar houses collectors were already published in 1976 (5). They are related to the 1967 collector-type whose characteristics are :

- vertical wall of coarse reinforced concrete (ρ = 2200Kg/m^3, λ = 1,75W/m.°C, Cp = 0,22Kcal/kg.°C) 0,60m thick, painted with black acrylo-vinylic;

- double glazing 2 x 3mm in a steel frame at 0,12m from the wall,

- height and width of the collector : 4,38 x 1,27m,

- distance between top and bottom openings for circulation: 3,50m.

- dimension of the openings : 0,565 x 0,110m (621,5cm^2)

Figures 9 and 10 are typical recordings of data during the winter time. The heat balances for the thermocirculation and the wall are computed separately from the hourly average data and added during 24 hours to determine the daily balance and, with respect to the daily incident solar energy, the daily global efficiency (Figure 11). To avoid the variations of the daily values (because of the inertia of the system) monthly balances and global efficiencies in terms of the incident solar energy were also calculated (figure 12).

The global efficiency of the system in terms of the incident solar energy is about 35% in winter and decreases to about 15% in summer because of the increasing incidence angle of the direct solar radiation on the south-facing wall between March and June. This phenomena which is accompanied of a reduction of the incident solar energy on the south-facing wall is interesting because it corresponds to a reduction of the needs for heat in buildings. So, the choice of this surface for the collectors seems to be justified for such a passive system.

Some visualisation of the flow in such solar collectors were realized using methaldehide particules. The photographs 1,2 and 3 are related with, respectively, the inlet and outlet zones and the whole collector for 12cm between plate and cover. The theoratical approach of such disturbed convective thermosiphon will be probably more complicated than the study of the solar chimney. Its experimental investigation will be undertaken in 1978.

BIBLIOGRAPHY
(1) French patent n° 1-152-129 à 1/04/1956

(2) French patent n° 2-144-066 - 29/06/1971
    Addition n° 2-189-686 - 20/06/1972

(3) US patent n° 3-832-992 - 26/06/1972

(4) F.TROMBE, A et M.LE PHAT VINH - Etude sur le chauffage des habitations par utilisation du rayonnement solaire. Revue Générale de Thermique. Dec 1965.

(5) F.TROMBE, J.F.ROBERT, M.CABANAT, B.SESOLIS. Caractéristiques de performance des insolateurs équipant les maisons à chauffage solaire du CNRS.
Colloque AFEDES "Echanges thermiques entre un batiment et son environnement". INSA Lyon - 6 - 8 Avril 1976

(6) S. OSTRACH - Internal viscous flows with body forces. Boundary layer research; IUTAM symposium Freiburg . Springer Verlag 1958.(see also NACA TN 3141, 1954 and NACA TN 3458, 1955 by S. OSTRACH).

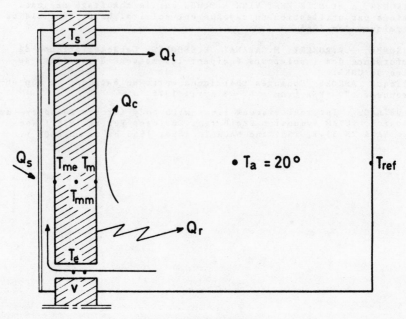

Fig.1 HEATING

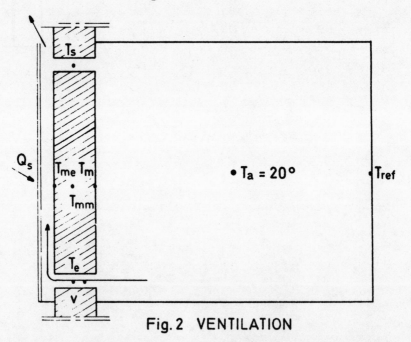

Fig. 2 VENTILATION

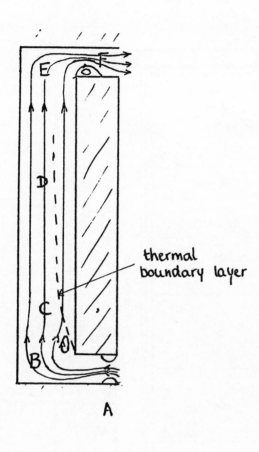

thermal
boundary layer

A

3a

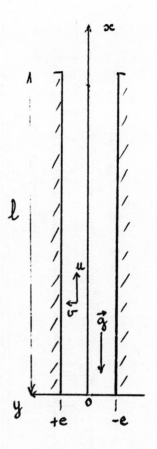

3b

- Figure 3 -

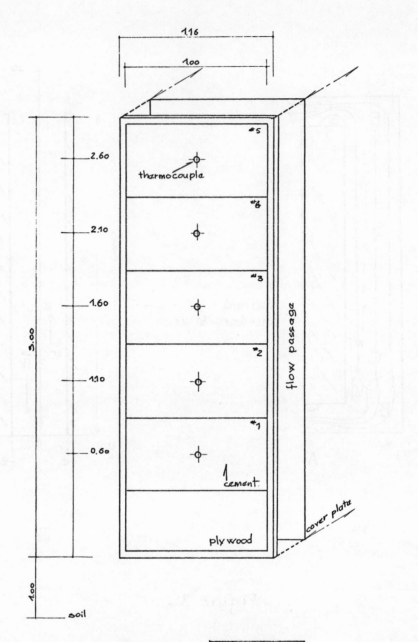

1.16

1.00

#5

2.60 — thermocouple

2.10 — #4

1.60 — #3

flow passage

1.10 — #2

3.00

0.60 — #1

cement

cover plate

ply wood

1.00

soil

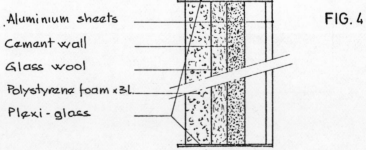

Aluminium sheets
Cement wall
Glass wool
Polystyrene foam x3l.
Plexi-glass

FIG. 4

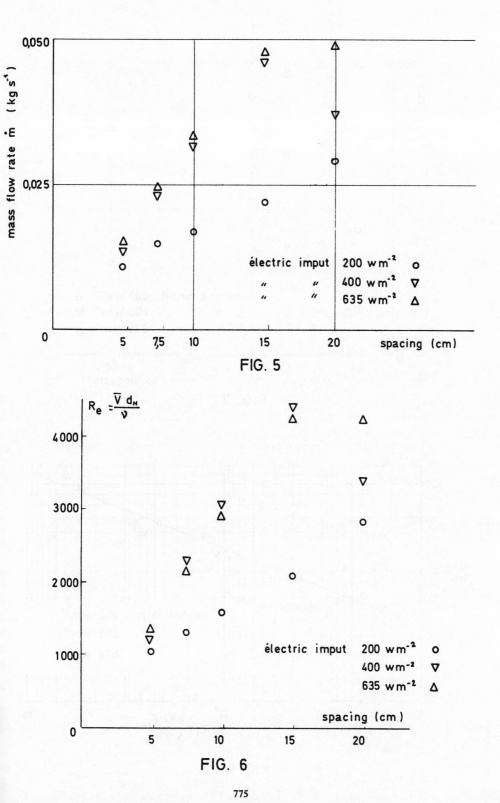

FIG. 5

FIG. 6

775

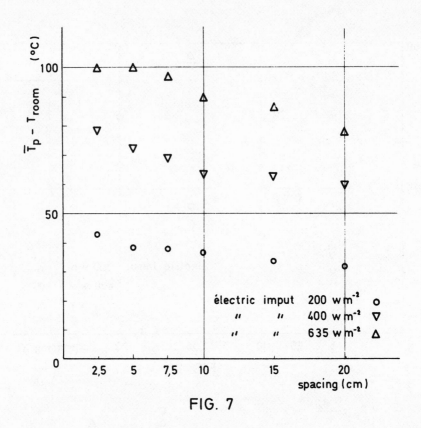

FIG. 7

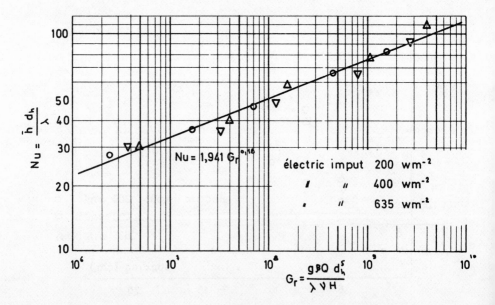

FIG. 8

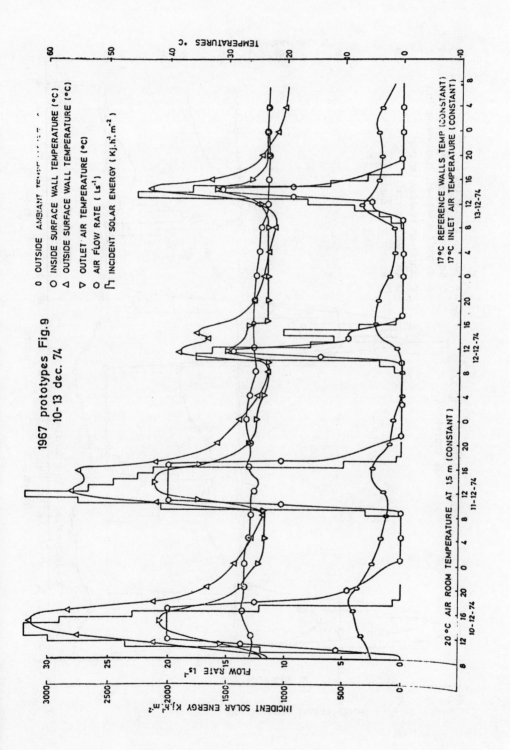

1967 prototypes Fig.9
10- 13 dec. 74

O OUTSIDE AMBIANT TEMPERATURE (°C)
O INSIDE SURFACE WALL TEMPERATURE (°C)
△ OUTSIDE SURFACE WALL TEMPERATURE (°C)
▽ OUTLET AIR TEMPERATURE (°C)
O AIR FLOW RATE (Ls⁻¹)
⌐ INCIDENT SOLAR ENERGY (Kj.h⁻¹.m⁻²)

17°C REFERENCE WALLS TEMP (CONSTANT)
17°C INLET AIR TEMPERATURE (CONSTANT)

20 °C AIR ROOM TEMPERATURE AT 1,5 m (CONSTANT)

TEMPERATURES °C

FLOW RATE Ls⁻¹

INCIDENT SOLAR ENERGY Kj.h⁻¹.m⁻²

10-12-74   11-12-74   12-12-74   13-12-74

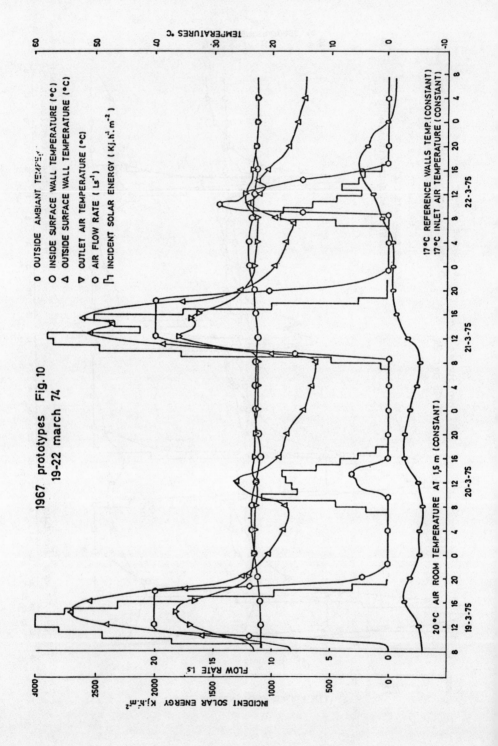

1967 prototypes Fig.10
19-22 march 74

O OUTSIDE AMBIANT TEMP.°C
O INSIDE SURFACE WALL TEMPERATURE (°C)
△ OUTSIDE SURFACE WALL TEMPERATURE (°C)
▽ OUTLET AIR TEMPERATURE (°C)
O AIR FLOW RATE (Ls⁻¹)
⊓ INCIDENT SOLAR ENERGY (KJ.h⁻¹.m⁻²)

TEMPERATURES °C

17°C REFERENCE WALLS TEMP. (CONSTANT)
17°C INLET AIR TEMPERATURE (CONSTANT)
20°C AIR ROOM TEMPERATURE AT 1,5 m (CONSTANT)

FLOW RATE Ls

INCIDENT SOLAR ENERGY KJ.h⁻¹.m⁻²

22-3-75   21-3-75   20-3-75   19-3-75

778

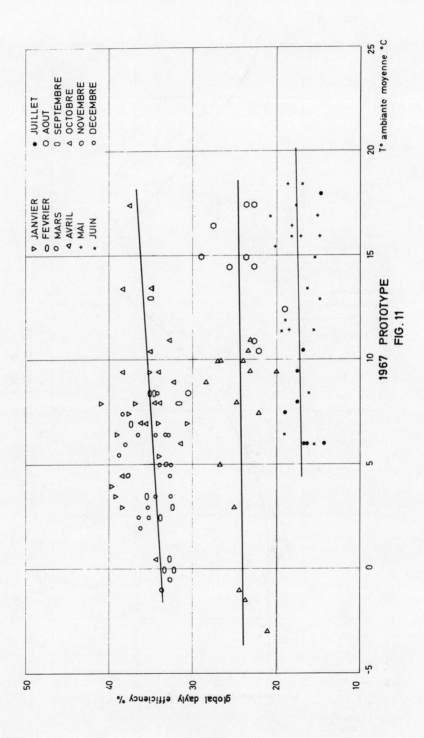

1967 PROTOTYPE
FIG. 11

T° ambiante moyenne °C

global dayly efficiency %

△ JANVIER
○ FEVRIER
○ MARS
△ AVRIL
+ MAI
× JUIN

● JUILLET
○ AOUT
○ SEPTEMBRE
△ OCTOBRE
○ NOVEMBRE
○ DECEMBRE

779

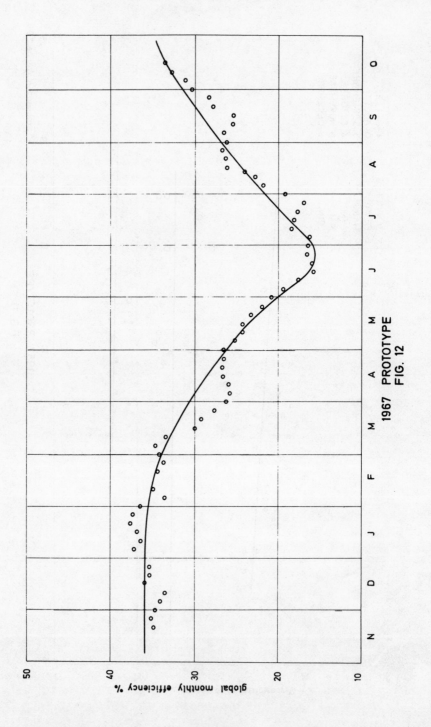

1967 PROTOTYPE

FIG. 12

global monthly efficiency %

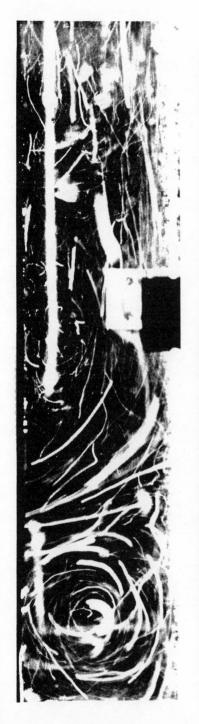

PHOTO 1   PHOTO 2   PHOTO 3

DIGITAL SIMULATION OF A SOLAR WATER HEATING SYSTEM UNDER NATURAL CIRCULATION CONDITIONS

JOSE A. MANRIQUE

Instituto Tecnologico y de Estudios Superiores de Monterrey
Monterrey, Mexico

ABSTRACT

One of the most serious problems with the simulation or design of a solar water heating system under natural circulation conditions is the lack of experimental radiation data. This is particularly true in developing countries where solar energy technology is most needed.

This work presents a mathematical model for the design of a solar water heating system in a location where only the monthly average of total radiation on a horizontal surface is available or may be estimated. The model is therefore a valuable design tool for estimating the most appropiate collector-storage tank geometry.

The computer program has been tested with an experimental prototype and the results are very satisfactory.

NOMENCLATURE

D outside tube diameter (m)

D_i inside tube diameter (m)

d fin thickness (m)

g acceleration of gravity (m/s^2)

$\bar{H}_T$ radiation on the collector (W/m^2)

h_{fi} heat transfer coefficient (W/m^{2}K)

k_m fin thermal conductivity (W/mK)

k_t tube thermal conductivity (W/mK)

L tube length (m)

$\dot{m}$ mass flow rate (kg/s)

q   heat flow (W)

R_c contact thermal resistance defined by Eq. (8)

R_t thermal resistance defined by Eq. (12)

S   radiation on absorber surface (W/m^2)

T   temperature (K)

T_a ambient temperature (K)

T_b absorber base temperature (K)

T_f fluid temperature (K)

U_L heat loss coefficient (W/m^2K)

W   distance between tubes (m)

x   coordinate across the collector (m)

y   coordinate along the collector tubes (m)

Subscripts

i   inlet

o   outlet

Greek Letters

β   coefficient of volume expansion (K^{-1})

ρ   density (kg/m^3)

INTRODUCTION

        As curtailments and prices of natural gas and other
fuels increase, a large portion of energy demands will have to be
satisfied by solar energy.

        In the particular case of Mexico there are only estima-
tes from related meteorological data on the monthly average daily
total radiation on a horizontal surface {1,2}. These estimates
also indicate that nearly 40% of the territory benefits with more
than 20 MJ/m^2day, on an annual basis, value which guarantees
adequate results in future solar energy applications {3}.

        The objective of this paper is to study the average
performance of a solar water heating system under natural circulation conditions, with emphasis on a simplified technique which
permits a rapid evaluation of the process. By using a parametric
analysis, the model under consideration allows the performance
of a solar water heating system to be appraised. Therefore, the
system geometry may be readily designed with a certain degree of

confidence.

Although several authors have investigated the performance of a solar water heating system under natural circulation conditions {4,5,6}, their works contemplate the transient behavior of the system under certain daily insolation characteristics. On the other hand, this paper intends to evaluate the performance of a system under average conditions throughout a larger period of time, such as a month, resorting only to estimates of solar radiation.

THEORETICAL ANALYSIS

The basic geometry of the flat plate solar collector absorber under consideration is shown in Figure 1. An energy balance in the absorber plate per unit length leads to the following differential equation and boundary conditions:

$$\frac{d^2T}{dx^2} - \frac{U_L}{k_m d}(T-T_a - \frac{S}{U_L}) = 0 \tag{1}$$

$$\frac{dT}{dx} = 0 \qquad \text{at} \qquad x = 0 \tag{2}$$

and

$$T = T_b \qquad \text{at} \qquad x = \frac{W-D}{2} \tag{3}$$

In deriving Equation (1) it has been assumed the collector operating under steady state conditions. For most practical systems, the effect of the heat capacity of the collector on the system performance is negligible {7}. The temperature distribution across the absorber plate may be obtained by solving Equations (1) to (3). Thus,

$$\frac{T - T_a - S/U_L}{T_b - T_a - S/U_L} = \frac{\cosh mx}{\cosh m\frac{W-D}{2}} \tag{4}$$

where $m^2 = U_L/k_m d$.

A squematic diagram for the plate temperature distribution for a constant value of y is also shown in Figure 1.

The heat flow per unit length entering the fin base may be calculated by using Fourier's law of heat conduction. This is,

$$q_c' = (W-D)\{S-U_L(T_b-T_a)\} \frac{\tanh\ m(W-D)/2}{m(W-D)/2} \tag{5}$$

or, in terms of the fin efficiency, F,

$$q_c' = (W-D)F\{S-U_L(T_b-T_a)\} \tag{6}$$

where

$$F = \frac{\tanh\ m(W-D)/2}{m(W-D)/2} \tag{7}$$

This heat flow is conducted through the thermal resistance between the absorbing plate and the tube surface. Referring to Figure 2,

$$q_c' = \frac{T_b-T_t}{R_c} \tag{8}$$

The base temperature T_b may be obtained from Equations (6) and (8). This is,

$$T_b = \frac{(W-D)FSR_c + (W-D)FU_LR_cT_a + T_t}{1 + (W-D)FU_LR_c} \tag{9}$$

By substituting Equation (9) into (8) the heat flow q_c' may be expressed in terms of the tube surface temperature,

$$q_c' = \frac{(W-D)F\{S-U_L(T_t-T_a)\}}{1 + (W-D)FU_LR_c} \tag{10}$$

Furthermore, since a fraction of the tube surface is also exposed to solar radiation,

$$q_r' = D\{S-U_L(T_t-T_a)\} \tag{11}$$

The useful energy gained by the fluid may be obtained analysing Figure 2. Thus,

$$q_u' = q_c' + q_r'$$

$$= \frac{T_t-T_f}{\frac{1}{h_{fi}\pi D_i} + \frac{\ln\ D/D_i}{2\pi k_t}} = \frac{T_t-T_f}{R_t} \tag{12}$$

Substituting Equations (10) and (11) into (12),

$$T_t = \frac{(W-D)F(S+U_L T_a) + \delta(DS+DU_L T_a+T_f/R_t)}{(W-D)FU_L + \delta(DU_L+1/R_t)} \qquad (13)$$

where

$$\delta = 1 + (W-D)FU_L R_c \qquad (14)$$

It is observed from the above expression that δ should have a value as close to unity as possible in a collector design. On the other hand, the useful heat flow in terms of the fluid temperature may be determined by combining Equations (12) and (13). Thus,

$$q_u' = F'W\{S-U_L(T_f-T_a)\} \qquad (15)$$

where the collector efficiency factor F' is defined by,

$$F' = \frac{(W-D)F + \delta D}{W\{(W-D)FU_L R_t + \delta U_L DR_t + \delta\}} \qquad (16)$$

The fluid temperature is described by the following differential equation:

$$\dot{m}_t c_p \frac{dT_f}{dy} - q_u' = 0 \qquad (17)$$

where $\dot{m}_t$ and c_p are the tube mass flow rate and specific heat of the working fluid, respectively. Furthermore, the above first order differential equation should satisfy the following boundary condition:

$$T_f = T_{fi} \quad \text{at} \quad y = 0 \qquad (18)$$

Solving Equations (17) and (18) it is obtained that

$$T_{fo} = (T_{fi}-T_a-S/U_L)\exp(-U_L WF'L/\dot{m}_t c_p) + T_a + S/U_L \qquad (19)$$

The mass flow rate which passes through the collector tubes depends on the buoyancy forces and pressure drop throughout the system. Figure 3 represents, schematically, the layout of a typical solar water heating system under natural circulation conditions. It is well known that the water temperature within the storage tank varies throughout the day and remains stratified when the flow rates are relatively small {8}. In order to estimate the buoyancy forces for an average radiation flux, a

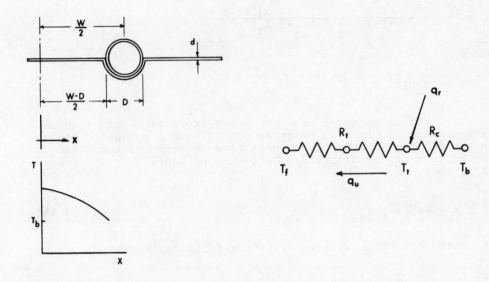

Fig. 1. Basic absorber geometry   Fig. 2. Thermal resistances

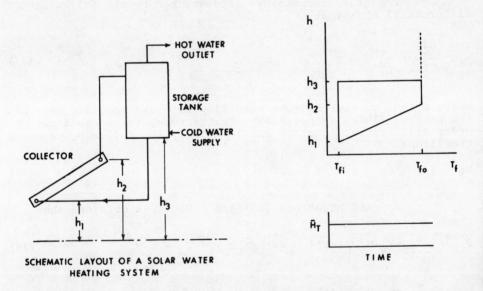

Fig. 3. Collector-storage tank   Fig. 4. Temperature distribution
         geometry   within the storage tank

temperature profile such as the one depicted in Figure 4 is
assumed. This is certainly a simplified model where all the water
in the storage tank is supposed to be warmed up. Since the diagram area is proportional to the thermosyphon head causing
flow {4}, this assumption will give an estimate of the system
capacity to heat water in an average day, as well as an indication of the storage tank capacity required.

Under these limit conditions, the thermosyphon head
causing flow is given by

$$h_T = \frac{1}{2}g(\rho_1 - \rho_2)f(h) \tag{20}$$

where $f(h) = 2h_3 - h_2 - h_1$.

On the other hand, if the water density is supposed to vary with
temperature according to the relation

$$\rho = AT^2 + BT + C$$

where A, B, and C are constants, the thermosyphon head is given
by

$$h_T = \frac{1}{2}\overline{\rho}g\overline{\beta}(T_{fo} - T_{fi})f(h) \tag{21}$$

where $\overline{\rho}$ and $\overline{\beta}$ are evaluated at the arithmetic average temperature. Furthermore, the pressure head opposing flow due to friction
and other losses may be represented by the expression

$$h_P = (f\frac{L}{D_i} + K)\frac{\overline{\rho}u^2}{2} \tag{22}$$

where f is the Darcy-Weisbach friction factor, u is the water
velocity, and K is a loss factor for losses associated with
bends, tees, and other restrictions in the tubing. For a flow
circuit involving a series of tubes, headers, and connecting
pipes, the total pressure loss may be represented by the
expression

$$h_P = h_{P1} + h_{P2} + h_{P3}$$

where subscript 1 refers to the collector tubes, 2 to the
headers, and 3 to the connecting pipes. Thus, for a system with
N tubes per collector and n collectors in parallel,

$$h_P = \frac{8\dot{m}^2}{\overline{\rho}\pi^2 D_1^4 N^2 n^2}(f_e\frac{L_1}{D_1} + K_e) \tag{23}$$

where

$$f_e = f_1 + f_2 N^2 \left(\frac{L_2}{L_1}\right) \left(\frac{D_1}{D_2}\right)^5 + f_3 N^2 n^2 \left(\frac{L_3}{L_1}\right) \left(\frac{D_1}{D_3}\right)^5 \tag{24}$$

and

$$K_e = K_1 + K_2 N^2 \left(\frac{D_1}{D_2}\right)^4 + K_3 N^2 n^2 \left(\frac{D_1}{D_3}\right)^4 \tag{25}$$

Noting that $f = 64/Re$ for laminar flow, and equating Equations (21) and (23),

$$\dot{m} = \frac{N n D_1^4 \bar{\rho}^2 g \bar{\beta} (T_{fo} - T_{fi}) f(h)}{256 \bar{\mu} L_1 \left\{ 1 + N \left(\frac{L_2}{L_1}\right) \left(\frac{D_1}{D_2}\right)^4 + N n \left(\frac{L_3}{L_1}\right) \left(\frac{D_1}{D_3}\right)^4 + \frac{K_e \dot{m}}{16 \bar{\mu} \pi N n L_1} \right\}} \tag{26}$$

Equations (19) and (26) permit an evaluation of the outlet fluid temperature and mass flow rate.

DISCUSSION OF RESULTS

One of the collector units under consideration consisted of eight type L copper tubes (OD=15.88 mm, ID=13.84 mm) brazed longitudinally at 10 cm pitch across a 20 SWG copper sheet 1.2 m long by 0.8 m wide. The parallel tubes were joined at the ends by type L copper tube headers (OD=28.58 mm, ID=26.04 mm). The entire plate was painted with a flat black paint with high solar absorptance, and placed inside an aluminum box insulated at the bottom and sides with a 5 cm thick fiberglass wool insulation. The top of the box was glazed with a 3.2 mm thick window glass plate. The collector unit was inclined at about 25.7° facing south and connected to an insulated 200 l storage tank.

Radiation estimates {1} for Monterrey, Mexico (25 40 N/100 18 W) are shown in Table 1. On the other hand, the total radiation on the collector surface was computed by resorting to the correlation of Liu and Jordan {9}. Heat transfer coefficients were calculated by using standard heat transfer techniques {10,11}.

A program to simulate the monthly performance of a given collector was written on an IBM 370 digital computer. Figure 5 shows the monthly water temperature increment above inlet conditions. As it is expected, the maximum outlet fluid temperature occurs during summer months. Similarly, Figure 6 depicts the collector efficiency variation throughout the year. Water load as well as total energy gained throughout the

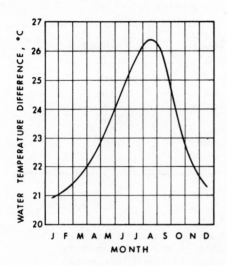

Fig. 5. Water temperature difference between inlet
and outlet

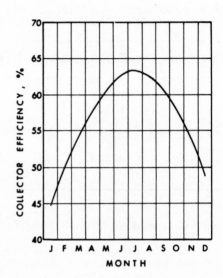

Fig. 6. Collector efficiency

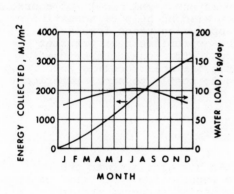

Fig. 7. Energy collected and
water load

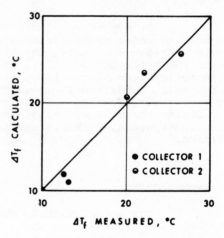

Fig. 8. Calculated and measured
results for two different
collectors

Table 1. Monthly average of total radiation on a horizontal
surface located in Monterrey, Mexico.

Month	$\bar{H}$ (kJ/m^{2}day)	Month	$\bar{H}$ (kJ/m^{2}day)
January	11302	July	21893
February	12139	August	19716
March	14149	September	17497
April	15070	October	13437
May	17079	November	11595
June	20218	December	10465

year are shown in Figure 7.

In order to compare the analytic results with experimental data, Figure 8 shows the monthly average water temperature, as measured and as calculated, for two solar water heating systems. It is observed that the agreement between both results is satisfactory.

CONCLUDING REMARKS

It may be concluded that a solar water heating system operating under natural circulation conditions may be designed with confidence by using the above mathematical model. Of course that the collector model itself can also be helpful to estimate the performance of a solar system operating under forced convection conditions, provided the appropiate heat transfer correlations are used. Thus, the collector model can be used to design the most appropiate module for solar cooling and heating systems, without resorting to costly physical experimentation.

ACKNOWLEDGMENT

The author wish to express his gratitude to the School of Engineering at Instituto Tecnologico y de Estudios Superiores de Monterrey for the financial support, and to Dean Santiago Chuck for his kind interest in the development of this work.

REFERENCES

1. Almanza, R., López, S., Radiación Solar Global en la Repú-
    blica Mexicana Mediante Datos de Insolación, Instituto de
    Ingeniería, UNAM, 357, 1975.

2. Hernández, E., La Distribución de la Radiación Global en
    México Evaluada Mediante la Fotointerpretación de la Nubosi-
    dad Observada por Satélites Meteorológicos, Centro de Inves-
    tigación de Materiales, UNAM, 1976.

3. Mayer, E. R., Present State and Perspective of Solar Energy
    Applications in Mexico, Centro de Investigación de Materia-
    les, UNAM, 1976.

4. Close, D. J., The Performance of Solar Water Heaters with
    Natural Circulation, Solar Energy, 6, 33, 1962.

5. Ong, K. S., A Finite-Difference Method to Evaluate the
    Thermal Performance of a Solar Water Heater, Solar Energy,
    16, 137, 1974.

6. Ong, K. S., An Improved Computer Program for the Thermal
    Performance of a Solar Water Heater, 18, 183, 1976.

7. Klein, S. A., Duffie, J. A., Beckman, W. A., Transient
    Considerations of Flat-Plate Solar Collectors, ASME, J. Engr.
    Power, 96A, 109, 1974.

8. Ragsdale, R. G., Namkoong, D., The NASA-Langley Building
    Solar Project and the Supporting Lewis Solar Technology
    Program, Solar Energy, 18, 41, 1976.

9. Liu, B. Y. H., Jordan, R. C., Daily Insolation on Surfaces
    Tilted Toward the Equator, Trans. ASHRAE, 526, 1962.

10. Kern, D. Q., Process Heat Transfer, McGraw-Hill, 1950.

11. Duffie, J. A., Beckman, W. A., Solar Energy Thermal
    Processes, Wiley, 1974.

AN EXPLICIT ANALYSIS OF A NATURAL CIRCULATION LOOP WITH REFERENCE TO FLAT PLATE SOLAR COLLECTORS

P. K. SARMA*, A. S. AL-JUBOURI, AND S. J. AL-JANABI

Department of Mechanical Engineering
Military Technical College
P. O. Box 478
Baghdad, Iraq

ABSTRACT

A simple dynamic analysis is presented with special reference to a natural circulation loop of a solar collector for two possible thermal conditions viz; a variable heat flux and a variable wall temperature. Some of the solutions, which may find application in the design are obtained explicitly.

NOMENCLATURE

a_n, b_n	constants in equation (1a)
A	Area of cross section
C_p	Specific heat at constant pressure
f	Friction coefficient
g	Accelaration due to gravity
G	Discharge rate of the fluid
H^+	Dimensionless enthalpy
M	Mass of the water in the reservoir
q_w	Heat flux at the wall
t	Time
T	Temperature
Z	Spatial coordinate representing direction of flow
Z_o	Length of the hot leg
θ, θ_c	Dimensionless time & Dimensionless period of active solar radiation
ν	Kinematic viscosity
μ	Absolute viscosity
τ	Shear stress at the wall
Re	Reynolds number
Gr	Grashof number
Nu_m	Mean Nusselt number

Subscripts

h	Hot
c	Cold
s	Steady state
m	Mean

---

*On leave from the Department of Mechanical Engineering, Andhra University, India.

INTRODUCTION

In countries where solar radiation is in abundance, several methods are being employed to harness the thermal energy to advantage of mankind. For domestic
purposes very often solar energy is tapped either for air-conditioning or heating
the water by employing a natural circulation loop in which the flat plate collector
exposed to solar radiation, is a main constituent. There exist good number of investigations both of experimental and theoretical nature pertaining to the analysis
of solar collectors [1,2,3]. However, the theoretical investigations are implicit
in nature and the designer often faces the problem of redesigning the loop with
reference to specific local thermal conditions. Thus, the purpose of the technical article is to offer a simple analysis which is amenable for straight
forward calculations.

ANALYSIS

The physical configuration of the loop is shown in Fig. 1. The theoretical
analysis of the loop is facilitated under the limitations of the following
assumptions.

1. The water reservoir is considered to be a sink having a finite ac-
   cumlating thermal capacity.
2. The flow is fully developed both in the hot and cold legs; (i.e., flow
   field is not dependent on the spatial coordinates except time).
3. The slug flow model is assumed. The flow is laminar in the hot leg.
4. The motion of the fluid is induced mainly because of the difference between mean densities of the fluid in the cold and hot legs.
5. The friction losses due to bends and constrictions in the passages are
   not accounted for in the analysis.

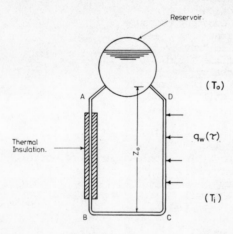

CD – HOT LEG.
AB – COLD LEG.
$q_w(\tau)$ – HEAT FLUX.
T_o, T_c, T_s – TEMPERATURES AT THE
               POINTS SHOWN.
Z_o – LENGTH OF THE HOT LEG.

Fig.-1-(PHYSICAL CONFIGURATION)

6. It is further assumed that in view of the very small velocities that are often met with, we have

$$\frac{dP}{dz}\Big|_{hot} = -\frac{dP}{dz}\Big|_{cold}$$

7. The solar radiation at the hot leg can be represented by the function

$$q_{w} = q_{m} \sum_{n=0}^{n} a_{n} \cos(n\pi\theta/\theta_{c}) + b_{n} \sin(n\pi\theta/\theta_{c}) \qquad (1a)$$

8. There is no phase transformation and the temperature potential $(T_{o}-T_{i})$ between the outlet and inlet of the hot leg is very small at any instant.

Thus, the equations of motion can be expressed as:

for downward flow

$$\frac{dG}{dt} = -A_{c}\frac{dP}{dz}\Big|_{cold} - \tau_{c}P_{c} + A_{c}\rho_{c}g \qquad (1)$$

for upward flow

$$\frac{dG}{dt} = -A_{h}\frac{dP}{dz}\Big|_{hot} - \tau_{h}P_{h} - A_{h}\rho_{h}g \qquad (2)$$

Where ρ_{h} and ρ_{c} are mean densities in the hot and cold legs. Combining eqs. (1) and (2) with the assumption (6.) we get

$$\frac{1}{A^{*}}\frac{dG}{dt} = g\Delta\rho - \frac{\tau_{h}P_{h}}{A_{h}}(1+\alpha) \qquad (3)$$

Where

$$\Delta\rho = \rho_{c} - \rho_{h}$$
$$\alpha = \tau_{c}P_{c}A_{h}/\tau_{h}P_{h}A_{c}$$
$$1/A^{*} = 1/A_{c} + 1/A_{h}$$

STEADY STATE ANALYSIS

For steady state conditions, the heat flux is time invariant i.e., $q_{w} \neq q_{w}(t)$ and $G \neq G(t)$. Equation (3) gives the balance between the body forces and the shear resistance. For a linear variation of density with respect to temperature we have

$$\Delta\rho = \beta\rho\Delta t_{o} \qquad (4)$$

Where $\Delta t_{o} = T_{h} - T_{c} = (T_{o} - T_{c})/2$. T_{h} is the mean bulk temperature of water in the hot leg. T_{c} is the temperature in the cold leg which is the same as T_{i}. Thus, for laminar flow conditions taking

$$\tau_{h} = \frac{1}{2}f\frac{G^{2}}{A_{h}^{2}\rho} \quad , \quad f = 16/Re \qquad (5)$$

Equation (3) can be reduced to the form shown below with the aid of eq. (4).

$$\beta\Delta t_{o} = \frac{1}{2}(1+\alpha)\,Re\left(\frac{\nu^{2}P_{h}^{3}}{g A_{h}^{3}}\right) \qquad (6)$$

Where

$$Re = 4G/P_{h}\mu \qquad (7)$$

Assumption (8.) gives the possibility to express eq. (6) as

$$\beta (T_o - T_i) = Re \ (1 + \alpha) \ (Ph^3 \ \nu^2/gA_h^3) \tag{8}$$

Equation (8) implies that $T_h = (T_i + T_o)/2$. At any instant, the energy balance on the hot leg side can be written as

$$G(T_o - T_i) \ C_p = q_w \ P_h \ Z_o \tag{9}$$

Combining eqs. (8) and (9) the Reynolds number can be expressed in terms of the heat flux as

$$Re = 2(g \ A_h^3/P_h^3\nu^2)^{1/2} \ (q_w \ Z_o\beta/C_p\mu)^{1/2} \ / \ (1 + \alpha)1/2 \tag{10}$$

Thus for the given geometry of the loop, as the heat flux increases the velocity of flow increases. Equations (6) and (10) yield

$$\beta\Delta t_o = (P_h^3 \ \nu^2/g \ A_h^3)^{1/2} \ (q_w \ Z_o\beta/C_p\mu)^{1/2} \ (1 + \alpha)^{1/2} \tag{11}$$

Thus, Reynolds number is given by

$$Re = 2 \ Gr*/(1 + \alpha)^{1/2} \tag{12}$$

Where $Gr* = g \ A_h^3\beta \ \Delta t_o/P_h^3 \ \nu^2$. In literature there exists good numbers of correlations to predict heat transfer rates in vertical conduits under laminar flow conditions which are generally of the type

$$Nu_m = Const. \ Re^p \ (Gr_f \ Pr_f)^q \tag{13}$$

The constants p and q are known, and subscript f refers to the mean temperature of the fluid as the reference value at which physical properties are to evaluated. Thus, the unknown value of Re in this particular case of the problem can be eliminated in terms of Gr* resulting in an expression for the mean heat transfer coefficient interms of known parameters. An example to illustrate the utility of the steady state analysis is given in the appendix.

UNSTEADY STATE ANALYSIS
    The unsteady state analysis is presented for two cases:
    1. Variable heat flux
    2. Variable wall temperature of the solar collector

    Realizing the fact that $\alpha << 1$ under natural working condition, eq. (3) can be put in the following form when the heat flux is a time variable

$$d/d_\theta \ (Re^2) = - Re^2 + Q^+ \tag{14}$$

Where

$$\theta = 4A*P_h^2t\mu/A_h^3\rho \tag{15}$$

and

$$Q^+ = 4Z_o\beta gA_h^3q_w(t)/C_pP_h^3\nu^3 \tag{16}$$

If Q^+ is a known function of time, solution of eq. (14) can be obtained. One important aspect which can be observed from eq. (14) is that whatever the type of time function Q^+ may be, the solution is exact and explicit. Even for the variation of the heat flux as per the assumption (7.) one can obtain an exact solution. In as much as the values of the coefficients a_n and b_n are dependent on longitude, latitude, solar time, and other local parameters [4,5] a simple type of heat flux variation with respect to time is assumed to illustrate the variation of circulation rate in the loop. For example, if the variation of the heat flux is

$$q_w = q_m (1 + \text{Sin } \pi\theta/\theta_c)$$

Such that $q_w = q_m$ when $\theta = 0$ and $\theta = \theta_c$ where q_m is the steady state component (i.e., Re = Re_s where Re_s is the steady state value of the Reynolds which can be anticipated under constant heat flux conditions). Dividing equation (14) by $Re_s^2 = 4 Z_o g A_h^3 q_m\beta/(C_p \rho P_h^3 \nu^3)$ the following can be arrived at.

$$d/d\theta \ (\varepsilon^2) = - \ \varepsilon^2 + Q^+/Re_s^2 \tag{18}$$

Where

$$\varepsilon = Re/Re_s \tag{19}$$

But we have

$$Q^+/Re_s^2 = 1 + \text{Sin } (\pi\theta/\theta_c) \tag{20}$$

Equation (18) with the help of eq. (20) can be written as

$$d/d\theta \ (\varepsilon^2) + \varepsilon^2 = 1 + \text{Sin } (\pi\theta/\theta_c) \tag{21}$$

The boundary condition for the linear first order differential equation is $\varepsilon = 1$ when $\theta = 0$. Thus, the solution for eq. (21) can be written as

$$\varepsilon^2 = 1 + (\pi\theta_c/\pi^2+\theta_c^2) \ [\text{Exp}(-\theta) + (c/\pi) \ \text{Sin } (\pi\theta/\theta_c) - \text{Cos } (\pi\theta/\theta_c)] \tag{22}$$

Further, the relative variation of the temperature potential between outlet and inlet of the solar collector, to the steady state value Δt_s can be obtained from eqs. (9), (17), and (19) as follows

$$\Delta t/\Delta t_s = [1 + \text{Sin } (\pi\theta/\theta_c)]/\varepsilon \tag{23}$$

Equations (20), (22), and (23) reveal the unsteady state behavior of the flow for stipulated thermal conditions.

VARIABLE WALL TEMPERATURE

It is further assumed that the wall temperature of the solar collector is not constant but time varying, i.e.,

$$T_w - T_h = \lambda f(\theta) \tag{24}$$

where λ is a constant $\lambda = P_h^2 Pr^{2/3} \nu^2/0.93 g A_h^2 \beta Z_o$. Equation (9) can be re-written as

$$G C_p (T_o - T_c) = h P_h Z_o f(\theta) \tag{25}$$

Assuming quasi-steady state conditions the unknown value of the heat transfer coefficient h can be evaluated by employing the equation of Sider and Tate [6] for laminar flow conditions. Thus,

$$Nu = 1.86 \ Re^{1/3} \ Pr^{1/3} \tag{26}$$

where $Nu = h (4 A_h)/k P_h$.

Thus, eqs. (25) and (26) can be combined to obtain

$$A_h (T_o - T_c)/(P_h Z_o) = 1.86 f(\theta)/Re^{2/3} Pr^{2/3} \tag{27}$$

Equation (3) can be transformed to the following form

$$d/d\theta (Re) = -1/2 (Re) + [f(\theta)/Re^{2/3}] \tag{28}$$

The boundary condition for the above differential equation is $Re = 0$ when $\theta = 0$. Equation (28) is very general in nature and its solution can be facilitated when $f(\theta)$ is a known function of time. Another important aspect is the evaluation of the change in enthalpy of the water in the reservoir with respect to time. If heat loss to the surroundings is neglected a simple energy balance can be written

$$d/dt [M C_p (T - T_R) = h P_h Z_o (T_w - T_h) \tag{29}$$

Combining eqs. (24), (27), and (29), and manipulating the same properly we obtain

$$dH^+/d\theta = Re^{1/3} f(\theta) \tag{30}$$

Where H^+ (Dimensionless enthalpy) $= [8 M C_p(T - T_R) g A^*\beta]/P_h^2 K Pr\nu$. Thus, eqs. (28) and (30) constitute a full set that are to be simultaneously solved for the given type of the function $f(\theta)$. To obtain the nature of the behavior of the eqs. (28) and (30), a sine function is chosen, i.e.,

$$f(\theta) = C \sin [-(\pi\theta/\theta_c)] \tag{31}$$

Where C may be constant. Thus, the solution of the differential eq. (28) consistent with the above assumption can be written as

$$Re^{5/3} = \left[\frac{C}{1+\left(\frac{5}{\pi}\frac{\theta_c}{}\right)^2}\right]\left[\frac{\theta_c}{\pi}\left\{Exp\left(-\frac{5}{6}\theta\right) - Cos\left(\pi\frac{\theta}{\theta_c}\right)\right\} + \frac{5}{6}\left(\frac{\theta_c}{\pi}\right)^2 Sin\left(\frac{\pi\theta}{\theta_c}\right)\right] \tag{32}$$

However, the solution of the equation can not be exact and numerical procedure is to be employed. Thus for a value of $C = 1$, solutions are obtained as a digital computer for different values of θ_c, the dimensionless time period of active solar radiation and the graphs are shown in Figs. 2 and 3.

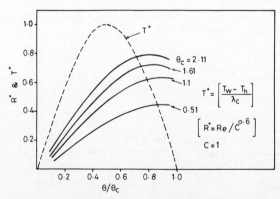

Fig.(2)   VARIATION OF CIRCULATION WITH TIME

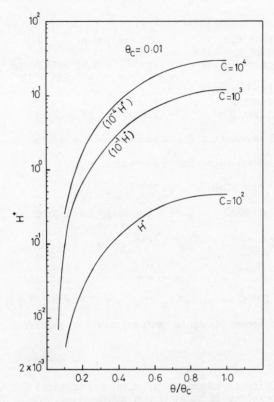

Fig.(3) VARIATION OF ENTHALPY OF
WATER W·R·T· TIME

CONCLUSIONS
    From the present analysis the following conclusions can be arrived at.
    1. The transient behavior of the flow conditions in the natural circulation
        loop can be arrived at by making use of the eqs. (22) and (32) for a
        variable heat flux and variable wall temperature respectively.
    2. Equation (23) gives the variation of the thermal potential between the
        outlet and inlet of the collector with respect to time for the assumed
        function of the heat flux variation. Evidently the thermal potential
        depends on the interaction between the heat flux and associated circu-
        lation rate.
    3. In the design of solar collectors for heating purposes a quasi steady
        state analysis is presented and the procedure to arrive at the number
        of tubes is indicated by an example in the appendix.

ACKNOWLEDGMENTS
    The authors thank the authorities of the Military Technical College, Baghdad,
for having supported this work.

APPENDIX
    An example for the steady state analysis to find the number tubes in a flat
plate collector is given:

Data Assumed

Mean effective radiant flux	=	350 Kcal/M^2 hr
Initial temperature of the water	=	$15^{\circ}C$
Final temperature of the water	=	$45^{\circ}C$: $\mu = 8.01 \times 10^{-4}$ Kgm/M.sec at $30^{\circ}C$
Quantity of water	=	100 Kgs: $\nu = 0.805 \times 10^{-6}$ M^2/sec
Duration of effective heating due to solar radiation	=	8 hrs
Diameter of tube in hot leg	=	2.2 cm

Length of tube (Z_o) connecting the headers to be restricted to a maximum value of 100 cm.

The result obtained using the present analysis

Mean temperature of the fluid	=	$30^{\circ}C$
Prandtl for water at $30^{\circ}C$	=	5.42
Reynolds for water at $30^{\circ}C$	=	2116 (calculated from eq. 10)
Heat flux parameter $[(4q_m z_o \beta)/\mu C_p]$	=	1.6
also $(g A_h^3/\nu^2 P_h^3)$	=	2.8×10^6

Since

$\Delta t_s = 2 \Delta t_o = 0.224^{\circ}C$ we get (G $C_p \Delta t_s$) n = $(100 \times 30)/8 = 375$ kcal/hr.

Thus, the number of tubes (n) to be used to meet the requirement is 16.

REFERENCES

[1] E. Lumsdine "Criteria transient solution and criteria for achieving maximum fluid temperature in solar energy application." Solar Energy, vol. 13, no. 1, April, 1970.
[2] C. L. Gupta, "On generatizing the dynamic performance of solar energy systems." Solar Energy, vol. 13, 1971.
[3] H. H. Safat & A. F. Souka, "Design of new solar heated house using double exposure flat plate collectors." Solar Energy, vol. 13, no. 1, April, 1970.
[4] S. J. Reddy, "An empirical method for the estimation of total solar radiation." Solar Energy, vol. 13, no. 2, May, 1971.
[5] P. Petheridge,"Sunpath diagrams and overall solar heat gain calculations." Building Research Station, 1969.
[6] J. P. Holman, "Heat Transfer" McGraw-Hill, 1976. (U.S.A.)

DIMENSIONING OF A SOLAR HEATING STATION

CARL AXEL SVENSSON

SUNVEX ENERGI AB
Växjö, Sweden

1. Introduction

The sun is the most powerful energy source mankind knows.
Almost all the energy we use today has its origin from the
sun. The oil which we use up in an increasing tempo these
days is nothing else than solar energy stored during millions
of years. Also other kinds of energy like wood, waterpower,
wind etc derive its origin from the sun.

Today we all are aware of the fact that we can't go on using
up the fossil fuels in the tempo we are used to. In other
words we can't go to the bank and borrow energy especially
longer, because the bank will very soon be empty. We have to
learn to play equal with production and use of energy. That
means that we have to learn to use the solar energy directly.
This will give rise to a far-reaching change of our community,
but the sooner we start the better we will be prepared, when
the conventional energy sources are finished.

This paper deals with one of the most obvious use of solar
energy - for heating purposes. In Sweden SUNVEX ENERGI AB
plans to build a solar heating station for a group of
50 houses (see figure 1).

The concept solar heating station implies a plant where solar
energy is collected and stored. Distribution of energy from
the solar heating station to the users is done with a conventional district heating system.

The reason to build a solar heating station common to several
houses, is of course that storing of solar energy is more
economically if it is done in a common storage instead of
in separate storages for each house. And storing of solar
energy is more or less neccessary in all the world. In Sweden
where we have extreme cold winters with very little sunshine,
we have to store solar energy for six months. Still solar energy
can be a technical and economical solution to our heating
problems in only a couple of years. The fastest way to benefit
from the cheap solar energy is, however, to build plants with
storages common to several users, that is to build solar heating
stations. (see figure 2)

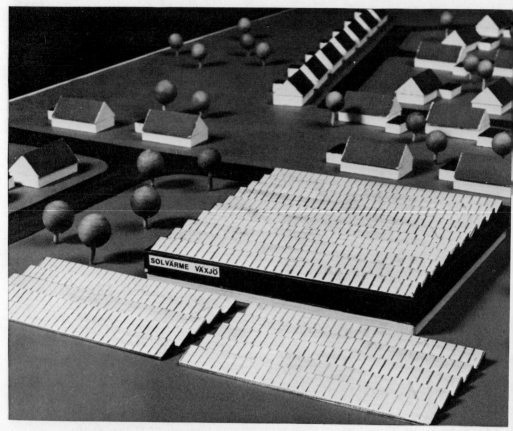

Figure 1. Model of a solar heating station for 50 houses in Ingelstad, Sweden.

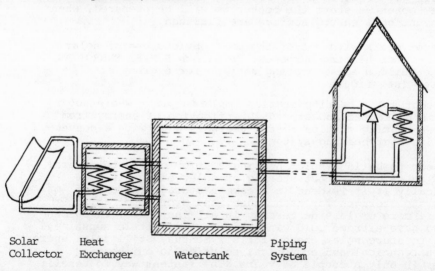

Solar   Heat   Piping
Collector   Exchanger   Watertank   System

Figure 2. Principle function for a solar heating station.

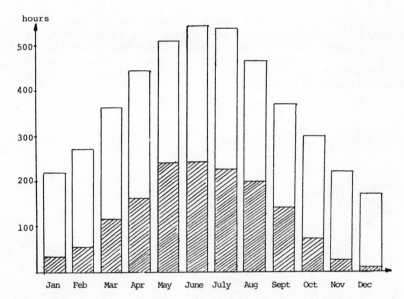

Figure 3. Distribution of sunny hours in Ingelstad, Sweden.

2. Solar Heating Conditions

The main condition when dimensioning a solar heating system
is, of course, the amount and the distribution of solar energy.
Most European contries has about 2000 hours of sunshine which
mainly are distributed to the summer half of the year.

In Ingelstad in Sweden where we are going to build our solar
heating station there are only 1.500 hours of sunshine which
are distributed according to figure 3.

The need of energy for heating is however, biggest during the
winter. A well insulated one family house of 140 square meters
in Sweden, needs about 18.000 kWh pro year for heating and
warmwater.

Table 1. Distribution of heating demand for one family house
         in Sweden.

Month	Heating demand	Warmwater demand
January	2.250	450
February	1.890	450
March	1.890	450
April	1.170	450
May	270	450
June	0	450
July	0	450
August	0	450
September	360	450
October	1.080	450
November	1.620	450
December	2.070	450
Total	12.600	5.400

As can be seen from table 1 above, 13.500 kWh or 75% of the
total heating demand is required during the time October till
March. The number of sunshine hours during the same time is
only about 300 or 20% of the total number. This implies that
the surplus of solar energy during the summer must be accumu-
laded to the winter.

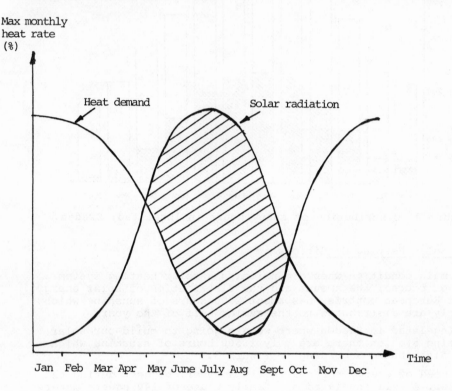

Figure 4. Seasonal variations in heat demand and solar radiation.

But is it possible to store solar energy from summer to winter?
The answer to this question is yes, _if_ the stored amount of
water is big enough.

3. Long Range Storage of Heated Water

The heat losses from a tank with heated water is determined
by the area and insulation of the tank, the temperature difference between the medium inside and outside the tank and the
time that the water is stored.

$$Q_1 = A \cdot k \cdot (T_w - T_{out}) \cdot (t_2 - t_1) \qquad \textcircled{1}$$

Q_1 = heat losses from the tank (Wh)

A = the area of the tank (m^2)

k = heat transfer coefficient $(W/m^2 c^o)$

T_w = temperature inside the tank (^oC)

T_{out} = temperature outside the tank (^oC)

$(t_2 - t_1)$ = storing time (h)

Due to the heat losses through the tank, the water temperature will decrease according to following formula;

$$Q_2 = V \cdot c \cdot (T_{w2} - T_{w1}) \qquad \textcircled{2}$$

Q_2 = heat losses from the water (Wh)

V = watervolume (m^3)

c = specific heat of water $(Wh/m^3 \,^oC)$

$(T_{w2} - T_{w1})$ = temperature decrease during the storing time. (^oC)

As it must be equilibrium between the heat transfer through the tank and the heat losses in the water, expression 1 and 2 must be equal.

$$Q_1 = Q_2$$

or

$$A \cdot k \cdot (T_w - T_{out})(t_2 - t_1) = V \cdot c \, (T_{w2} - T_{w1})$$

$$(T_{w2} - T_{w1}) = \frac{A}{V} \cdot k \cdot \frac{1}{c} (T_w - T_{out}) \cdot (t_2 - t_1) \qquad \textcircled{3}$$

As already mentioned the storing time of warm water for a solar heating station in Sweden must be 6 months, that is $24 \cdot 30 \cdot 6 = 4.320$ hours.

The temperature in the tank shall vary between 90^oC and 40^oC during the actual storing period. Higher water temperature than $100 \,^oC$ is inappropriate, because the water must be under pressure, which raises the price of the tank considerably. On the other hand it is not wise to let the water in the tank be colder than 40^oC, because then the water is not hot enough for bathing and other purposes. The average temperature in the tank is therefore $\frac{90 + 40}{2} = 65 \,^oC$.

The temperature outside the tank, T_{out}, can be found in existing statistics of climate. For the actual area, the southern part of Sweden, the average air temperature during October – March is $\pm \, 0^oC$.

The specific heat of water, c, is constant and equal to 1.160 Wh/m^3 $^\circ C$.

When putting these values into equation (3) , the expression for the temperature decrease during the storing time reduces to following formula;

$$(T_2 - T_1) = \frac{A}{V} \cdot k \cdot \frac{1}{1160} \cdot (65 - 0) \cdot 4320$$

$$(T_2 - T_1) = \frac{A}{V} \cdot k \cdot 251,03$$

The remaining parameters of the tank, that is the surface/volume ratio and the insulation, can consequently be choosen freely.

For small tanks the surface/volume ratio is relatively big, but decreases rapidly for bigger tanks. Let us study some diffrent types of tanks to see this fact more clearly.

Spherical tank

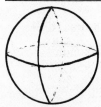

$$A = 4 \pi r^2$$
$$V = \frac{4}{3} \pi r^3$$

$$\frac{A}{V} = \frac{3}{r}$$

Cubical tank

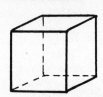

$$A = 6 a^2$$
$$V = a^3$$

$$\frac{A}{V} = \frac{6}{a}$$

Cylindrical tank

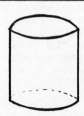

$$A = 2 \pi r^2 + 2 \pi rh$$
$$V = \pi r^2 h$$

$$\frac{A}{V} = 2 \left(\frac{1}{r} + \frac{1}{h} \right)$$

Volume (m^3)	Spherical tank	Cubical tank	Cylindrical tank
100	1,04	1,29	1,19
1.000	0,48	0,60	0,55
10.000	0,22	0,28	0,26
100.000	0,10	0,13	0,12

Table 2. Surface/volume ratio ($\frac{A}{V}$) for different types and sizes
of tanks.

The optimal shape of the tank is obviously a sphere. However,
to build a spherical tank is expansive. In practice it is therefore more realistic to build cylindrical or cubical tanks.

Assume that the water tank is insulated with 1 meter of insulation
material. The heat transfer coefficient, k, will then be as
small as 0.04 (W/m^2 $^{\circ}$C). The decrease in temperature for the
different types of tanks which we have studied, will then be
according to table 3 below.

Volume m^3	Spherical tank	Cubical tank	Cylindrical tank
100	10,4 $^{\circ}$C	13,0 $^{\circ}$C	12,0 $^{\circ}$C
1.000	4,8 $^{\circ}$C	6,0 $^{\circ}$C	5,5 $^{\circ}$C
10.000	2,2 $^{\circ}$C	2,8 $^{\circ}$C	2,6 $^{\circ}$C
100.000	1,0 $^{\circ}$C	1,3 $^{\circ}$C	1,2 $^{\circ}$C

Table 3. Decrease of watertemperature depending on heat losses
from the tank.

The results in table 3 above, is of fundamental significance.
It gives us the key to technical and economical use of solar
energy. Because table 3 shows that it is possible to store
solar energy during long periods with very high efficiency,
if it is done in big tanks, common to several users. This implies
that solar energy will be a real alternative to conventional
fuels in only a couple of years, if the solar systems are
designed as solar heating stations.

4. Dimensioning of a Solar Heating Station

The starting point when dimensioning a solar heating station
is to determine the size of the water tank. In the southern
part of Sweden, where Sunvex Energi AB plans to build the solar
heating station mentioned, there is a deficiency of solar energy
from the 1st of October to the 1st of April (see figure 4). This
implies that the water tank must be designed to store all the
energy needed during this period. From table 1, it is evident
that each house need 13.500 kWh for heating and warmwater during
the time October to April. As the pilot plant consists of 50
houses the required need of accumulated energy is 50x13.500 =
675.000 kWh.

The water tank must, however, be dimensioned for a bigger amount of energy, depending on heat losses in the tank and in the distribution pipe system. The latter are wellknown from conventional district heating systems and are approximately 10% of the energy delivered.

The heat losses from the tank on the other hand, can easily be calculated from equation ③ . Assume that the tank is insulated with 1 meter of insulation material, e.g. k = 0.04, and that the surface/volume ratio ($\frac{A}{V}$) is 0.4. The heat losses from the tank is then;

$$T_2 - T_1 = 0.4 \cdot 0.04 \cdot 251,03$$
$$T_2 - T_1 = 4,02 \ ^\circ C$$

As the total temperature descrease during the accumulation period is 50 $^\circ$C, the heat losses from the tank will be $\frac{4,02}{50} \cdot 100 = 8,04\%$.

Now it is very easy to calculate the volume of the heat storage.

Energy required for the houses	675.000 kWh
Heat losses in the pipe system (10%)	67.500 kWh
Heat losses in the tank (8,04%)	54.270 kWh
Total requirement for accumulated energy	796.770 kWh

The tank volume can now be calculated by menas of equation ② .

$$Q_2 = V \cdot c \ (T_{w2} - T_{w1})$$
$$796.770 = V \cdot 1,16 \cdot (90-40)$$
$$V = \frac{796.770}{1,16 \cdot 50}$$
$$\underline{V = 13.737,4}$$

With the surface/volume ratio ($\frac{A}{V}$) fixed to 0.4, the total area of the tank must be 13.737,4 $\cdot$ 0,4 = 5.495 m^2.

As it is decided to build the water tank for this pilot plant above ground and rater close to the houses it is desirable that the tank is not too high, because then it might hav a negative influence on the surroundings. The tank dimensions has therefore been fixed to 46 x 46 x 6,5 meters.

The second step when dimensioning a solar heating station is to determine the neccessary area of solar collectors. As Sunvex Energi has been developing and testing solar collectors for several years, this step in the calculation has been very easy to do. Let us first establish the fact that concentrating, tracking solar collectors shall be used in this system. The reasons for this standpoint are several:

a)   Concentrating, tracking solar collectors are very suitable for north european countries with their long sunny summer days. (In June the night in Sweden is only 3-4 hours) Solar energy can be collected from 6 am to 6 pm with tracking collectors!

b) The average air temperature during the summer is not
    especially high (15-19 $^{\circ}$C), why a concentrating collector,
    with its small heat losses, is to prefer to a flat plate
    collector.

c) It is very easy to produce heated water of 90°C with
    concentrating, tracking collectors, while the maximum
    temperature for flat plate collectors in Sweden is about
    70°C. And the higher the water temperature the smaller the
    water tank can be.

The total area of solar collectors is then determined by the
amount of energy needed, the distribution of solar energy and
the performance of the solar collectors.

Energy_demand_during_the_winter_(see page 9)   796.770 kWh

Energy_demand_during_the_summer

Required energy for the houses (see figure 4)   225.000 kWh
Heat losses in the pipe system (10%)   22.500 kWh
Heat losses in the tank l)   52.000 kWh

                                                  299.500 kWh

Total demand of energy   1.096.270 kWh

During the time April to October there is about 1.200 hours
of sunshine in Ingelstad, where the pilot plant is to be built.
This implies that we can calculate with 90 sunny days during
this period.

The concentrating, tracking solar collectors, which we are
going to use, has a capacity of 5 kWh/m^2 and sunny day. This
implies that we need $\frac{1.096.270}{90 \cdot 5} \sim 2.437$ m^2 of solar collectors.

l) The heat losses in the tank during the summer are a little
   smaller than during the winter, depending on the higher air
   temperature during the summer.

5. Cost Analysis

The total need of investment for this pilot plant is 1,7 millions
of dollars and are distributed according to the table below.

	M$
Solar collectors (2.600 m^2 á 270$)	0,70
Primary pipe system	0,12
Water tank (13.800 m^3)	0,60
Secondary pipe system	0,04
Projecting	0,12
	1.58
Unexpected costs	0,12
Total_need_of_investment	1,70

This investment might be regarded as high, compared to the costs for conventional energy today, but considering that this is an initial small scale plant, the future for solar heating stations must be estimated as very good.

We already know that the costs for concentrating, tracking solar collectors will fall to about 115 $/m^2 in a couple of years. In the same way the costs for water tanks will be reduced to less than 20 $/m^3 in the future.

This implies that in a couple of years the need of investment for a solar heating plant like this, might be only a third of what it is today. With 10% yearly instalment, this implies a cost of just about 5 cents pro kWh! That is, solar heating stations will make solar energy competitive to conventional energy, even in areas with very little sunshine.

6. Summary

Solar energy can be a real technical and economical alternative to existing forms of energy in just a couple of years. One way to make the best use of solar energy is to build solar heating stations where energy is collected and stored. The merit of solar heating stations is that the large scale factor is used both technically and economically, which reduces the cost considerably.

The particular solar heating station discussed in this paper is situated in Sweden, where the conditions for solar heating are hard. In spite of this, there is an obvious tendency that solar heating stations will be able to make the use of solar energy competitive to conventional energy, even in north european contries like Sweden. Most other contries in the world has better solar energy conditions than Sweden. I am therefore quite convinced that solar heating stations will in a near future be an everyday occurrence all over the world.

THERMAL ANALYSIS OF A FLAT-PLATE SOLAR COLLECTOR: SHORT- AND LONG-TERM EFFECTS OF A THIN-FILM COATED COVER PLATE

D. L. SIEBERS AND R. VISKANTA*

Heat Transfer Laboratory
School of Mechanical Engineering, Purdue University
West Lafayette, Indiana, USA 47907

ABSTRACT

A detailed two-dimensional, steady-state, nodal, heat transfer analysis is developed for a flat-plate solar energy collector to study designs for improving its thermal performance. The long-term effects on collector performance of the most promising of these designs are then examined in some detail. The analysis accounts for the temperature gradients in the fluid flow and vertical directions in the collector, the physical and thermodynamic properties of the materials in the collector, the collector location, its orientation and dimensions, the number of cover plates, and any thin-film selective coatings on the cover plates or absorber. Also accounted for are the time dependent variations in meteorological conditions and the beam (collimated) and diffuse solar irradiation.

The results of the study indicate that very little improvement in collector efficiency can be obtained by altering collector design to reduce heat loss out the insulation on the back of the collector or to increase heat transfer between the working fluid and the absorber beyond what is already technically and economically feasible. The results do indicate that suppressing thermal radiation heat loss from the absorber such as with selective surfaces is one of the most promising means for improving collector performance. Improvements between 10% and 50% in long-term collector efficiency are noted.

NOMENCLATURE

b = thickness of a collector component
CLC = cloud cover (in tenths)
c_p = specific heat
$\underline{G}$ = incident radiant flux
$\overline{h}$ = average heat transfer coefficient
k = thermal conductivity or index of absorption
$\tilde{k}$ = effective thermal conductivity of the gap
$\dot{m}$ = fluid mass flow rate
n = number of cover plates or index of refraction

Q = heat transfer rate
q = heat flux
T = temperature
U = effective heat transfer coefficient
V = wind speed
α = effective absorptance
δ = distance between absorber and cover plate
ε = emittance
κ = absorption coefficient
ρ = reflectance
σ = Stefan-Boltzmann constant
ϕ = relative humidity

*Presently at Sandia Laboratories, Livermore, California, USA.

Subscripts

a = ambient conditions   o = outlet
b = back of collector   s = solar
i = inlet   in = insulation

INTRODUCTION

The dwindling supply of oil and natural gas has stimulated great interest
in solar energy as an alternate energy source. Collection of solar energy
using flat-plate collectors provides a means for converting solar into thermal
energy. Solar thermal processes which can provide low temperature heat for
application for heating and cooling of buildings, for drying and other evaporative processes, and for industrial process heat can become technically and economically viable. Flat-plate solar energy collection systems, consisting of a solar
collector and an energy storage device, are discussed in great detail in the
literature [1-8]. The collector consists of an absorber surface, thermal insulation on the sides and back of the collector, one or more semitransparent cover
plates on the front of the collector spaced two to three centimeters from each
other and from the absorber surface, and a working fluid circulating through a
fluid flow structure in contact with the absorber surface.

The objective of this work is to examine several possible areas for improvement in flat-plate collector design. Then, with consideration of what is already
technically and economically feasible, determine which designs provide significant improvements in collector performance and examine the effects of meteorological conditions on long-term flat-plate collector performance. The design
changes considered are increased insulation on the back of the collector to
decrease the heat loss from the back, designs to increase the effective absorber-
to-working-fluid heat transfer coefficient to improve heat transfer from the
absorber to the working fluid such as discussed by Whillier and Saluja [9],
variations in working fluid properties and flow rate, and designs to reduce the
heat loss from the top of the absorber to the atmosphere such as selective
surfaces on a cover plate or absorber [6, 10-12], honeycomb structures in the
air gap between the absorber and cover plate [13-18], or evacuation of the air
gap between the absorber and cover plate [19, 20].

ANALYSIS

Physical Model

The mathematical model for the thermal analysis of a flat-plate solar energy
collector must take into account many factors. These can be divided into two
groups, environmental factors (directional and spectral nature of incident solar
radiation, air temperature, wind speed, cloud cover, etc.) and collector factors
(collector materials and their properties, collector orientation, and collector
operating conditions). A detailed discussion of these factors is given elsewhere [6, 8, 21].

Incorporating some of the factors in very great detail in the mathematical
model of a solar collector is not warranted, as well as difficult. As a result,
to make the analysis mathematically tractable and at the same time realistic,
several assumptions have been made:

1. Solar irradiation can be divided into direct and diffuse components.
    The diffuse component is uniform across the sky's hemisphere.
2. A two-band spectral model for the analysis of radiation heat transfer,
    surface radiation properties, and overall collector radiation character-
    istics is adequate. The wavelength cutoff between the two bands,

solar (short-wave) and thermal (long-wave), is about 3.0 μm.
3. The glass cover plates are opaque to thermal radiation.
4. Emission of radiation from the collector surfaces is diffuse. Surfaces are assumed gray in each band.
5. The absorber surface is either a specular or a diffuse reflector. The cover plates are specular reflectors.
6. Nodal approximations of uniform temperature and uniform heat flux on a node are valid.
7. Collector is assumed quasi-steady, responding instantaneously to changes in such variables as solar flux or air temperature.

Several other assumptions common to solar collector analysis listed in Reference 6 are also used. The validity of these idealizations is discussed in detail elsewhere [21].

Mathematical Model

Developed in this section is a mathematical model for a flat-plate solar collector which accounts for the environmental and collector factors previously mentioned, and the temperature gradients in the fluid flow direction in the collector and the vertical direction in the collector (insulation to absorber, absorber to working fluid, absorber to cover plate, cover plate to cover plate, and cover plate to atmosphere). These two are the most important temperature gradients. Other temperature gradients such as would occur between the tubes in Figure 1 can be accounted for by fin efficiency factors [6].

Since only the temperature gradients in the fluid flow and vertical directions are considered, the heat transfer is two-dimensional for the thermal analysis of the collector. The two-dimensional mathematical model for the thermal analysis of a collector is derived by performing an energy balance on a component of the collector. Figure 1 shows the various modes of heat transfer occurring in a section of a collector in the fluid flow direction. There is direct and diffuse solar energy absorbed by the cover plates and absorber, thermal radiation heat exchange between the cover plates and absorber, thermal radiation heat exchange between the top cover and the atmosphere, convection of energy from the absorber to the working fluid, advection of energy by the fluid conduction in the flow direction, and finally, energy lost out the back of the collector.

For a steady state analysis the energy balance derived for each component of an n-cover plate collector are the following:

Cover Plate (n):

$$b_n k_n \frac{d^2 T_n}{dx^2} + \tilde{k}_{n-1,n}(T_{n-1}-T_n)/\delta_{n-1,n} + \varepsilon_n^-(G_n^- - \sigma T_n^4) + q_{s,n}$$

$$+ \bar{h}_{n,a}(T_a-T_n) + \varepsilon_n^+\sigma(T_{sky}^4 - T_n^4) = 0 \tag{1}$$

Cover Plates (2 → n-1):

$$b_j k_j \frac{d^2 T_j}{dx^2} + \tilde{k}_{j-1,j}(T_{j-1}-T_j)/\delta_{j-1,j} + \varepsilon_j^-(G_j^- - \sigma T_j^4) + q_{s,j}$$

$$+ \tilde{k}_{j,j+1}(T_{j+1}-T_j)/\delta_{j,j+1} + \varepsilon_j^+(G_j^+ - \sigma T_j^4) = 0 \tag{2}$$

$$j = 2, 3, \ldots n-1$$

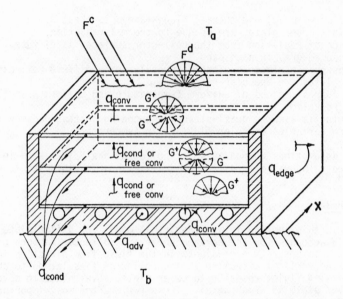

Figure 1   Schematic of a Section of a Flat-Plate Collector in the
Flow Direction Illustrating the Modes of Heat Transfer

Cover Plate (1):

$$b_1 k_1 \frac{d^2 T_1}{dx^2} + \tilde{k}_{p,1}(T_p - T_1)/\delta_{p,1} + \varepsilon_1^-(G_1^- - \sigma T_1^4) + q_{s,1}$$

$$+ \tilde{k}_{1,2}(T_2 - T_1)/\delta_{2,1} + \varepsilon_1^+(G_1^- - \sigma T_1^4) = 0 \tag{3}$$

Absorber Plate (p):

$$b_p k_p \frac{d^2 T_p}{dx^2} + \tilde{k}_{p,1}(T_1 - T_p)/\delta_{1,p} + \varepsilon_p^+(G_p^+ - \sigma T_p^4) + q_{s,p}$$

$$+ U_{p,f}(T_f - T_p) + U_{p,in}(T_{in} - T_p) = 0 \tag{4}$$

Working Fluid (f):

$$\dot{m}_f c_{p_f} \frac{dT_f}{dx} + U_{p,f}(T_p - T_f) = 0 \tag{5}$$

Insulation (in):

$$b_{in} k_{in} \frac{d^2 T_{in}}{dx^2} + U_{p,in}(T_p - T_{in}) + U_{f,in}(T_f - T_{in}) + U_{b,in}(T_b - T_{in}) = 0 \tag{6}$$

In Eq. (1), for example, the first term accounts for conduction along the flow
direction of the nth-cover plate; the second term is convective-conductive heat
transfer across the air gap between cover plates n-1 and n; the third term
represents the net absorption and emission of thermal (longwave) radiation from
the bottom of cover plate n, respectively; the fourth term denotes absorption of

solar (short-wave) radiation by the cover plate n; the fifth term represents convection to the atmosphere from cover plate n; the final term is the radiation heat exchange between the sky and the top of the cover plate n. The solar energy absorbed by cover plate n ($q_{s,n}$) will be discussed in greater detail in the following paragraphs. The physical interpretation of terms in the remaining equations is similar.

The net solar radiation flux [the fourth term in Eqs. (1) through (3)] is represented as follows using Assumption 1:

$$q_{s,j} = \alpha_j(\theta)F^C \cos\theta + \alpha_j F^d(1+\cos s)/2 + \rho_g \alpha_j(F^C+F^d)(1-\cos s)/2,$$

$$j = (p, 1, 2 \ldots n) \quad (7)$$

The first term in Eq. (7) is the collimated or direct solar irradiation at an incident angle θ absorbed by absorber or cover plates. The second term is the diffuse solar energy flux absorbed by absorber or cover plates with the collector tilted at an angle s from the horizontal. The last term is the insolation that is reflected diffusely from the ground and is absorbed by the collector.

In Eq. (7) the incidence angle θ is given in terms of the collector location and orientation and the time of day [6]. The effective absorptances $\alpha(\theta)$ and α of the cover plates and absorber in a cover plate-absorber system are determined using the net radiation method [21]. The procedure is to first determine the surface radiation properties of a cover plate from its optical properties (n_λ, κ_λ, etc.) and dimensions by the analysis presented by Taylor and Viskanta [22]. Second, the effective radiation characteristics of each cover plate are determined from its surface radiation properties. Next, the radiation charac- teristics of a multiple cover plate system are developed from the radiation characteristics of each individual cover plate by combining them one at a time to the system until the desired number of cover plates is reached. Finally, the radiation characteristics of the cover plate system are combined with the radiation properties of the absorber to determine the effective radiation charac- teristics of the cover plate-absorber system. Each of the effective absorptances is then integrated over the solar part of the spectrum, using the spectral distribution of the solar radiation as a weighting function, to obtain total directional and total hemispherical absorptances of the cover plates and absorber for Eq. (7). A more detailed derivation of these radiation characteristics as well as a derivation of the n-cover plate radiation characteristics is presented elsewhere [21].

The boundary conditions on the edges of the cover plates, absorber, and insulation are,

$$dT/dx = 0 \qquad \text{at} \quad x = 0 \quad \text{and} \quad x = L \tag{8}$$

assuming the top surface area of the collector is large compared to the sides and the sides are well insulated [6]. The initial condition for Eq. (5) is

$$T_f = T_i \qquad \text{at} \quad x = 0 \tag{9}$$

Collector Performance Parameters

The parameters which will be used to compare the performance of various solar collectors are the outlet temperature of the working fluid, T_o, and the useful energy gain of the fluid, Q_u. The useful energy gain is the energy absorbed and carried away by the working fluid as thermal energy,

$$Q_u = \dot{m}c_p(T_o - T_i) \tag{10}$$

As is usually done in solar collector analysis, the useful energy is normalized with the insolation on the collector. This gives the efficiency of the collector, η, or the ratio of the energy gain of the working fluid to the solar energy incident on the collector. Both instantaneous efficiencies,

$$\eta = \int_A q_u dA / \int_A G_s dA \simeq \sum_{i=1}^{N} q_{u_i} \Delta A_i / \sum_{i=1}^{N} G_{s_i} \Delta A_i \simeq Q_u/Q_i \tag{11}$$

and accumulated efficiencies

$$\overline{\eta} = \int_{t_i}^{t_f} Q_u dt / \int_{t_i}^{t_f} Q_i dt \simeq \sum_{t=t_i}^{t_f} Q_u \Delta t / \sum_{t=t_i}^{t_f} Q_i \Delta t \tag{12}$$

are used to compare collector performance. The nodal approximations of uniform heat flux and temperature on a node have been used to write Eqs. (11) and (12).

RESULTS AND DISCUSSION

Collector Parameters

The results in this work point out the design changes in flat-plate collectors which result in the most significant improvements in collector performance. The effects of meteorological conditions on long-term collector performance of the most beneficial design changes are also presented. A large number of design improvements, insolation, and meteorological conditions have been considered and the results are detailed elsewhere [21]. Here, it is possible to include some selected results.

Before presenting the results, a brief description of the collector design used in the analysis to obtain the results is warranted. The assumed basic collector design is one cover plate collector with an absorber surface consisting of two parallel aluminum sheets each 2.0 mm thick with the 0.5 cm gap between them divided into channels in the flow direction through which the working fluid, water, flows. Each channel is a 0.45 cm by 0.5 cm rectangle. Reynolds numbers between 10 and 1000 were calculated for the range of fluid flow rates used in these channels which means the flow is laminar. The Nusselt number correlation needed for laminar flow in the channel just described can be obtained from the literature [23].

The orientation and dimensions of the collector are:

$$s = 0° \qquad\qquad \delta_{p,1} = 2.0 \text{ cm}$$
$$L = 2.0 \text{ m} \qquad\qquad b_{in} = 6.0 \text{ cm}$$
$$W = 1.0 \text{ m} \qquad\qquad b_j = 0.32 \text{ cm} \quad (j = 1, 2, \ldots n)$$

The spectral optical properties of the glass cover plates are:

$$n_\lambda = 1.5 \qquad \text{for all } \lambda$$
$$\kappa_\lambda = 0.04 \text{ cm}^{-1} \text{ for } \lambda < 3.0 \text{ μm}$$
$$\kappa_\lambda = 10.0 \text{ cm}^{-1} \text{ for } \lambda > 3.0 \text{ μm}$$

The thin film selective surface used in some cases on the cover plate surface facing the absorber is In_2O_3, which has a thickness of 0.24 μm and the complex index of refraction ($\tilde{n}_\lambda = n_\lambda + ik_\lambda$) obtained from weighted averages of spectral optical properties [11]:

$$n = 1.7 \atop k = 0.01 \Big\} \lambda < 3.0 \text{ } \mu m \qquad\qquad n = 7.25 \atop k = 13.5 \Big\} \lambda > 3.0 \text{ } \mu m$$

The absorber which is painted with a diffuse black coating has the following spectral, hemispherical solar and thermal radiation properties:

$\alpha_\lambda = 0.98$ and $\rho_\lambda = 0.02$ for $\lambda < 3.0$ μm; $\varepsilon_\lambda = 0.89$ and $\rho_\lambda = 0.11$ for $\lambda > 3.0$ μm

The computed total, hemispherical thermal radiation properties for a 0.32 cm thick glass cover plate with the optical properties given above are calculated to be [21]: $\varepsilon = 0.91$ and $\rho = 0.09$ for both sides. The computed total hemispherical thermal radiation properties using the same analysis for the same cover plate, only this time with the In_2O_3 thin-film selective surface on the side facing the absorber, are: $\varepsilon = 0.91$ and $\rho = 0.09$ for top side; $\varepsilon = 0.17$ and $\rho = 0.83$ for bottom side coated with a thin film of In_2O_3.

The total, directional solar radiation characteristics of a collector with one glass cover plate computed from the analysis are given in Table 1. The computed total, hemispherical solar radiation characteristics are: $\alpha_p = 0.84$ and $\alpha_{cp_1} = 0.02$.

The total, directional solar radiation characteristics of a collector with one glass cover plate with the In_2O_3 thin-film on the side facing the absorber was computed [21] and is given in Table 2. The computed total, hemispherical solar radiation characteristics are: $\alpha_p = 0.77$ and $\alpha_{cp_1} = 0.05$.

The Effects of Design Parameters on Collector Performance

Figure 2 is a plot of the collector efficiency versus the effective heat transfer coefficient between the absorber and working fluid. The effective absorber-to-fluid heat transfer coefficient, h_{p-f}, is determined from the

Table 1.   Total, Directional Radiation Characteristics of
a Collector with One Cover Plate

θ (degrees)	0.8	10.4	28.4	50.6	71.4	85.7
α_p (θ)	0.91	0.91	0.91	0.88	0.69	0.21
α_{cp_1} (θ)	0.01	0.01	0.01	0.02	0.02	0.02

Table 2.   Total, Directional Radiation Characteristics of a
Collector with One Thin-Film Coated Selective
Cover Plate

θ	0.8	10.4	28.4	50.6	71.4	85.7
α_p (θ)	0.84	0.84	0.84	0.81	0.64	0.19
α_{cp_1} (θ)	0.05	0.05	0.05	0.06	0.05	0.03

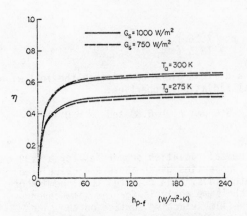

Fig. 2   Variation of Collector
         Efficiency with the Effective
         Absorber-to-Fluid Heat Trans-
         fer Coefficient; One Cover
         Plate, $\dot{m}$ = 0.005 kg/s, T_i =
         25 C, T_a = T_b, ϕ = 80%, V =
         4.0 m/s, CLC = 0, θ = 0°

Fig. 3   Variation of Collector Effi-
         ciency with the Effective
         Insulation Conduction Resis-
         tance: One Cover Plate, $\dot{m}$ =
         0.005 kg/s, T_i = 25 C, T_a = T_b,
         ϕ = 80%, V = 4.0 m/s, CLC = 0,
         θ = 0°

knowledge of the sum of the resistances to heat transfer between the absorber
and working fluid for an absorber node i. The figure shows that up to an effective heat transfer coefficient of about 120 W/m² K, significant improvements in
collector performance result from increases in the heat transfer coefficient
for all environmental conditions. Above this value, very little is gained by
improving collector design to increase the effective absorber-to-fluid heat
transfer coefficient.

    To visualize where a conventional flat plate collector design would fit on
the curve in Figure 2, consider the bonded tube absorber design shown in Figure
1 with water as a working fluid. This type of design with a poor bond between
tube and absorber could have an effective heat transfer coefficient below 10 W/
m² K. One with a good bond between absorber and tube which is fairly easily
achieved could lie above 60 W/m² K [9].

    The effect of reducing heat loss from the back of the absorber on collector
performance is examined by varying the ratio of insulation thickness to the
thermal conductivity of the insulation on the back of the collector as shown in
Figure 3. The results indicate that increasing the resistance beyond 1.0 to
1.5 m²K/W has no effect on the collector performance for the various environmental conditions and collector designs considered. This is because heat loss
from the back of the collector, even with only a small thickness of insulation
or a relatively high thermal conductivity insulation, is very small compared to
front surface heat losses from the collector as noted by Tabor [24].

    The effects of the fluid flow rate, specific heat of the fluid, and the
surface area of the top of the collector on collector performance can be seen
by determining the effects of the flow factor ($\dot{m}c_p/A$) on collector performance.
Figure 4 shows that as the flow factor increases, the efficiency of the collector increases until a flow factor of about 60 W/m²K is reached. After this the
efficiency becomes rather insensitive to the flow factor. However, the fact
that the collector efficiency increases with increasing flow rate can be mis-
leading since it would indicate that a large flow factor is desirable for
collection of solar energy. This is not true as Figure 4 shows. A large flow
factor also results in a low fluid outlet temperature. This means that there
is no optimum flow rate when considering only the thermal performance of a

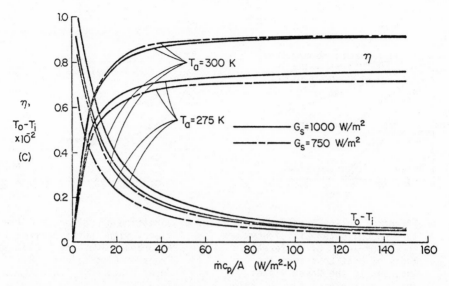

Figure 4   Variation of Collector Efficiency and Outlet-Inlet Temperature
Difference with the Flow Factor ($\dot{m}c_p/A$): One Cover Plate, T_i =
25 C, $T_a = T_b$, ϕ = 80%, V = 4.0 m/s, CLC = 0, θ = 0°

flat-plate collector. As with most heat exchange equipment, a trade-off in
collector design between outlet fluid temperature and collector efficiency must
be made.

The Effects of Reducing Thermal Radiation, Conduction, and Free
Convection Losses from the Absorber Surface

Collector efficiencies resulting from the elimination of each mode of heat
transfer in the air gap (conduction, free convection or thermal radiation) one
at a time or from combinations of different modes of heat transfer between the
absorber and cover plate are shown in Figure 5 as a function of the inlet temperature of the working fluid. The figure shows that suppressing thermal radiation heat transfer is the most effective means of improving collector performance
when trying to eliminate only one mode of heat transfer. This is followed by
conduction and then free convection. For example, one cover plate collector,
first with no radiation, then with no conduction (no free convection), then with
no free convection in the air gap, and a one cover plate collector with no
suppression of heat transfer in the air gap, respectively, have efficiencies of
57%, 37%, 33%, and 28% at T_i = 80 C. In other words, Figure 5 shows that suppression of free convection and elimination of conduction are less effective
means of improving collector efficiency when compared with thermal radiation
suppression.

The ratio of the effective thermal conductivity of the gap and the gap
spacing can be used to examine the effects of free convection and conduction
suppression in the air gap on collector performance in a more detailed manner,
since suppression of either of these modes of heat transfer is the result of
some design change which lowers the effective conductance (k/δ). Figure 6 shows
that the effective conductance decreases, the efficiency of the collector
increases. But for most collector designs the effective conductance is less
than 10 W/m²K to start with, so the maximum increase in efficiency obtainable
by lowering the ratio is apparently about 20% for winter-like environmental

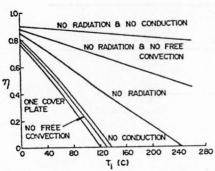

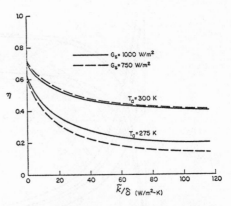

Fig. 5 Effect of Inlet Fluid Tempera-
        ture on Collector Efficiency
        for Suppression of Different
        Modes of Heat Loss from the
        Absorber to the Cover Plate:
        One Cover Plate, $\dot{m}$ = 0.005 kg/
        s, G_s = 1000 W/m^2 , T_a = T_b =
        300 K, ϕ = 75%, V = 4.0 m/s,
        CLC = 0, θ = 0°

Fig. 6 Variation of Collector Effi-
        ciency with the Effective
        Absorber to Working Fluid Heat
        Transfer Coefficient: One
        Cover Plate, $\dot{m}$ = 0.005 kg/s,
        T_i = 25 C, T_a = T_b, ϕ = 80%,
        V = 4.0 m/s, CLC = 0, θ = 0°

conditions, G_s = 750 W/m^2 and T_a = 275 K, as shown in Figure 6.

Suppression of combinations of different modes of heat transfer shown in Figure 5 give the most significant increases in collector performance. At fluid inlet temperature of 80 C, suppression of thermal radiation and free convection heat transfer in the air gap yields a collector efficiency of 74%, while suppression of thermal radiation and conduction results in a collector efficiency of about 86%. These are large improvements over previously obtained efficiencies for suppression of any one mode of heat transfer in a one cover plate collector. This then demonstrates that once thermal radiation heat transfer in the air gap has been reduced, free convection or conduction suppression will result in significant increases in collector performance, but not before. Eaton et al. [20] called attention to this possibility in their experimental work. This is also examined in much more detail elsewhere [21].

The Short and Long Term Effects of Selective Surface on Flat-Plate Solar Collector Performance

This section examines the short and long term effects of selective cover plates on flat-plate collector performance. As seen in the previous section, selective surfaces for the purpose of suppressing thermal radiation heat loss from the absorber would be one of the single most effective design changes above what is already technically and <u>economically</u> feasible for improving collector performance.

Figures 7 and 8 show the effects of selective surfaces on two different flat-plate collectors. In Figure 7 the mass flow rate through the collector is held constant. In Figure 8 the outlet temperature of the collector is held constant. In both cases the selective cover plate gives a significant increase in flat-plate collector performance at higher operating temperatures (T_i or T_i-T_0 > 40 C). In the constant mass flow rate collector, the performance is reflected in a larger efficiency and a higher outlet temperature. These trends agree well with results presented by Taylor and Viskanta [11] for a one cover

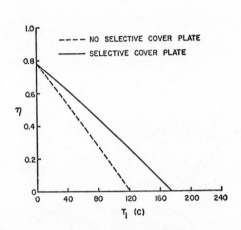

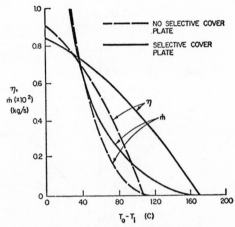

Fig. 7   Effects of the Number of Selective Surfaces on Collector
Efficiency: One Cover Plate,
$\dot{m}$ = 0.005 kg/s, G_s = 1000 W/m²,
T_a = T_b = 300 K, ϕ = 80%, V =
4.0 m/s, CLC = 0, θ = 0°

Fig. 8   Variation of Collector Efficiency with the Working Fluid
Outlet-Inlet Temperature Difference: One Cover Plate,
G_s = 750 W/m², $\dot{m}$ = 0.005 kg/s,
T_a = T_b = 300 K, ϕ = 75%, V =
4.0 m/s, CLC = 0, θ = 0°

plate collector. For a constant outlet temperature collector the improved
performance is reflected in a higher mass flow rate and a larger collector
efficiency.

The effects of a selective surface placed on a cover plate of a flat-
plate solar energy collector over a diurnal cycle are shown in Figures 9 and
10. These figures are for a typical diurnal cycle of the diurnal cycles that
were simulated. Figure 9 shows that the efficiency of a collector is increased
most during early morning and late afternoon hours, but the amount of energy
collected and the outlet temperature of the fluid are increased the most around
noon.

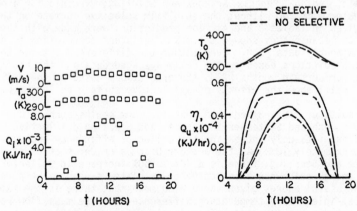

Figure 9   Effects of a Selective Surface on Collector Performance Over
a Diurnal Cycle (June 16, 1964, Sterling, Va.): One Cover
Plate, $\dot{m}$ = 0.004 kg/s, T_i = 300 K, T_b = 297 K.

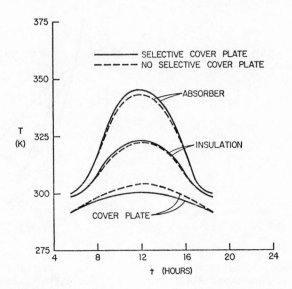

Figure 10   Effects of a Selective Surface on Collector Temperatures
Over a Diurnal Cycle (June 16, 1964, Sterling, Va.):   One
Cover Plate, ṁ = 0.004 kg/s, T_i = 300 K, T_b = 297 K.

Figure 10 shows that a selective surface causes the largest temperature changes in the collector at noon with almost no change in the early morning and late afternoon.   The figure also indicates two reasons why the selective cover plate increases the performance of the collector.   First, the cover plate has a steady state temperature closer to the ambient air temperature.   Therefore, less heat loss occurs from the cover plate to the atmosphere.   Second, the absorber temperature is higher so more energy is transferred to the working fluid.   The insulation temperature as seen in Figure 10 is almost unchanged resulting in very little change in heat loss from the back of the collector.

Collector performance predictions for March, June, September, and December for Lafayette, Indiana, Phoenix, Arizona and Sterling, Virginia for a one cover plate collector with and without the thin film selective surface on the cover plate are given in Tables 3 and 4.   The predictions were made with the daily averaged data because they showed the most consistent correlation to predictions based on the hourly data [25].   Even though the differences between the collector performance predictions based on the daily averaged data and the hourly data were significant for certain months [25], the predictions based on the daily averaged data should still show the effects of a selective surface on long term collector performance.

The results of Table 3 indicate that for warm months such as June in Lafayette and June and September in Phoenix and Sterling a selective surface is of marginal benefit, only increasing the collector efficiency and useful energy gain by approximately 10%.   As the weather becomes colder such as in December in Lafayette and Sterling, there is a 40% to 50% increase in collector efficiency and useful energy gain when a selective cover plate is used.   The percentage increases in useful energy given above also mean that there is the same increase in the outlet-inlet fluid temperature difference when the mass flow rate and inlet temperatures are constant.   This is clear from Eq. (10).   As a result, when there is an increase in the amount of energy collected by using a selective cover plate, there is also an increase in the "quality" of the energy collected.

Table 3. The Long Term Effects of a Selective Cover Plate on Collector Performance: Collector Design Conditions; One Cover Plate, $\dot{m}$ = 0.004 kg/s, T_i = 300 K.

		Lafayette, Indiana (1971)			Phoenix, Arizona (1964)			Sterling, Virginia (1964)		
		No Selective Surface	Selective Surface	Diff. (%)	No Selective Surface	Selective Surface	Diff. (%)	No Selective Surface	Selective Surface	Diff. (%)
March	$\bar{\eta}$	0.29	0.41	41	0.50	0.57	14	0.37	0.49	32
	$\overline{Q}_u \times 10^{-6}$ (KJ/month)	0.26	0.37	42	0.74	0.86	16	0.40	0.52	30
June	$\bar{\eta}$	0.54	0.60	11	0.60	0.64	7	0.53	0.59	11
	$\overline{Q}_u \times 10^{-6}$ (KJ/month)	0.68	0.75	10	0.92	0.98	7	0.66	0.74	12
Sept.	$\bar{\eta}$	0.46	0.56	22	0.59	0.65	10	0.52	0.58	12
	$\overline{Q}_u \times 10^{-6}$ (KJ/month)	0.42	0.51	21	0.78	0.85	13	0.58	0.65	12
Dec.	$\bar{\eta}$	0.26	0.37	42	0.44	0.54	23	0.27	0.40	48
	$\overline{Q}_u \times 10^{-6}$ (KJ/month)	0.13	0.18	38	0.39	0.48	23	0.16	0.23	44

Table 3. The Long Term Effects of a Selective Cover Plate on Collector Performance: Collector Design Conditions; One Cover Plate, $\dot{m}$ = 0.004 kg/s, T_i = 300 K.

		Lafayette, Indiana (1971)			Phoenix, Arizona (1964)			Sterling, Virginia (1964)		
		No Selective Surface	Selective Surface	Diff. (%)	No Selective Surface	Selective Surface	Diff. (%)	No Selective Surface	Selective Surface	Diff. (%)
March	$\bar{\eta}$	0.29	0.41	41	0.50	0.57	14	0.37	0.49	32
	$Q_u \times 10^{-6}$ (KJ/month)	0.26	0.37	42	0.74	0.86	16	0.40	0.52	30
June	$\bar{\eta}$	0.54	0.60	11	0.60	0.64	7	0.53	0.59	11
	$Q_u \times 10^{-6}$ (KJ/month)	0.68	0.75	10	0.92	0.98	7	0.66	0.74	12
Sept.	$\bar{\eta}$	0.46	0.56	22	0.59	0.65	10	0.52	0.58	12
	$Q_u \times 10^{-6}$ (KJ/month)	0.42	0.51	21	0.78	0.85	13	0.58	0.65	12
Dec.	$\bar{\eta}$	0.26	0.37	42	0.44	0.54	23	0.27	0.40	48
	$Q_u \times 10^{-6}$ (KJ/month)	0.13	0.18	38	0.39	0.48	23	0.16	0.23	44

Table 4. The Long Term Effects of a Selective Cover Plate on Collector Performance Based on Data for Sterling, Va. (1964): Collector Design Conditions; One Cover Plate, $\dot{m}$ = 0.002 kg/s, T_i = 300 K.

		No Selective Surface	Selective Surface	Difference (%)
March	$\bar{\eta}$	0.27	0.37	37
	$Q_u \times 10^{-6}$ (KJ/month)	0.28	0.40	43
June	$\bar{\eta}$	0.37	0.46	24
	$Q_u \times 10^{-6}$ (KJ/month)	0.46	0.58	26
September	$\bar{\eta}$	0.36	0.46	28
	$Q_u \times 10^{-6}$ (KJ/month)	0.40	0.51	28
December	$\bar{\eta}$	0.20	0.30	50
	$Q_u \times 10^{-6}$ (KJ/month)	0.11	0.17	55

If the flow rate is lowered as was done to obtain the results in Table 4, a higher collection temperature and a lower collector efficiency result as shown in Figure 4. For the months presented in Table 4 the flow rate was reduced to one-half of the flow rate for the months in Table 3. This resulted in a 40% increase in the mean outlet-inlet temperature difference and a 30% decrease in useful energy collected for each month due to greater heat loss at the higher collector temperatures.

For March, June, September and December the selective cover plate causes a significant increase in collector performance even during the warmer months as seen in Table 4. This is because the thermal radiation heat loss from the absorber, which the selective surface suppresses, is more important when the collector is operating at a higher temperature with a larger temperature difference between the collector and the atmosphere.

The results obtained indicate that for cold weather conditions or for high temperature operation of the collector a selective surface will improve collector performance. For warm weather and low temperature operation of the collector the selective surface only gives marginal improvement in collector performance.

CONCLUSIONS

Several conclusions can be drawn from these results. First, collector efficiency is improved by increasing the effective conduction resistance of the insulation, b/k, up to a value of 1.0 to 1.5 m^2K/W or by increasing the effective absorber-to-working-fluid heat transfer coefficient up to a value of 120 W/m^2K. If either parameter is increased above its respective value given above, the collector performance is insensitive to the change. Also, since values given above for each of those parameters are economically and rather easily achieved, there is little need for improvements in flat-plate collector design in these areas.

Second, collector efficiency can be improved significantly by increasing the flow factor, $\dot{m}c_p/A$, up to a value of 60 W/m^2K, but the working fluid outlet temperature decreases at the same time. As a result, a trade-off must be made between collector efficiency and the "quality" of the energy collected when considering the thermal performance of a flat-plate collector. Above a flow factor of 60 W/m^2K the collector is relatively insensitive to changes in the flow factor. Another conclusion which can be drawn is that suppressing thermal radiation heat loss from the absorber by a selective surface on the absorber or on a cover plate is the most effective means of improving collector performance when suppressing only one mode of heat loss. A thin-film coated selective cover plate gives significant improvements in collector performance when insolation is at its highest value for a diurnal cycle. For a period of a month a selective surface gives significant improvements during cold weather operation of the collector or for high temperature operation of the collector. Only marginal improvement in collector performance is realized for warm weather and low temperature operation of the collector when a selective cover plate is used. As a result, since effective, durable and economical selective surfaces have not been developed as yet, it is one area where significant improvements in collector performance can be achieved for certain operating conditions.

ACKNOWLEDGMENTS

The authors wish to acknowledge the editorial assistance and expert typing of the manuscript by Mrs. Debbie Drake.

REFERENCES

1. Tybout, R. A., and Löf, G. O. F., Natural Resources Journal 10, 270 (1970).

2. Freeman, S. D., Energy, The New Era, Vintage Books, New York (1974).

3. Daniels, F., Direct Use of the Suns' Energy, Yale University Press, New Haven, Conn. (1964).

4. Hottel, H. C., and Erway, A. D., in Introduction to the Utilization of Solar Energy (A. M. Zarem and A. D. Erway, Eds.), McGraw Hill, New York (1963), pp. 87-106.

5. Williams, J. R., Solar Energy: Technology and Applications, Ann Arbor Science Publishers, Mich. (1974).

6. Duffie, J. A., and Beckman, W. A., Solar Energy Thermal Processes, John Wiley, New York (1974).

7. Whillier, A., in Low Temperature Engineering Application of Solar Energy (R. C. Jordan, Ed.), ASHRAE, New York (1967), pp. 27-40.

8. Meinel, A. B., and Meinel, M. P., Applied Solar Energy, Addison-Wesley

Publishing Company, Menlo Park, Calif. (1976).

9. Whillier, A., and Saluja, G., Solar Energy $\underline{9}$, 21 (1965).

10. Tabor, H., in Low Temperature Engineering Applications of Solar Energy (R. C. Jordan, Ed.), ASHRAE, New York (1967), pp. 41-52.

11. Taylor, R. P., and Viskanta, R., in Application of Solar Energy (T. S. Wu, et al., Eds.), UAH Press, Huntsville, Alabama (1975), pp. 189-207.

12. Goodman, R. D., and Menke, A. G., Solar Energy $\underline{17}$, 207 (1975).

13. Hollands, K. G. T., Solar Energy $\underline{9}$, 159 (1965).

14. Charters, W. S., and Peterson, L. F., Solar Energy $\underline{13}$, 353 (1972).

15. Chun, K. R., and Crandall, W. E., ASME Paper No. 74-WA/HT-11 (1974).

16. Buchberg, H., Catton, I., and Edwards, D. K., ASME Paper No. 74-WA/HT-12 (1974).

17. Tien, C. L., and Yuen, W. W., Intern. J. Heat Mass Transfer $\underline{18}$, 1409 (1975).

18. Marshall, K. N., Wede, R. K., and Dammann, R. E., LMSC/D462879, Lockheed Missiles and Space Co., Inc., Palo Alto, Calif. (1976).

19. Speyer, E., J. Eng. Power $\underline{87}$, 270 (1965).

20. Eaton, C. B., and Blum, H. A., Solar Energy $\underline{17}$, 151 (1975).

21. Siebers, D. L., "Analytical Studies of Flat-Plate Solar Collector Performance: Realistic Models and Design Concepts," M.S. Thesis, Purdue University (1975).

22. Taylor, R. P., and Viskanta, R., Wärme-und Stoffübertragung $\underline{8}$, 219 (1975).

23. Kays, W. M., Convective Heat and Mass Transfer, McGraw-Hill, New York (1966).

24. Tabor, H., Bull. Res. Counc. Israel $\underline{6C}$, 165 (1958).

25. Siebers, D. L., and Viskanta, R., Solar Energy $\underline{19}$, 163 (1977).

EXPERIMENTAL AND THEORETICAL PERFORMANCE STUDY OF SOLAR COLLECTORS

A. E. DABIRI, G. GROSSMAN*, AND F. BAHAR

Arya Mehr University of Technology, P.O. Box 3406
Tehran, Iran

ABSTRACT

Performance studies of four types of solar collectors have been made to determine their adaptability for domestic water and space heating systems in Iran. Flat-plate collectors with various forms of plate design and a parabolic type concentrating collector were considered. A specially designed test facility has been erected, where the collectors could be tested under steady state conditions. Flow rates, temperature and solar radiation measurements allowed for calculation of collected energy and efficiency. A numerical model has been developed to calculate heat losses from the collectors and to predict their performance.

Experimental results are given and compared with the theoretical ones obtained from the numerical analysis. Performance characteristics, including collected heat and efficiency, are calculated.

NOMENCLATURE

A - area of collector surface, (m^2)

A_1 - area of glazing, (m^2)

A_e - heat transfer area of collector edges, (m^2)

A_w - heat transfer area of water passages, (m^2)

A_5 - reflector area, (m^2)

A_{50} - projected area of reflector perpendicular to solar radiation, (m^2)

c - specific heat of water, (Joules/kg-$^\circ$C)

h_{23} - heat transfer coefficient from water to absorber, (Joules/m^2-sec-$^\circ$C)

$\dot{m}$ - mass flow rate of water through collector, (kg/sec)

Q_a - amount of heat absorbed in the absorber plate, (Joules/m^2-sec)

Q_s - solar heat reaching the collector, (Joules/sec)

Q_{10} - heat transferred from glazing to atmosphere, (Joules/sec)

---

*Visiting Professor, Presently at Technion Haifa, Israel.

829

Q_{23} - heat transferred from absorber to water, (Joules/sec)

Q_{24} - heat transferred from absorber through insulation, (Joules/sec)

Q_{21} - heat transferred from absorber to glazing, (Joules/sec)

Q_{20} - heat losses through edges, (Joules/sec)

Q_{15} - heat transferred from glazing to reflector, (Joules/sec)

Q_{54} - heat transferred from reflector through insulator, (Joules/sec)

Q_{50} - heat transferred from reflector to atmosphere, (Joules/sec)

R - reflectivity of reflector surface, (dimensionless)

T_0 - ambient temperature, ($^{\circ}$C)

T_2 - temperature of absorber, ($^{\circ}$C)

T_3 - temperature of water, ($^{\circ}$C)

T_s - effective sky temperature, ($^{\circ}$C)

T_{in} - inlet water temperature, ($^{\circ}$C)

T_{out} - outlet water temperature, ($^{\circ}$C)

α - absorptivity of absorber surface, (dimensionless)

ε_1 - emissivity of glazing, (dimensionless)

ε_2 - emissivity of absorber, (dimensionless)

ε_5 - emissivity of reflector surface

τ - transmissivity of glazing, (dimensionless)

INTRODUCTION

Domestic water and space heating are attractive applications of solar energy in Iran, where the average annual insolation is about 4500 kcal/m^2-day (1) and January mean daily temperatures range from -10°C in the extreme northeast to 20°C in the extreme southeast part of the country (2). Solar cooling and air conditioning is another important application with summer temperatures reaching up to 40°C in most parts of Iran (2).

Several types of solar collectors may be considered for the above applications. Flat-plate collectors have been widely used to supply solar energy at low temperatures such as is usually required for water and space heating (3,4). Flat-plate collectors are relatively simple to build and install; they do not need to track the sun and can utilize both the direct and diffuse components of the solar radiation. With a good design of the plate from the heat transfer and flow point of view, and utilizing appropriate materials, these devices can collect solar heat at efficiencies up to 90% (3). One important limitation of flat-plate collectors is that they normally cannot supply heat at temperatures exceeding $100^\circ C$. While these temperatures are quite adequate for water and space heating, higher temperatures are required for the efficient operation of cooling and air conditioning systems. For these applications, concentrating collectors need to be used.

The present paper describes part of a solar energy program initiated to develop solar domestic applications in Iran. Several types of solar collectors, both flat-plate and concentrating, have been designed, developed and tested. A theoretical and numerical analysis of the collectors has been performed to aid in their design. A test facility has been erected, where the collectors could be experimented with under steady state conditions. Flow rate, tempera- ture, and solar radiation measurements allowed for calculation of collected energy and efficiency.

THEORETICAL MODEL

In order to calculate heat losses from the collectors and predict their performance, it was necessary to develop a heat transfer and flow model. Such a model, implemented on a computer, provides an important tool for evaluating the geometrical dimensions and other design parameters for an efficient and economical operation of the collector.

Several theoretical analyses are described in the literature, particularly on heat transfer in flat-plate collectors. Among the most important ones is the work of Whillier (5), who made a comprehensive study on design factors influencing solar collector performance, following earlier studies by Hottel and Woertz (6). Whillier's calculations were based on a one-dimensional model with average heat transfer coefficients. Bliss (7) used a similar approach and calculated plate efficiency factors for rectangular and fin-and tube type geometries. More detailed analyses based on two- and three-dimensional models have been performed by others (8-10).

In our numerical model, the one-dimensional approach was adopted. For the flat-plate collectors, heat balances similar to those of Whillier (5) have been used, with a detailed calculation of the heat transfer coefficients. The lat- ter include the effect of heat losses by radiation and convection, each as a function of surface properties, geometry, thermal properties, and other influ- encing parameters. For the concentrating collector a similar model was deve- loped which also takes into account the heat losses from the reflector (Fig.1).

At this point, the details of the calculation is omitted and we only pre- sent some curves describing typical results of analysis. The efficiencies of a concentrating collector.and of a flat-plate collector versus an important dimensionless parameter, $\frac{mc}{h_{23} A_w}$, for different values of ε_2, are plotted in Figs. 2 and 3, respectively. Water temperature difference between outlet and inlet is shown as well. Typical values of the other para- meters have been used for plotting these figures. The collector dimensions used in the calculations are equal to those of the collectors built and tested in the present work. Other parameters have been taken as follows:

$$T_0 = 298.15^\circ K \; ; \; T_{in} = 293.15^\circ K \; ; \; T_s = 293.15^\circ K$$

$$\varepsilon_1 = 0.1 \qquad \alpha = \varepsilon_2$$

for the concentrating collector:

$$\varepsilon_5 = 0.2 \qquad ; \; \tau = 0.65 \qquad ; \; R = 0.8$$

$$A_1 = 2513 \; cm^2 \; ; \; A_5 = 24000 \; cm^2 \; ; \; A_{50} = 1520 \; cm^2$$

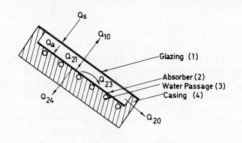

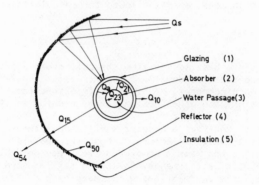

Fig. 1   Schematic of collectors

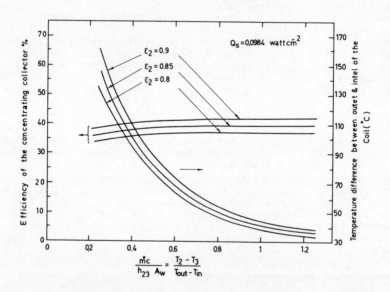

Fig. 2   The efficiency of a concentrating collector and also
temperature difference between outlet and inlet of the
coil, versus $\dfrac{\dot{m}c}{h_{23}\,A_w}$ for different values of ε_2

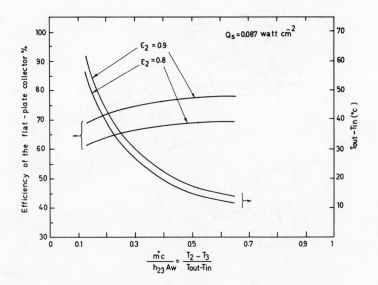

Fig. 3   The efficiency of a flat-plate collector and
         also temperature difference between outlet
         and inlet of the coil versus, $\dfrac{\dot{m}c}{h_{23}\,A_w}$ for
         different values of ε_2.

For the flat-plate collector,

$\tau = 0.9$

$A = 14365 \text{ cm}^2$; $A_e = 2545 \text{ cm}^2$

The results show that the efficiency of flat-plate collectors is usually
higher than that of concentrating collectors, but on the other hand, the water
temperature difference in a concentrating collector is higher than that of a
flat-plate collector. The efficiencies of both collectors drop at low flow
rates. This is due to the fact that at low flow rates, high fluid temperatures
are encountered which increase the heat losses.

CONSTRUCTION OF COLLECTORS

Three types of flat-plate collectors and one concentrating collector were
designed and built.

All three flat-plate collectors had the same construction of the casing
and transparent cover and differed only in the absorber plate. The plate of
collector 1 (Figure 4a) was fabricated from a grill of seven parallel ½" gal-
vanized steel pipes interconnected with two 1" headers. The pipes were spaced
4" apart. Straps of 19 gauge black iron were welded between the pipes to form
a continuous surface. The plate of collector 2 (Figure 4b) was made from the
same grill of pipes, but instead of welding iron straps between the pipes, a
sheet of 19 gauge black iron was laid under the grill. Seven longitudinal

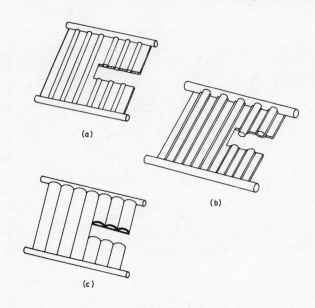

(a)

(b)

(c)

Fig. 4   Three different plates of flat-plate collectors

indentations had been pressed into the sheet to fit around the pipes, as shown.
The pipes were fastened to the sheet by means of steel clamps. The plate of
collector 3 was fabricated from a flat 19 gauge black iron sheet and a corrugated 16 gauge black iron sheet. The two sheets were connected together by
spot welding them along the lines of contact, and formed a plate with seven
channels (Figure 4c). Two headers, made of 1'' black steel pipe, were welded
at both ends of the plate.

Each of the three absorber plates was contained in a rectangular casing
fabricated from pressed wood. The inside dimensions of the casing were 180 x
88.5 cm and it was painted on the outside with water resistant paint. The
absorber plates, painted flat black, were laid in the casing on the bottom
insulation which consisted of a 2'' thick glass wool mattress. The edges were
insulated with 1'' thick polysterene coated with reflecting aluminum foil.
Each collector was glazed with one layer of 4 mm thick glass, with an option
to add a second layer of the same glass. Each of the three flat-plate collectors thus had an absorbing area A of 170 x 84.5 cm^2.

The concentrating collector is described in Figure 5. The reflector, in
the shape of a parabolic cylinder, was fabricated from chrome-plated steel
sheets supported on a wooden frame. The dimensions of its opening, facing the
sun, were 100 x 160 cm. The absorber was fabricated from a coil of $\frac{1}{4}$'' copper
tubing wound on a 1'' copper pipe and painted flat black. The absorber was
mounted inside an evacuated glass tube, 5 cm in diameter, and placed along
the line focus of the reflector. The area of the absorber was around 2010 cm^2.
The supporting structure carrying the collector was positioned along the east-
west direction and the tilt angle of the south facing reflector could be
changed by pivoting it on bearings. Additional details are given in Figure 5.

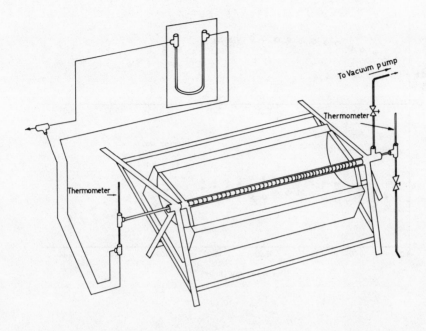

Fig. 5 Schematic of concentrating collector

EXPERIMENTAL RESULTS AND DISCUSSION

Tests according to the aforementioned procedure were conducted during the period June to October in Tehran (latitude 36°). Temperature, flow and radiation measurements were taken every 15 minutes from 09.30 to 16.00 hours.

Typical temperature profiles are shown in Figure 6 for a flow of 710 cm^3/ min. for the three types of flat-plate collectors. The radiation intensity variation is shown as well. The temperature increase is found to be largest for collector 3, where the plate design had the best heat transfer properties. Second best was collector 1, which had better thermal contact between the absorbing surface and the water passages than collector 2.

Fig. 7 describes the efficiencies of the three flat-plate collectors. The efficiency seems more or less constant, with somewhat lower values in the early morning and late afternoon, arising probably from the fact that a greater portion of the direct radiation is reflected from the glazing when the sun shines on the collector at a large angle from the normal. Part of this effect is compensated for by the fact that heat losses are smaller in the morning and afternoon, due to the lower temperature reached in the collector at these hours.

Fig. 8 shows characteristic curves for three flat-plate collectors descri- bing the variation of efficiency and water temperature difference versus the flow rate. In this figure, we have also plotted the curves based on our ana- lysis for two different values of ε_2. The heat transfer area of water passage

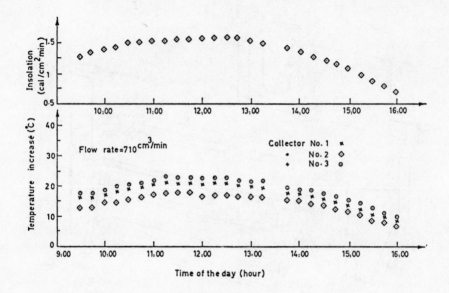

Fig. 6   Temperature profiles for three types of
         flat-plate collectors

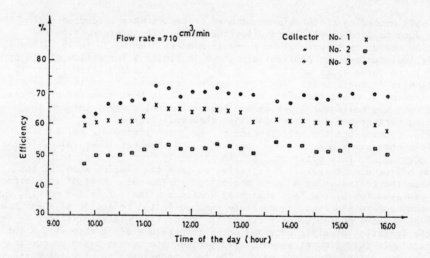

Fig. 7   The efficiencies of three flat-plate collectors,
         versus time of the day

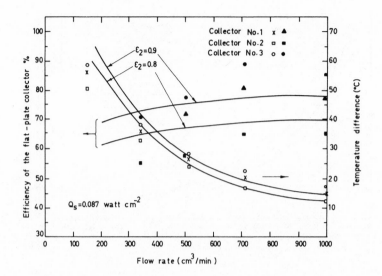

Fig. 8 Variation of efficiency and water temperature
difference versus the flow rate for three flat-
plat collectors

for plotting these curves is $A_w = 4673$ cm^2. We notice that the efficiencies
of three flat plate collectors are scattered around the theoretical curves
and the model predicts higher temperature differences than that of the experimental results for flow rates.

Experimental data obtained with the concentrating collector are plotted
in Figure 9 for two typical days. The mean radiation intensity during the
experiment was 984 watt/m^2. A temperature increase of up to 120°C has been
obtained. Efficiencies were lower than those obtained with the flat-plate
collectors. The curves based on our analysis are also plotted in this figure
for a heat transfer area of water passage of $A_w = 2010$ cm^2. The comparison
between the analysis and the experiment shows that experimental results fall
in between the theoretical curves for $\varepsilon_2 = 0.9$ and $\varepsilon_2 = 0.8$ assuming other
parameters constant.

CONCLUSIONS

A numerical model based on theoretical heat transfer analyses has been
developed to calculate design parameters of solar collectors. A test facility
has been erected which allowed testing of the collectors under steady state
conditions, and an experimental procedure was developed to perform the tests.
Three types of flat-plate collectors and a parabolic concentrating collector
have been designed and tested for application in domestic solar heating and
cooling systems. Good results have been obtained with a good agreement between
the numerical and experimental data. Both the numerical model and the test
facility have proved to be important tools in the efficient design of solar
collectors.

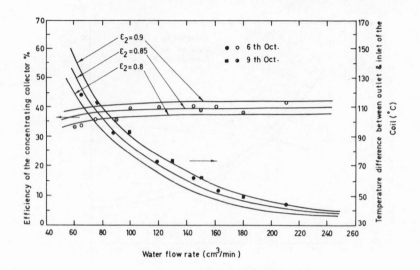

Fig. 9 Variation of efficiency and water temperature
        difference versus the flow rate for concentrating
        collector, for two typical days

REFERENCES

1. G.O.G. Lof, J.A. Duffie, and E.O. Smith,1960 , World distribution of
    solar energy. Solar Energy, 10; 27.
2. Meteorological Year Book,1965, Iranian Meteorological Department, Tehran,
    Iran.
3. R.C. Jordan (editor),1967, Low temperature engineering applications of
    solar energy, American Society of Heating, Refrigerating and Air Condi-
    tioning Engineers, Inc., New York
4. J.I. Yellott: Utilization of sun and sky radiation for heating and
    cooling of buildings, ASHRAE Journal, 15: 13-42, (December 1973).
5. A. Whillier,1967, Design factors influencing solar collector performance.
    In "Low Temperature Engineering Applications of Solar Energy", R.C. Jordan,
    editor, ASHRAE Journal, 27-40.
6. H.C. Hottel and B.B. Woertz,1942, The performance of flat-plate solar
    collectors. Trans. ASME, 64: 91-105.
7. R.W. Bliss,1959, The derivation of several "Plate efficiency factors"
    useful in the design of flat-plate solar heat collectors. Solar Energy,
    3, No. 4: 55-64.
8. M. Iqbal,1966, Free convection effects inside tubes of flat-plate
    collectors. Solar Energy, 10: 207.
9. D.J. Close,1962, The performance of solar water heaters with natural cir-
    culation. Solar Energy, 6: 33-40.
10. B.V. Petukhov,1967, Heat transfer in a tubular type solar water heater.
    Geliotekhnika, 3: 37-41.

COOLING OF A FLUID BY A NOCTURNAL RADIATOR

B. BOLDRIN AND G. SCALABRIN

Laboratorio per la Tecnica del Freddo del CNR
Padova, Italy

R. LAZZARIN AND M. SOVRANO

Facoltà di Ingegneria, Istituto di Fisica Tecnica
Università di Padova
Padova, Italy

ABSTRACT

Atmospheric infrared radiative heat exchange has been investigated with regard to its cooling effect on a flat-plate radiator using an i.r. transparent wind screen.

A computer program has been developed to determine the heat exchange available to cool a fluid circulating within the collector for a variety of operational and clear night-sky conditions.

Along with the theoretical analysis, experiments were carried out by employing one of the most common flat-plat collector types and an automatic station according to the national solar energy program was built up. In this station, besides the usual actinometric data, downward atmospheric flux, ambient temperature, relative humidity and wind speed were continuously recorded and transferred to a computer via punch tape. The simultaneous knowledge of these parameters and of the cooling effect obtained by the collector allows to find out the efficiency of the cooling system.

NOMENCLATURE

A   collector area (m^2)

C   constant in Eq. (3c)

c_p specific heat at constant pressure (J/kg.K)

d   diameter (m)

F   fin efficiency

F'   collector efficiency

F_R collector heat removal factor

G   atmospheric radiation (W/m^2)

h   convection coefficient ($W/m^2 {}^\circ C$)

k   thermal conductivity ($W/m^\circ C$)

L   duct length (m)

n   number of ducts

p   distance between ducts (m)

Q   heat flux (W/m^2)

R   thermal resistance $(m^2 °C/W)$

s   thickness   (m)

T,t   temperature $(K, °C)$

U   overall loss coefficient $(W/m^2 °C)$

W   flow rate   (kg/h)

x,y   cartesian coordinates   (m)

ε emittance

σ Stefan-Boltzmann constant

τ trasmittance

φ relative humidity

Subscripts

a   ambient, air

b   back-insulation

c   cover

f   fluid

H   hydraulic

i   inlet

m   mean

o   outlet

p   plate

s   sky

w   wind

INTRODUCTION

   It is common knowledge that, especially under clear-sky conditions and with transparent atmosphere, a rapid cooling of the
ground and of the atmospheric layers closer to it takes place
soon after sunset. This cooling is the result of heat loss by
infrared radiation to the cold night sky, about 300-500 W/m^2; the

main factor opposing nocturnal radiative cooling is the atmospheric radiation. This radiative downward flux coming from the atmosphere depends on concentration of the emitting gases, such as
water vapour, carbon dioxide, and ozone, as well as on their temperatures and on the degree of cloudiness.

In order to determine the atmospheric radiation a variety of
instruments has been designed and in recent years a series of
measurements were made during night-time hours by employing a differential radiometer. /1,2,3/.

Theoretical analyses have also been attempted by several investigators who solved the radiative heat transfer equation for a
model atmosphere, and correlated the values obtained by means
of a simple equation as a function of air temperature and humidity at ground level /4,5,6,7/.

The atmospheric downward fluw G can be conveniently expressed
in either of the following ways:

(a) by considering the atmosphere as a grey body at the ambient
temperature T_a, whose equivalent emissivity ε is:

$$\varepsilon = \frac{G}{\sigma T_a^4} \qquad (1)$$

(b) by regarding the atmosphere as a black-body at the equivalent
radiative temperature T_s of the sky:

$$T_s = \left(\frac{G}{\sigma} \right)^{1/4} \qquad (2)$$

Figure 1 shows typical spectra of atmospheric emission (a) and of
the blackbody radiative flux (b) at the equivalent temperature t_s,
for which the two areas are equal to each other, and Fig. 2 represents t_s as a function of relative humidity at different values
of ambient temperature. From Fig. 2 one can see the remarkable
difference between the values of the ambient temperature and the
equivalent blackbody temperature of the sky, and the influence of
relative humidity on it.

The difference between the upward and the downward radiative
fluxes gives the net radiative flux exchanged between the ground
and the atmosphere. Under clear night sky conditions the net flux
is normally about 100 W/m^2.

RADIATIVE HEAT FLOW MODEL

The nocturnal radiation cooling of a metal flat-plate, even
under the most favourable clear night sky conditions, is not considerable because of heat transfer by convection from the ambient
air to the exposed radiator surface /8/. It is necessary, therefore, to protect the radiator plate from the wind by means of a
plastic film which should be highly transparent in the infrared

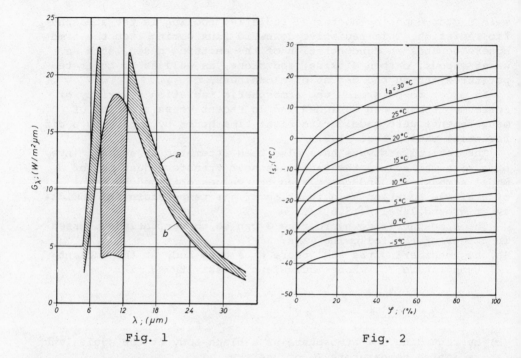

Fig. 1   Fig. 2

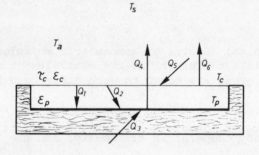

Fig. 3

region of the spectrum. The energy balance equations for the flat-
plate radiator and wind screen assembly shown in Fig. 3, assuming
that:
1) performance is steady state;
2) properties are independent of wavelength and temperature;
3) heat transfer is one-dimensional, losses through the sides are
ignored and are conveniently increased those through back insulation; by a constant C;
4) external surface temperature of insulation is equal to ambient;
5) wind convection coefficient is given by McAdams's formula /9/

are:

$$Q_1 = \varepsilon_p \, \varepsilon_c \, \sigma(T_c^4 - T_p^4) \tag{3a}$$

$$Q_2 = \frac{k_a}{s_a} \, (T_c - T_p) \tag{3b}$$

$$Q_3 = \frac{k_b}{s_b} \, (T_a - T_p) \, C \tag{3c}$$

$$Q_4 = \tau_c \, \varepsilon_p \, \sigma(T_p^4 - T_s^4) \tag{3d}$$

$$Q_5 = h_w \, (T_a - T_c) \tag{3e}$$

$$Q_6 = \varepsilon_c \, \sigma(T_c^4 - T_s^4) \tag{3f}$$

Under given environmental conditions the wind screen temperature t_c is obtained from the balance equations (3) for every set
flat-plate temperature t_p below ambient temperature. This screen
temperature is higher than the radiator temperature and a little
lower than ambient temperature.

In order to determine the net flux exchanged Q_o:

$$Q_o = Q_4 - (Q_1 + Q_2 + Q_3) \tag{4}$$

a computer program has been carried out making use of the above
equations (3) and making the following assumptions: the radiator
plate emissivity is $\varepsilon_p = 0.9$; the wind screen cover emissivity
and transmissivity are $\varepsilon_c = 0.07$ and $\tau_c = 0.85$ respectively;
the back-side is insulated by 5 cm polyurethane foam. The spacing
between radiator plate and screen cover is enough to minimize
heat flow by conduction but not so wide as to favour heat exchange
by convection.

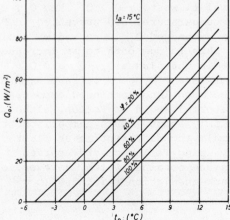

Fig. 4

Figures 4 illustrates the results relative to the net cooling
effect of the radiator as a function of plate temperatures and
for different relative humidities. One can see how Q_o decreases
with a decreasing plate temperature. It is possible to obtain
very low plate temperatures as compared to ambient, but that would
yield an insignificant cooling effect.

One can also realize the strong influence of humidity which
makes the practical application of the process more convenient
in dry climates.

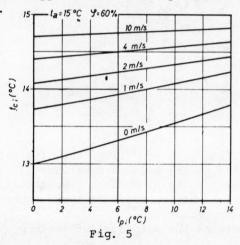

Fig. 5

Values of cover temperature t_c for a given ambient temperature
and relative humidity are plotted in Fig. 5 for different values
of wind velocity as a function of plate temperature t_p. It can
be seen that cover temperature is always very near ambient temperature and increases slightly with t_p and with wind velocity. For
this reason wind velocity was not taken into account in Fig. 4,
where its influence would have been limited. On the contrary if
the plate had not been provided with screen cover, wind velocity
would have been of great importance /8/.

Examples of thermal fluxes Q_1, Q_2, Q_3, and Q_4, calculated as
a function of plate temperature, are presented in Fig. 6. One can
see how the net flux exchanged Q_o is raised when plate temperature t_p is increased: this is due both to an increase in the flux
Q_4 and to a decrease in the fluxes Q_1, Q_2 and Q_3.

Polyethylene films have very good infrared transmittance but
are soon deteriorated upon long-continued exposure. A more suitable cover plastic material would be Teflon, although its infrared
transparency can be only about 70%. A comparison between the results obtained by employing different cover plastic films is presented in Fig. 7. By using Teflon covers, curves b and c, less
satisfactory results are obtained which, however, can be improved
by increasing the back-insulation thickness (b = 8 cm and c = 5cm).

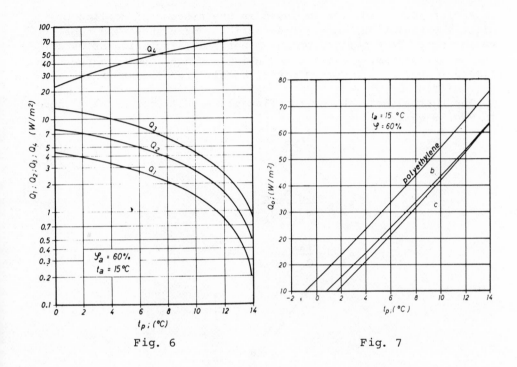

Fig. 6

Fig. 7

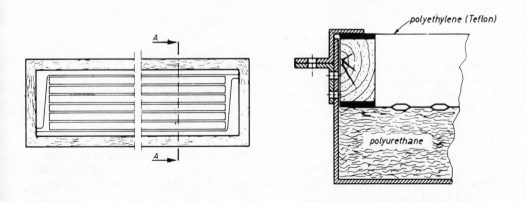

Fig. 8

PERFORMANCE OF FLAT-PLATE RADIATOR IN THE COOLING PROCESS

A flat-plate radiator employed in the process of cooling by night-sky radiation, consists essentially of a sheet-and tube radiating surface through which an antifreeze solution of water can be circulated. Above the radiating surface, spaced a few centimeters apart, is a screen cover, highly transparent in the infrared (Fig. 8).

The thermal network for the flat-plate collector is represented in Fig. 9 where:

$$R_1 = \frac{T_a - T_p}{Q_3} \tag{5a}$$

$$R_2 = \frac{T_c - T_p}{Q_1 + Q_2} \tag{5b}$$

$$R_3 = \frac{T_a - T_c}{Q_5 - Q_6} \tag{5c}$$

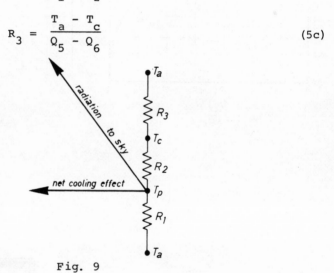

Fig. 9

A number of simplifying assumptions are made:
1. Steady-state performance.
2. Temperature gradients around ducts neglected.
3. Temperature gradients in the direction of flow and between ducts treated independently.
4. Uniform flow in ducts.

Figure 10 shows the region between two ducts and, at any location y, the temperature distribution in the x direction (Fig. 10a) and, at any location x, the temperature distribution in the y direction (Fig. 10b).

By operating in the same way as with calculations from solar collectors, according to Bliss's theory /10/, one can write the

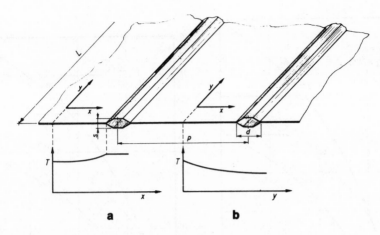

Fig. 10

following equations:

$$F' = \frac{1/U}{P\left\{\dfrac{1}{U\left[d + (p-d)F\right]} + \dfrac{1}{\pi d_H h_f}\right\}} \tag{6}$$

$$F_R = \frac{Wc_p}{A U}\left\{1 - \exp\left(-\frac{U F'A}{Wc_p}\right)\right\} \tag{7}$$

$$T_{f,m} = T_{f,i} - \frac{Q_o}{UF_R}\left(1 - \frac{F_R}{F'}\right) \tag{8}$$

$$T_{p,m} = T_{f,m} - \frac{Q_o A}{h_f \pi d_H nL} \tag{9}$$

It is, therefore, possible to calculate the temperature $t_{f,o}$ of the fluid at the outlet for a given fluid inlet temperature $t_{f,i}$ as a function of flow rate.

Figures 11 and 12 show these temperature variations $t_{f,o}$ under two different climate conditions; the better results are obtained when the atmosphere is more transparent.

The dashed curves in Figs. 11 and 12 refer to the performance of a teflon cover radiator.

EXPERIMENTAL RESULTS

The experimental apparatus employed consists of a 'roll-bond' flat-plate collector, whose area is $2.94 \cdot 0.54 \ m^2$. The radiator plate is dull black and back insulated by 5 cm polyurethane foam.

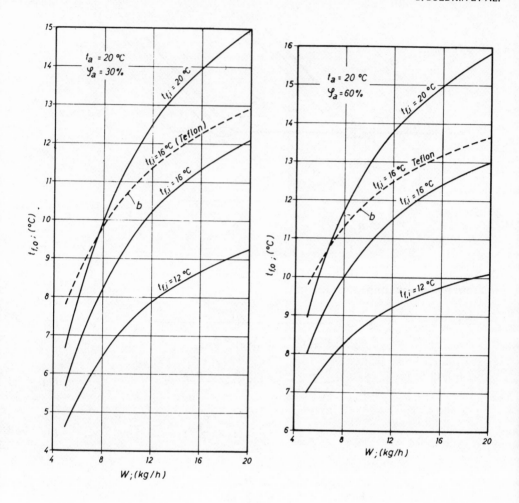

Fig. 11 Fig. 12

The plate is crossed by only a few tubes in order to reach a fluid velocity sufficient to obtain a reasonable thermal exchange between fluid and plate in spite of the low flow rate of fluid. A polyethylene screen cover is set above the radiating plate, spaced 5 cm apart. The collector is fed by a constant level tank in order to keep a practically steady flow (Fig. 13).

The evaluation of results is made easier by the presence nearby of a solar station built up according to the national solar energy program /11/.

In this station, besides the usual actinometric data, the follo

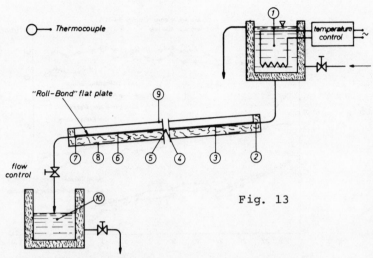

Fig. 13

wing data are continuously recorded on punched tape:
1. downward atmospheric flux;
2. ambient temperature;
3. relative humidity;
4. wind speed.

Investigations were carried out last summer and the results relative to two nights are reported in Figs. 14 and 15. The following quantities, recorded all night long, are represented for a given flow rate:

Fig. 14

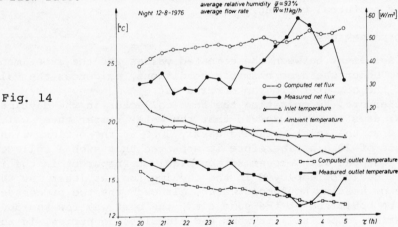

1. inlet and outlet temperatures;
2. ambient temperature;
3. net thermal flux.

In spite of the humidity, the net exchange is always about 40-50 W/m^2 for a 5°C cooling below ambient temperature. The high humidity often caused the formation of condensate on the screen, which was at a temperature 1-2°C below ambient temperature, thus

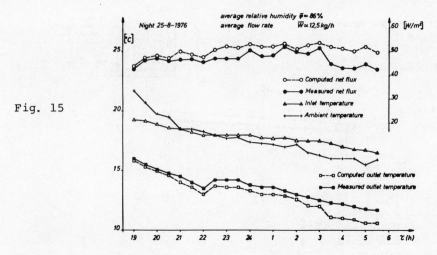

Fig. 15

reducing the cooling effect of the system.

The values computed according to the above theoretical study
and referring to clear night sky conditions appear in dashed line
in Figs. 14 and 15. There is a good agreement between these values
and the highest experimental data.

Further investigations will be developed for steady flow and
different inlet temperatures in order to determine the performance
of the collector for a large range of conditions. This knowledge
will make it possible to define in detail some practical applications of nocturnal cooling.

CONCLUDING REMARKS

The agreement between the computed values and the experimental
results, under the same climatic conditions, evidences the validity of the computer program carried out.

Furthermore, by operating the same collector in the daytime
on clear days it was observed that the water temperature, which
was 20°C at the inlet, reached easily 50°C at the outlet, with a
flow rate of 20kg/h. Therefore it appeared that such a collector,
though equipped with a cover screen highly transparent in the infrared, could be employed in the daytime too, at least for the
purpose of heating household water. Although the two processes
cannot be optimized at the same time, the best way for the double
utilization, especially in warm and cloudless countries, is now
being investigated.

REFERENCES

M. SOVRANO and B. BOLDRIN, Investigation on nocturnal radiation
measured by a new differential radiometer, Fifth Int. Heat Transfer
Conference, Tokyo (1974).

2. R.M. GOODY, Atmospheric Radiation, Vol. 1 Clarendon Press, Oxford (1964).

3. W.M. ELSASSER, Heat Transfer by infrared radiation in the atmosphere, Harvard Meteor. Studies 6, 1-107 (1942).

4. K.Ya.KONDRATYEV, Radiative Heat Exchange in the Atmosphere (trans. from Russian), 111-129. Pergamon Press, Oxford (1965).

5. W.C. SWINBANK, Long-wave radiation from clear skies, Quart. J.R. Met. Soc. 89, 339-348 (1963).

6. B. BOLDRIN, R. LAZZARIN and M. SOVRANO, A numerical model of infrared atmospheric emission over a clear sky, Report No.44, Fisica Tecnica, Ingegneria, Padova (1973).

7. B. BOLDRIN, R. LAZZARIN e M. SOVRANO, A study of infrared radiation from the atmosphere, Report n. 55, Fisica Tecnica, Padova 1974.

8. B. BOLDRIN, G. SCALABRIN e M. SOVRANO, On the utilisation of radiative cooling in the atmosphere, Int. Inst. of Refrigeration XIVth Int. Congress Moscow, 1975.

9. J.A. DUFFIE and W.A. BECKMAN, Solar energy thermal processes, Wiley, 1974.

10. R. BLISS, The derivations of several plate- efficiency factors, useful in the design of flat plate solar collectors, Solar Energy, III, 4, 55-64, 1959.

11. G. BETTI, An automatic self-sufficient station for the recording of solar, radiative and climatic data, 3rd Course on "Solar Energy Conversion, Alghero (Italy), 1976.

THE EQUIVALENT DEGREE-DAY CONCEPT FOR RESIDENTIAL SPACE HEATING ANALYSIS

J. T. ROGERS, J. B. FINDLAY, M. C. SWINTON, J. D. MOIZER, AND R. H. H. MOLL

Energy Research Group, Carleton University
Ottawa, Ontario, Canada K1S 5B6

ABSTRACT

A model has been developed to represent the residential space-heating demand in Ontario, so as to permit better forecasting of energy consumption in this sector and the evaluation of various policy options. The Ontario housing stock is represented by four typical residences whose structural and thermal characteristics were established by an extensive survey. To provide the accuracy desired, the conventional degree-day method is replaced by the equivalent degree-day method, which allows for passive solar heat and for wild heat.

Typical results obtained by use of the model are given.

NOMENCLATURE

A area, m^2

C_p specific heat of air, kJ/kgK

D_d nominal degree (K)-days, based on the conventional North American balance temperature of $65^{\circ}F$ ($\sim18^{\circ}C$), K-days

D_{de} equivalent degree-days, K-days

H_S^* heating-season average solar heat gain rate, kW

H_W^* heating-season average wild heat gain rate, kW

K overall heat loss factor for a building, kW/K

k extinction coefficient for solar radiation in glass, cm^{-1}

l thickness of glass, cm

n average number of air changes per hour

Q annual net building heat loss, kJ

$SHGF$ solar heat gain factor, W/m^2

$\overline{SHGF}$ heating-season average solar heat gain factor, W/m^2

T_i internal design temperature, K

T_o mean of maximum and minimum ambient temperatures on a given day, K

T^* balance temperature, the temperature below which the heating system is required to operate, $^{\circ}C$ or K

$\overline{T}_o$ yearly average of T_o, K

$T_{o\ max}$ yearly mximum of T_o, K

$T_{o\ min}$ yearly minimum of T_o, K

t   time, sec., hr. or days

U   overall heat transfer coefficient, kW/m^2K

V   internal volume of building, m^3

ρ   interior air density, kg/m^3

ψ   latitude angle, degrees

Subscripts

i   house type i

w   windows

A bar over a symbol indicates an average value.

INTRODUCTION

To improve the forecasting of energy consumption and to permit the evaluation of the effects on energy demand of specific policy options in the residential and commercial sectors, the Ministry of Energy of the Province of Ontario
commissioned the development of forecasting models for these sectors which would
include technical as well as conventional economic factors. The Energy Research
Group of Carleton University has been responsible for developing the technical
aspects of the models.

In the residential sector, about 65 to 70% of the total energy demand is
for space heating, so that any model of energy consumption in this sector
must represent the space-heating demand realistically if it is to be useful
for forecasting overall demands. The development of a model to represent
the residential space-heating energy demand of Ontario was, therefore, a
major task in this study.

To accomplish its purpose, the model must fulfill two major requirements.
First, it must represent the essential physical features of the losses and gains
of heat in the entire Ontario housing stock in sufficient detail to permit
accurate computation of the present energy consumption for space-heating without
being too unwieldy nor requiring excessive computer time or storage capacity.
Second, it must permit the investigation of energy-saving options (e.g. the
addition of below-grade insulation in basements) and new energy sources (e.g.
solar heating, district heating) in sufficient detail to enable realistic
forecasts to be made of future space-heating energy consumption resulting
from appropriate and economically-justified improvements in the housing stock
and equipment.

To fulfill these requirements, it was necessary a) to define a manageable
but realistic representation of the present Ontario housing stock, with respect
to its thermal characteristics, and b) to establish equations representing the
heat losses and gains of this stock in a realistic manner.

This paper concentrates on the second necessity, but provides a brief
outline of the housing stock model as well as some of the specific results of the
study.

REPRESENTATION OF ONTARIO HOUSING STOCK

In addition to the requirement that the housing stock model did not become too
unwieldy, an important constraint was obviously the information that was available on the Ontario housing stock. The model was established from a study of
references 1, 2, 3 and 4, as well as from many discussions with representatives
of government bodies, fuel suppliers and utilities. Of course, several itera-
tions using the heat loss and gain model described later were required to arrive
at the final housing stock (and heat loss) model.

First of all, the Ontario housing stock was classified as to its space-
heating demand characteristics by the matrix shown in Table 1. The classification as to type follows that of references 1 and 2 and is fairly obvious, but
the other classifications require a little explanation. The vintage category
allows for changes in the insulation standards in the National Building Code
(3,4). The regional category allows for significantly-different degree-day
areas which are reflected in varying building code requirements (3,4). The
energy-supply category allows for the different insulation standards required
for electrically-heated as opposed to non-electrically heated houses (3,4).

Table 1. Classification of Ontario Housing Stock for Space-Heating Forecasts

Type, i	Single family, Detached	Semi-detached and Duplex	Row	Apartment
Vintage, j	Pre-1965	1966-71	1972-76	1977 & later
Region, l	Eastern (Ottawa)	Central (Toronto)	South-western (Windsor)	Northern (Thunder Bay)
Energy Supply, n	Non-electric	Electric		

For each type of residence listed in Table 1, weighted-average sizes and
typical building layouts were established. In each category, typical wall,
window, roof, and basement characteristics were established, using information
from several sources but mainly references 1 to 5. Since the detailed description of each category of residence in Table 1 would be quite lengthy and since
description of the residence types is not the primary purpose of this paper,
these descriptions are not given here. Essentially, the housing stock model is
oriented to modern frame construction. Full details can be found in reference
6.

From the foregoing information, using the space-heating model described
later, the present space-heating energy demand can be calculated. With further
assumptions as to fuel types and furnace efficiencies, the total input energy
demand for space heating can be calculated and compared to actual values. Of
course, re-iterative methods were used until satisfactory agreement was achieved.

The model, which can predict the present Ontario space-heating energy
consumption with adequate accuracy, provides a basis for reasonable forecasts
of energy consumption for this purpose. By using various econometric approaches
to forecast changes in housing stock, fuel types and other factors (7) and by
accounting for technical improvements and changes as described earlier, forecasts
of energy demand can be made and the impacts of various technological changes can
be established. The model has not yet been used for these latter purposes.

HEAT LOSS MODEL

The degree-day method for calculating annual heating requirements for residential buildings is well known (e.g. 8). However, the degree-day method, when
applied to modern, well-insulated residences with low infiltration rates generally results in a significant over-prediction (typically 25 to 30%) of the actual

annual heating load. While such a conservative over-prediction may be acceptable for design purposes for individual residences, a more accurate method was desired for the present study.

The annual net building heat loss according to the degree-day method is given in principle by

$$Q = 24 \ K \ \sum_n (T^* - T_o)n \tag{1}$$

$$= 24 \ K \ D_d \tag{2}$$

The overall heat loss factor for the building, K, is given by:

$$K = \sum_k U_k \ A_k + \frac{n\rho C_p V}{3600} \tag{3}$$

where the first term represents the perimeter or fabric heat loss through the various building perimeter components, k, and the second term represents the infiltration or ventilation heat loss.

Values of U_k for the various components, including the effects of internal and external film coefficients and allowances for varying insulation levels were taken from reference 9. Effective values of U_k for the basement walls below grade and the basement floor were taken from reference 5. The computer program developed for the analysis permits the independent specification of any level of insulation for the various components. A complete listing of all values used is contained in reference 6.

The greatest uncertainty in determining K is the number of air changes per hour, n. This depends on many factors including window and door crack lengths, type of windows and doors, house structure, wind velocity and direction, internal to external temperature differences, existence or not of fireplaces, type of heating (electric or non-electric), furnace operation, ventilation fan operation and habits of residents (9,10,11,12,13). Thus the determination of a heating-season average value of n characteristic of the various residence types is very difficult. Values eventually chosen from a study of available data ranged between 0.2 to 0.4 air changes per hour for electrically-heated houses and about 0.5 air changes per hour for oil- or gas-heated houses. However, because of the large uncertainties in these values, and since older houses may have higher values of n, assessments of the sensitivity of residential space-heating demand to variations in n were made.

Conventionally in North America, the balance temperature, T^*, is taken as $65^\circ F$ ($18.3^\circ C$) and values of degree-days, D_d, based on this balance temperature and determined from meteorological data averaged over many years, are available for most localities. The choice of $65^\circ F$ as the balance temperature recognizes that solar gains and "wild heat" from appliances, equipment and occupants play a role in establishing the heating requirements of a building, since the normal internal heating season design temperature is about 70 to $72^\circ F$ ($\sim 21\text{-}22^\circ C$).

However, it appears that the use of an arbitrary fixed balance temperature of $65^\circ F$ is not an adequate approximation for modern, well-insulated "tight" residential buildings, equipped with a conventional range of appliances. The over-estimation of the annual heating load by the conventional degree-day method can be attributed mainly to this assumption of a fixed balance temperature. To provide the accuracy desireable in the present forecasting model, a method has been developed which, in effect, permits the calculation of an appropriate balance temperature allowing for the heating season average solar gain for the housing stock classification of Table 1, and for the heating season average wild heat.

In this method, the actual heating season average balance temperature is calculated by:

$$T^* = T_i - \frac{H_S^*}{K} - \frac{H_W^*}{K} \tag{4}$$

Knowing the actual balance temperature, a modified value of degree-days which allows for solar gains and wild heat, the equivalent degree-days, D_{de}, can be calculated, from which a reasonably accurate annual heat loss may be calculated. The calculations of H_S^*, H_W^* and D_{de} are described in the following sections.

HEATING-SEASON AVERAGE SOLAR HEAT GAIN RATE, H_S^*

Calculations of solar contributions to space heating or cooling are greatly facilitated by the use of tables of solar heat gain factors (SGHF) (9,14). Solar heat gain factors are the instantaneous rates of solar heat transfer on a clear day through a unit area of one layer of unshaded double-strength sheet glass at a given location, month of year, time of day and orientation of surface.

In the present study, solar heat gain factors from reference 14 were used. The SHGF's from reference 14 are given for an average clear day (Clearness Number = 1.0(15)), and allow for direct, scattered and ground-reflected radiation. The direct component allows for annual variations of the solar constant and the atmospheric extinction coefficient. The diffuse component allows for the seasonal variation of scattering in the atmosphere and accounts for the diffuse radiation incident on a vertical surface by the empirical equation of Threlkeld (16). The ground-reflected component assumes a ground reflectivity of 0.2. The window glass assumed in the table has a value of kl, i.e., the product of extinction coefficient and glass thickness, equal to 0.05, which is typical of window glass used in North America (14). The glass is also assumed to have an index of refraction of 1.52.

Note that the model used in the study assumes that all solar heat gains occur through windows. While this is not rigorously true since solar radiation will increase the outer surface temperatures of opaque walls and roofs, thus reducing the heat loss rate, the magnitudes of such effects are considerably less than those of solar heat gains through windows (17) and are ignored in this study.

The heating-season average solar heat gain rate was determined in the following manner. For a given latitude the tables of reference 9 give daily total SHGF's for a typical day of each month on vertical surfaces oriented in eight directions (N, NE, E, SE, S, SW, W, NW). For a given latitude, the daily totals were averaged over the day and over the eight directions, it being assumed that the housing stock in Ontario is randomly oriented. The resulting directional-average value of SHGF was then assumed to represent also a time-average value for the month in question. This process was repeated for each month of the heating season, which, in Ontario may be assumed to extend from the end of September to the end of March, and the resulting values averaged to find the heating-season average solar heat gain factor for a given latitude. This procedure was repeated for a number of latitudes over the range of $43^{\circ}N$ to $49^{\circ}N$, which covers the latitudes of interest in this study, that is, those within which the majority of the population of Ontario lives. The resulting heating-season average solar heat gain factor was found to be a linear function of latitude and an equation representing it was established by the least-squares technique:

$$\overline{SHGF} = 159.4 - 1.517\psi \tag{5}$$
$$43^{\circ}N \leq \psi \leq 49^{\circ}N$$

Since almost all of the Ontario housing stock is equipped with double-glazed windows, the model assumes that double-glazing is standard. Therefore, the above heating-season average solar heat gain factor must be modified to allow for double-glazed windows. The conventional approach to allow for such factors is to employ shading coefficients (9,14). For example, the shading coefficient for a double-glazed window of ordinary window glass is 0.9 (9). That is, to calculate the solar heat transfer through a double-window, the appropriate solar heat gain factor is multiplied by 0.9. However, the shading coefficient is designed to establish the <u>maximum</u> heat gain rate resulting from the particular configuration, since the main use of shading coefficients is to establish design heat loads for cooling systems, while what was needed for our purposes was a heating-season average value.

To establish a value for the heating-season average shading coefficient, the following procedure was used. The incidence angles for direct solar radiation for vertical windows with various orientations at various times of the year were established as a function of time of day using solar altitude and azimuth angles given in reference 14. Then, taking values of transmissivity of ordinary window-glass as a function of incidence angle from tables in reference 18, the reduced solar heat gain because of the second sheet of glass was calculated as a function of time of day by multiplying the solar heat gain factor for the chosen orientation from reference 14 by the appropriate transmissivity. The resulting reduced daily total solar heat gain factor was divided by the normal daily total solar heat gain factor from reference 14 to establish an average shading coefficient for the day. By repeating this procedure for various window orientations and times of year, a heating-season average shading co-efficient for double-glazed windows was established as 0.8.

An allowance must also be made for the presence of blinds or curtains on the windows. Values of shading coefficients for blinds of various types and for curtains are available (9), but again these would give the maximum solar heat loads for such components rather than heating-season average values. For example, for light-coloured venetian blinds on a double-glazed window, the shading co-efficient is 0.51 (9). Reducing this value in proportion to the reduction in the value for a double-glazed window alone, an estimated heating-season average value is about 0.45. Of course, there is no way of actually determining the number of windows in the Ontario housing stock that will be equipped with blinds and curtains nor the percentage of time during the heating season that the blinds and curtains will be closed. Therefore, there is no justification for attempting to develop a very elaborate model for the windows. It was thus assumed that one-third of the windows would have closed blinds or curtains at any one time, so that using the above estimated values, the overall correction factor for the heating-season-average solar heat gain factor was estimated to be 0.68. This correction factor was assumed to apply in each of the regions of the Province used in the model.

An additional correction to the heating-season average solar heat gain factor is required for the effects of cloud cover. A correlation for the overall effect of cloud cover on direct and diffuse solar radiation is given by Kimura and Stephenson (19), but this correlation could not be used because of lack of information on typical cloud covers in the various regions of the province. However, data extending over a period of 23 years were available for the monthly-average solar irradiation on a horizontal plane in the Ottawa region (20). These data were compared to the monthly-average values of solar insolation for a clear day on a horizontal plane calculated from data given in reference 14, and the ratios of the two values were taken as representing the monthly-average effects of cloud cover. The monthly-average values were then averaged to determine the heating-season average value. The heating-season average correction factor for cloud cover determined in this way was 0.67. Lacking any other information, it was assumed that this correction factor applied in all the regions of the Province.

Therefore, the modified value of solar heat gain factor which allows for double-glazed windows, blinds and curtains and cloud cover is given by:

$$\overline{SHGF}_c = (0.68)(0.67)(159.4 - 1.517\psi)$$
$$= 72.6 - 0.691\psi \tag{6}$$

It can be easily shown that the heating-season average solar heat gain factor is related to the heating-season average solar heat gain rate, H_S^*, by:

$$H_S^* = (\overline{SHGF}_c) \, A_{wi} \tag{7}$$

Therefore, the heating-season average solar heat gain rate is given by:

$$H_S^* = (72.6 - 0.691\psi) \, A_{wi} \tag{8}$$

There are several other factors which can affect the heating-season average solar heat gain rate which are not accounted for in equation 8. These factors include the effects of adjacent buildings, trees, topographical features and other objects on solar radiation. These effects include reductions in solar heat gains because of interference with direct, diffuse and reflected sunlight as well as increases resulting from increased reflection. Obviously, it is impossible to estimate accurately the overall impact for Ontario, but it is expected that such effects would result in a net reduction in the heat gain rate. Another factor is the effect of snow cover on ground reflectivity, which increases the reflectivity greatly (14,21). In the heating season in Ontario solar heat gain factors may be about 20% higher for ground freshly covered with snow than for bare ground (14). Again, the overall impact of snow cover on solar heat gains in Ontario is very difficult to estimate accurately. However, since the overall impacts of surrounding objects and of snow cover are opposed to each other, it was assumed that they would compensate and that the net impact could be ignored. Obviously, this point, including possible regional variations, requires further investigation.

Another factor that is ignored in equation 8 is the effect of heat storage. Again, this is done on the basis that heat storage effects will balance out over the heating season.

Because of the several uncertainties in the various factors affecting the heating season average solar heat gain rate, the sensitivity of the annual space heating requirements to variations in this factor was examined as will be reported later.

HEATING-SEASON AVERAGE WILD HEAT GAIN RATE, H_W^*

The heating-season average wild heat gain rate for the Ontario housing stock was established from the results of several surveys by various government and utility bodies, (22,23,24,25,26). For the most common electrical appliances, including lighting, the average capacities, average usages and penetrations have been estimated. Similar estimates are available for non-electric household equipment. In the model, appliances and equipment have been grouped into three categories:

a) major appliances with 100% penetration (lighting, stoves, refrigerators, radios, television sets)

b) major appliances with less than 100% penetration (automatic clothes washers, clothes dryers, dishwashers, room air-conditioners)

c) minor appliances

A separate category is used for non-electric (gas) appliances.

The average annual appliance energy consumption per household has been established for the Province as a whole; no information is available on regional variations but these are not expected to be significant. Similarly, there was not sufficient information to establish any seasonal variation of appliance usage. Also, no variation with house type or age was apparent.

In addition to energy generation by appliances, an allowance was made for the contribution of the inefficiency of water heating to wild heat. Although there is some dependence of this contribution on housing type, since demand for hot water varies directly with the number of persons in a household, and the average number of persons varies with housing type (6), this variation has been ignored in the model, since its overall effect will be small and within the uncertainties of the data.

Finally, an allowance for the heat generated by occupants can be included in the wild heat contribution.

From a survey of the available information, the heating-season average wild heat gain rate, H_W^*, was established as 0.94 kW per household, which was assumed constant for all the categories given in Table 1. Because of the many uncertainties in this factor, the sensitivity of space-heating demand to wild heat gain rate was determined by considering variations in this factor of $\pm$ 25%, i.e. from about 0.71 kW to about 1.18 kW.

EQUIVALENT DEGREE DAYS, D_{de}

The balance temperature, T^*, for any category given in Table 1 can now be calculated from equation 4 since H_S^*, H_W^* and K are now known. An internal design temperature, T_i, of approximately $22^{\circ}C$ was assumed in the study, but this value is left as a variable in the program so that parameter studies may be done.

Knowing T^*, a modified value of degree-days, the equivalent degree-days, D_{de}, may be calculated by:

$$D_{de} = \int_{HS} (T^* - T_0)dt \qquad (9)$$

where the integral is taken over the heating season. To calculate the equivalent degree-days for any balance temperature, it is assumed that the yearly variation of mean daily temperatures can be simulated by a sine function, i.e.,

$$T_0 = \overline{T}_0 + \left| \frac{T_{o\,max} - T_{o\,min}}{2} \right| \sin \frac{2\pi t}{365} \qquad (10)$$

Obviously, T_0 will vary from region to region within the Province. To validate this procedure, the conventional degree-days, based on $65^{\circ}F$ ($18.3^{\circ}C$), were calculated from equations 9 and 10 and compared to the tabulated values. This comparison is shown in Table 2, where the largest error is seen to be about 1.5%.

Table 2. Comparison of Tabulated and Calculated Degree-Days

Region	Eastern (Ottawa)	Central (Toronto)	South-Western (Windsor)	Northern (Thunder Bay)
$\overline{T}_0$, $^{\circ}C$	5.7	8.6	8.9	2.3
$T_{o\,max}$, $^{\circ}C$	20.6	21.7	21.7	17.2
$T_{o\,min}$, $^{\circ}C$	-11.1	-4.4	-3.9	-15.0
D_d, calculated K-d (Equation 9)	4763	3736	3611	5841
D_d, tabulated, K-d	4829	3793	3655	5780

RESULTS

To illustrate the use of the model and to assess the importance of various factors, we will apply the method to some specific cases for the Eastern Ontario (Ottawa) region.

For a single-family detached house in the Ottawa region, equipped with an oil-fired furnace, the selected average values of the major parameters, as discussed in the previous sections, are an infiltration rate, n, of 0.5 air changes per hour, a heating-season average solar heat gain rate per unit window area, H_s^*/A_{wi}, of 41 W/m^2, and a heating-season average wild heat rate, H_w^*, of 0.94 kW. The total heating-season heat losses and gains as functions of the vintage (insulation standard) of the residence are given in Table 3.

Table 3 shows that even with the relatively poor insulation specified by the National Building Code prior to 1965, solar and wild-heat gains combined reduce the total heating-season average heat losses by about 17%, while with present-day insulation standards, the reduction is 25%, with the solar contribution being about half of this value. Solar and wild-heat gains are seen to be of the same magnitude for all the insulation levels considered in Table 3. Therefore, the importance of these factors is evident, even with pre-1965 insulation standards. Older houses tend to have higher infiltration rates than newer ones (11), but this possibility is ignored in Table 3. We will consider the sensitivity of net heat losses to infiltration rate later.

The influence of the type of residence on heat gains and losses is shown in Table 4, for residences with oil-fired furnaces in the Ottawa region and for insulation standards of the period 1972-1976. The significant reduction in net heat losses from single-family detached houses to multi-unit residences is evident, with an apartment unit having a net heat loss of only about 20% of that of a single-family detached house. Therefore, the trend towards multi-unit residences which is expected to continue in Ontario (27), should yield considerable benefits in the reduction of energy demand. Table 4 also shows that infiltration losses become relatively very important in row-houses and apartment buildings, as the peripheral heat losses are reduced. We also see that solar heat gains are somewhat less important relatively in apartments than in detached houses, while wild heat gains tend to become very significant in apartment units, since the mix of appliances is a function of family income rather than residence type.

Table 5 shows that electrically-heated detached houses, because of their higher insulation standards and lower infiltration rates, have lower net heat losses than do oil- or gas-heated detached houses. However, the difference in heat losses between electric- and non-electric heated houses is considerably less for 1972-76 standards than it would be for houses insulated to previous standards, because the standards for non-electrically-heated residences have recently approached much more closely to those for electrically-heated residences than used to be the case (6).

Tables 3 and 4 also give the balance temperature calculated by equation 4. We see from Table 3 that for detached houses insulated to pre-1965 standards, the calculated balance temperature is 18.1°C, almost equal to the conventional balance temperature of 18.3°C (65°F) upon which the degree-day tables are based. Therefore, the use of the conventional degree-day approach will give reasonably accurate results for detached houses insulated to earlier standards, as might be expected. However, as insulation levels are improved, the balance temperature for detached houses is lowered significantly, which accounts for the increasing inaccuracy of the conventional degree-day approach. The equivalent degree-day method is, of course, designed to account for this effect.

Table 4 shows that the balance temperature drops as low as 11°C for reasonably well-insulated apartments. The conventional degree-day method would be very

Table 3. Effect of Residence Vintage (Insulation Standards) on Heating-Season Performance

Region: Eastern Ontario (Ottawa)
House Type: Single-family detached
Heating System: Oil-fired furnace
Residence Vintage: Variable

$n = 0.5$ air changes/h
$H^*_S/A_{wi} = 41$ W/m^2
$H^*_W = 0.94$ kW

Vintage of Residence	Peripheral Losses	Infiltration Losses	Solar Gains	Wild Heat Gains	Net Heat Losses	Balance Temp, °C
Pre-1965	48.3 (90)	5.51 (10)	4.16 (7.7)	4.78 (8.9)	44.9 (83)	18.1
1966-71	36.0 (87)	5.51 (13)	4.16 (10)	4.78 (11)	32.6 (79)	16.8
1972-76	33.7 (86)	5.51 (14)	4.16 (11)	4.78 (12)	30.3 (77)	16.4
1977-81	30.2 (85)	5.51 (15)	4.16 (12)	4.78 (13)	26.8 (75)	15.9

Table 4. Effect of Residence Type on Heating-Season Performance

Region: Eastern Ontario (Ottawa)
Residence Type: Variable
Heating System: Oil-fired furnace
Residence Vintage: 1972-76

$n = 0.5$ air changes/h
$H^*_S/A_{wi} = 41$ W/m^2
$H^*_W = 0.94$ kW*

Type of Residence	Peripheral Losses	Infiltration Losses	Solar Gains	Wild Heat Gains	Net Heat Losses	Balance Temp, °C
Single-family detached	33.7 (86)	5.51 (14)	4.16 (11)	4.78 (12)	30.3 (77)	16.4
Attached & Duplex	23.0 (82)	4.98 (18)	2.79 (10)	4.78 (17)	20.4 (73)	15.4
Row-house	14.4 (76)	4.45 (24)	2.11 (11)	4.78 (25)	12.0 (64)	13.1
Apartment	7.56 (71)	3.05 (29)	0.79 (7)	4.01* (38)	5.81 (55)	11.1

* Wild heat for apartments does not include hot water system contribution, as hot water tank is part of central heating system, remote from individual units. $H^*_W = 0.8$ kW

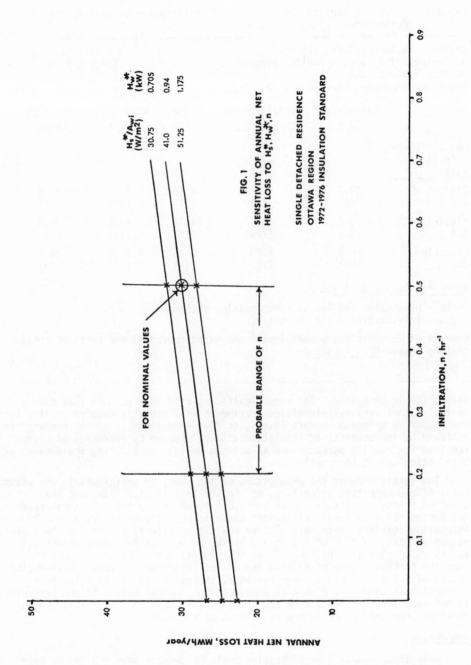

FIG. 1

SENSITIVITY OF ANNUAL NET
HEAT LOSS TO H_s^*, H_w^*, n

SINGLE DETACHED RESIDENCE
OTTAWA REGION
1972-1976 INSULATION STANDARD

H_s^*/A_{wi} (W/m²)	H_w^* (kW)
30.75	0.705
41.0	0.94
51.25	1.175

Table 5. Effect of Heating System Type and Infiltration on Heating Season
Performance

Region: Eastern Ontario (Ottawa) n = Variable
Residence Type: Single-family Detached H_S^*/A_{wi} = 41 W/m^2
Heating System: Variable H_W^* = 0.94 kW
Residence Vintage: 1972-76

Type of Heating System & Infiltration Rate	Peripheral Losses	Infiltration Losses	Solar Gains	Wild Heat Gains	Net Heat Losses
Oil/gas 0.5	33.7 (86)	5.51 (14)	4.16 (11)	4.78 (12)	30.3 (77)
Electric 0.4	30.8 (88)	4.39 (12)	4.16 (12)	4.78 (14)	26.3 (75)
Electric 0.2	30.8 (93)	2.20 (7)	4.16 (13)	4.78 (14)	24.1 (73)

Notes for Tables 3, 4 and 5

Units - Heat gains and losses - MWh/heating season
- Infiltration - air changes/h.

Numbers in brackets below heat losses and gains represent per cents of total
heating-season heat losses.

---

inaccurate in this case. The very significant wild heat gain in this case
indicates that very well-insulated apartments will probably require cooling in
the summer to maintain comfort levels, so that reductions in energy consumption
achieved by improvements in insulation will be opposed by increases of energy
required for cooling purposes and an optimum design, considering the demands of
both seasons, must be established.

The sensitivity of the predictions of the model to variations in the assumptions of the important parameters, H_S^*, H_W^* and infiltration rate, has been
examined. Some results are shown in Figure 1, where the total heating-season
net losses for a detached, oil-heated house in the Ottawa area with 1972-76
insulation standards have been plotted against infiltration rate for the nominal
values of H_S^*/A_{wi} = 41 W/m^2 and H_W^* = 0.94 kW, as well as for combinations of
values 25% higher and 25% lower than the nominal ones. Figure 1 shows that,
over the probable range of heating-season infiltrations for modern well-built
houses of 0.2 to 0.5 air changes per hour, the variation in predicted total
seasonal heating demand from that predicted for nominal values of the parameters
is not very large, ranging from about +6% to -17%, at the most. This conclusion
provides some confidence in the usefulness of the model.

CONCLUSIONS

Many other comparisons of heating performance have been and can be made
using the model described here. Preliminary estimates have been made of the
present total space-heating demand in Ontario using this method (6,7) but these
are not reported here. However, it appears that passive solar heat and wild
heat combined contribute about 10 to 15% at present to the total Ontario space-
heating demand. The utility of the equivalent degree-day concept, and the

importance of passive solar heating and wild heat in properly assessing the space-heating demands of modern well-insulated buildings has been shown.

It is anticipated that the model will be used in forecasting future space-heating demands in Ontario and in the evaluation of various energy-conservation policy options. It is hoped that it will play a role in the important task of reducing the growth rate of demand for energy in the residential sector.

APPENDIX

The heating-season average solar heat gain rate was determined by two independent methods. In one method, solar heat gains on an hour-by-hour basis throughout the heating season were determined by fundamental principles, based on the work of Stephenson (14) and Threlkeld (16), allowing for the effects of double-glazing and for heating-season average sun-hours in the Ottawa area. In this model, the residence window area was assumed to be randomly oriented in four directions (N,E,S,W). A computer program developed by one of the authors (JBF) was used in this analysis.

The other method used is a more approximate method, but since it emphasizes the various assumptions and estimates used in the analysis, it has been described in detail in the paper. Both methods yield essentially the same value of heating-season average solar heat gain rate.

REFERENCES

1. -
    Housebuilders' Survey. Central Mortgage and Housing Corporation,
    Government of Canada   1969

2. -
    Housebuilders' Survey. Central Mortgage and Housing Corporation,
    Government of Canada   1974

3. -
    Residential Standards, Canada (1965 & 1970). Associate Committee on
    the National Building Code, National Research Council of Canada
    NRCC 11562   1970

4. -
    Residential Standards, Canada (1975). Associate Committee on the
    National Building Code, National Research Council of Canada NRCC 13991 1975

5. Latta, J.K. Walls, Windows and Roofs for the Canadian Climate.
    Division of Building Research, National Research Council of Canada
    NRCC 13487

6. -
    Analysis of Technical Factors Affecting Residential Energy Consumption
    Section 2 of Report to Ontario Ministry of Energy.
    Energy Research Group, Carleton University   March, 1977
    To be published by Ontario Ministry of Energy

7. -
    The Residential and Commercial Sectors. Specification and Base
    Period Values. Section 1 of Report to Ontario Ministry of Energy.
    Informetrica Ltd., and Energy Research Group, Carleton University, Mar.1977
    To be published by Ontario Ministry of Energy

8. -
    ASHRAE Handbook: 1976 Systems Volume
    American Society of Heating, Refrigerating and Air-Conditioning
    Engineers   1976

9. -
    ASHRAE Handbook of Fundamentals. American Society of Heating,
    Refrigerating and Air-Conditioning Engineers   1972

10. Jordan, R.C., G.A. Erickson and R.R. Leonard. Infiltration
    Measurements in Two Research Houses. ASHRAE Journal   May,1963

11. Coblentz, C.W., and P.R. Achenbach. Field Measurements of Air
    Infiltration in Ten Electrically-heated Houses. ASHRAE Journal July,1963

12. Tamura, G.T., and A.G. Wilson. Air Leakage and Pressure
    Measurements on Two Occupied Houses. ASHRAE Journal   Dec.,1963

13. Tamura, G.T. Measurement of Air Leakage Characteristics of
    House Enclosures. Division of Building Research, Research Paper
    No.653, National Research Council of Canada NRCC 14950   1975

14. Stephenson, D.G. Tables of Solar Altitude, Azimuth, Intensity
    and Heat Gain Factors for Latitudes from 43 to 55 Degrees North.
    Division of Building Research, Technical Paper No.243, National
    Research Council of Canada NRC 9528   April,1967

15. Threlkeld, J.L., and R.C. Jordan. Direct Solar Radiation
    Available on Clear Days. Transactions of ASHRAE, 64, 45   1958

16. Threlkeld, J.L. Solar Irradiation of Surfaces on Clear Days.
    Transactions of ASHRAE, 69, 24   1963

17. Mitalas, G.P. Net Annual Heat Loss Factor Method for Estimating
    Heat Requirements of Buildings. Division of Building Research,
    Building Research Note No.117, National Research Council of
    Canada   Nov.,1976

18. Mitalas, G.P., and D.G. Stephenson. Absorption and Transmission
    of Thermal Radiation by Single and Double Glazed Windows.
    Division of Building Research, Research Paper No.173, National
    Research Council of Canada NRC 7104   Dec.,1962

19. Kimura, K., and D.G. Stephenson. Solar Radiation on Cloudy Days.
    Division of Building Research, Research Paper No.418, National
    Research Council of Canada NRC 11111   Nov.,1969

20. Webb, M.M. Personal communication. Atmospheric Environment
    Services   June,1977

21. Stephenson, D.G. Equations for Solar Heat Gain Through Windows.
    Division of Building Research, Research Paper No.255, National
    Research Council of Canada NRC 8527   June,1965

22. -
    Energy Application Survey - 1974. Regional and Marketing Branch
    Ontario Hydro   1975

23. -
    Household Facilities by Income and Other Characteristics, 1972
    Statistics Canada 64-202   1974

24. -
    Household Income Facilities and Equipment, Micro Data File - 1971
    Income. Consumer Income and Expenditure Division, Statistics Canada

25. -
    How You Use the Electricity You Use. Conserve Energy Publication
    Ontario Hydro   1976

26. -
     Energy Utilization and the Role of Electricity. Ontario Hydro
     Submission to the Ontario Royal Commission on Electric Power
     Planning (Porter Commission)   April,1976

27. -
     An Energy Policy for Canada. Phase 1. Vol.II Appendices.
     Dept. of Energy, Mines and Resources, Canada   1973

ACKNOWLEDGEMENTS

The work described in this paper was part of a project on residential and
commercial energy demand in Ontario for the Ontario Ministry of Energy. The
project was undertaken by Informetrica Ltd., Ottawa, which was responsible for
the economic aspects, and by the Energy Research Group of Carleton University,
which was responsible for the technical aspects.

The authors thank the Ontario Ministry of Energy for permission to publish
this paper. We are grateful to the many people who provided information which
assisted us greatly in the project. Among those who helped particularly with
the work described in the paper were:

D.G. Stephenson, Division of Building Research, NRCC

A. Konrad, Division of Building Research, NRCC

A.C.S. Hayden, Dept. of Energy, Mines and Resources

M.M. Webb, Atmospheric Environment Services, Downsview, Ont.

M. Robinson, Canadian Standards Association, Toronto

D. Brayshaw, Consumer's Gas, Toronto

C.E. Snow, Dept. of Consumer and Corporate Affairs

A. Levasseur, Dept. of Energy, Mines and Resources.

We apologize to anyone whose name has been inadvertently omitted.

SOLAR RADIATION TRANSMISSION AND HEAT TRANSFER THROUGH ARCHITECTURAL WINDOWS

R. VISKANTA AND E. D. HIRLEMAN

School of Mechanical Engineering
Purdue University
West Lafayette, Indiana, USA 47907

ABSTRACT

A detailed investigation of the thermal performance of architectural window systems has been undertaken. Unique aspects of the analysis included consideration of the spectral and directional nature of solar radiation transmission and thermal (longwave) radiation exchange processes of ordinary and radiation selective thin film coated glass window panes. Also, hourly meteorological and insolation data obtained from the U.S. Weather Bureau were used in realistic predictions of the long-term energy transfer through several window design configurations for both summer and winter conditions.

A number of window design parameters were investigated including multiple panes, gap spacings, filler gas thermal conductivity, window orientation and thin film coatings yielding high solar transmission but simultaneously high reflection in the thermal portion of the spectrum. The computations indicated that double glazed windows can cut window heat losses by more than 50% over conventional single panes for a typical winter month in Sterling, Virginia (suburb of Washington, D. C.). Somewhat smaller reductions in window heat gain are possible for a typical summer month in Phoenix, Arizona. Our calculations also predicted that the use of advanced window systems utilizing one thin film coating on a double glazed window results in approximately a 70% reduction in window losses over conventional single panes for the winter season.

INTRODUCTION

The growing disparity between the supply and demand for nonrenewable fossil fuels by the world market is an accepted fact. Also, a significant fraction of this demand (∿23% in the U.S.) is directly attributable to heating and cooling of buildings. As a result, there has been considerable interest generated during the past few years, as also attested by this Seminar, in detailed investigation of heat and mass transfer in buildings. The increased understanding would permit judicious building construction and remodeling activities to be effective in conserving energy and maximizing the use of renewable sources such as solar radiation.

Two beneficial ingredients of solar energy, light and heat, in the architectural and engineering design of buildings are recognized and are being considered in design solutions [1]. The recent study [2] has shown the architectural window to be an important factor in energy consumption for residential and commercial climate control. The study also revealed that for a typical home in the 40° - 42° North latitude in the United States, replacing of single pane windows by double pane windows would reduce the window heat losses by approximately 30% in the winter. Thus it appears that advanced window systems have potential for significant primary energy savings in certain climates.

A large number of studies dealing with transmission of solar radiation and heat transfer through architectural windows have been performed and a recent literature review [3] indicates that the problem is relatively well understood. However, a more realistic analysis of solar radiation transmission through and thermal radiation heat transfer between window panes coated with radiation selective films including spectral and directional effects is needed. The previous solar transmission and heat transfer computations have been performed using insolation and meteorological data averaged for the entire season [2]. The variation of solar gains and heat losses through windows during the diurnal cycle do not appear to have been reported in the open literature, and it remains to be determined whether calculations based on the averaged meteorological data are accurate.

The objective of this work is to gain improved understanding of solar energy transmission and heat transfer through double and multiple pane architectural windows on both diurnal and long-term bases. To this end, a realistic spectral analysis is presented to predict solar transmission through and thermal radiation exchange between window panes which are coated on one or both sides with radiation selective thin films. The total energy transmission through windows are then calculated using hourly meteorological data. Sample results are presented for single and double windows which are uncoated or coated with radiation selective thin-films that are highly transparent in the solar and highly reflecting in thermal parts of the spectrum.

ANALYSIS

This section summarizes the mathematical model for analyzing the thermal performance of architectural windows. The purpose, of course, is to calculate the long-term performance of the window system. In general, the factors which control the performance of a window can be divided into two groups: (1) the environmental, and (2) the window design. The environmental factors which affect the thermal performance of a window include the spectral, directional, diurnal and seasonal nature of solar radiation, and atmospheric conditions such as air temperature, wind speed and direction, relative humidity, cloud cover, haze and pollution. The window design parameters are such items as number, thickness, and composition of glass panes, window orientation and other factors such as coatings on the glass surfaces, multiple pane spacings and filler gas composition.

It was possible to account for the majority of these factors in the mathematical model. However, in order to make the analysis mathematically tractable, several idealizations were necessary but these introduced no appreciable errors. The details of the physical and mathematical model are summarized here.

Physical and Mathematical Model

The transport of energy through the window system is illustrated schematically in Figure 1. Total (direct plus reflected from ground or surrounding buildings) solar (short-wave) radiation on the outside surface of the window is partly reflected, partly absorbed by the glass panes and the remaining fraction is transmitted through the glass into the building. During the winter, for example, when the outside air temperature T_a is lower than the inside room temperature T_r, heat is transported by coupled convection-conduction-thermal (long-wave) radiation heat transfer, from the interior of the building to the exterior. Heat that is lost from the outside surface by convection and thermal radiation is under steady state conditions transported from the inside to the outside by

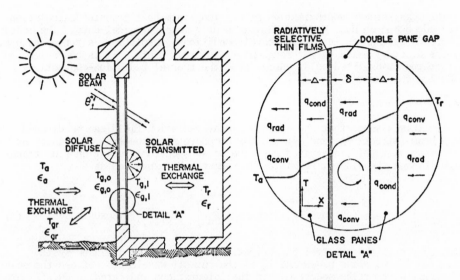

Figure 1. Schematic Diagram of a Window System

conduction through the glass. Between the inside glass surface and the air in the interior is exchanged by convection and by radiation with the surrounding walls.

Figure 1 shows schematically the various modes of heat transfer occurring in a double-pane window. The direction of the heat transfer indicated is for the winter case, i.e., $T_a < T_r$. For the "summer problem" when $T_a > T_r$, the direction of heat transfer is generally in, although during summer months the direction of heat transfer may be reversed during the diurnal cycle. On the other hand, the solar energy transport through the window is always positive. The heat exchange between the two panes occurs by free convection or pure conduction and radiation. Heat can also be transferred through the frame and by exfiltration through openings. This heat loss component must be replenished by an interior heat source to maintain comfortable interior temperatures. During the summer months there will be net heat gain through the windows, and the cooling loads would be increased.

The following assumptions are made in the analysis:

1. Heat transfer through the window is one-dimensional, i.e., temperature gradients parallel to the window and edge effects are ignored.
2. The glass is semitransparent to solar and opaque to thermal radiation.
3. A two-band spectral model for the analysis of radiation heat transfer, surface radiation properties and overall radiation characteristics is adequate.
4. The incident solar radiation field can be divided into a direct (beam) and diffuse components of flux. The diffuse component is uniform over the sky's hemisphere.
5. The interior walls are gray and diffuse emitters of thermal radiation, but for the purpose of calculating heat transfer through the glass panes, the walls are assumed not to reflect any short wave solar radiation back to the window.
6. The gas between the glass panes does participate in the transfer of radiation.
7. The system is assumed to be quasi-steady, responding instantaneously to changes in such parameters as insolation, air temperature or wind speed.

The quasi-steady approximation allows long-term heat transfer calculations to be made based on U.S. Weather Bureau hourly meteorological data. The model is "forced" by changes in environmental conditions (insolation, air temperature, wind speed, etc.) which vary throughout the diurnal cycle. It is convenient to separate the transmitted solar energy from the thermal radiation-convection-conduction components.

Solar Energy Input

If it is assumed that the incident solar radiation field can be divided into a beam (direct) $F_{b\lambda}$ and diffuse $F_{d\lambda}$ components of spectral flux incident on the window, total (integrated over the solar spectrum) radiation flux transmitted through the window, F_t, into the room can be expressed as

$$F_t = \int_0^{\lambda_s} T_{s(2),\lambda}(\theta°)\cos\theta° F_{b\lambda}(\theta°)d\lambda + \int_0^{\lambda_s} F_{d\lambda} \int_0^{\pi/2} T_{s(2),\lambda}(\theta)\cos\theta\sin\theta d\theta d\lambda \qquad (1)$$

where $T_{s(2),\lambda}$ is the spectral directional transmittance of a window system consisting of two glass panes and λ_s is the cut-off wavelength between the solar and thermal part of the spectrum. If the integrations indicated in Eq. (1) are performed, the total transmitted flux F_t can be expressed as

$$F_t = T_{s(2)}(\theta°)\cos\theta° F_b(\theta°) + T_{s(2)}F_d \qquad (2)$$

where the total directional transmittance $T_{s(2)}(\theta°)$ and the total hemispherical transmittance $T_{s(2)}$ are defined as

$$T_{s(2)}(\theta°) = \int_0^{\lambda_s} T_{s(2),\lambda}(\theta°)F_{b\lambda}(\theta°)d\lambda / \int_0^{\lambda_s} F_{b\lambda}(\theta°)d\lambda \qquad (3)$$

and

$$T_{s(2)} = \int_0^{\lambda_s} F_{d\lambda} \int_0^{\pi/2} T_{s(2),\lambda}(\theta)\cos\theta\sin\theta d\theta d\lambda / \int_0^{\lambda_s} F_{d\lambda}d\lambda \qquad (4)$$

The total solar flux absorbed by the first and second window panes of the two pane system can be expressed respectively as

$$F_{a,1} = A_{s(2),1}(\theta°)\cos\theta° F_b(\theta°) + A_{s(2),1}F_d \qquad (5)$$

and

$$F_{a,2} = A_{s(2),2}(\theta°)\cos\theta° F_b(\theta°) + A_{s(2),2}F_d \qquad (6)$$

where $A_{s(2),1}(\theta°)$ and $A_{s(2),1}$, for example, represent the total directional and hemispherical absorptances of the first pane in a two pane system. They are defined in a similar manner as the transmittances given by Eqs. (3) and (4). Spectral data (index of refraction and absorption coefficient) for clear float window glass was obtained from a manufacturer (PPG INDUSTRIES, Inc.) and used to predict spectral directional and then total directional and hemispherical radiation characteristics [5]. Thin film coated glass panes were analyzed following Taylor and Viskanta [6].

If it is assumed that the window is perpendicular to the earth surface, that the diffuse solar flux "sees" half of the sky's dome, that the diffuse radiation is distributed uniformly over the sky's dome, and that the ground and surroundings reflect diffuse the solar radiation incident on a horizontal surface, the beam and diffuse components of the solar radiation flux incident on the window are given by [4]

$$F_b = [1 + (\rho_{gr}/2)]F_b^o \tag{7}$$

and

$$F_d = \tfrac{1}{2}(1 + \rho_{gr})F_d^o \tag{8}$$

In this equation ρ_{gr} is the reflectance of the ground and $\cos\theta^o$ is defined as

$$\cos\theta^o = -\sin\delta\cos\phi\cos\gamma + \cos\delta\sin\phi\cos\gamma\cos\omega + \cos\delta\sin\gamma\sin\omega \tag{9}$$

where ϕ, δ, γ and ω are the latitude, declination, azimuth and hour angles, respectively. The spectral or total beam (F_b^o) and diffuse (F_d^o) components of the solar flux incident on the earth surface (horizontal) can be calculated following very simple [4, 7] and realistic [8] procedures available in the literature or determined from observed data on hourly, daily or monthly basis.

Heat Loss or Gain Through Window

If edge effects are neglected and the heat transfer through the window is assumed to be quasi-steady, energy balances on the various components of the system shown in Detail A of Figure 1 yields that the heat transfer rate from the outside surface of the glass to air (q_o), the heat transfer rate through the first glass pane (q_{g1}), across the gap (q_{gap}), through the second glass pane (q_{g2}), and from the inside of the room to the inside surface of the glass (q_i) must all be equal,

$$q_o = q_{g1} = q_{gap} = q_{g2} = q_i \tag{10}$$

An extension to triple and multiple pane glass windows is obvious. The different terms arising in Eq. (10) will be discussed below.

The heat transfer rate from the outside surface of the window is due to combined convection and radiation, and are expressed as

$$q_o = \bar{h}_o(T - T_{g1,o}) + \varepsilon_{g1,o}\sigma T_{g1,o}^4 - \alpha_{g1,o}\tilde{\varepsilon}_{e,o}\sigma \tilde{T}_{e,o}^4 \tag{11}$$

where $\varepsilon_{g1,o}$ and $\alpha_{g1,o}$ are respectively the emissivities and absorptivities of the outside glass surface and $\tilde{\varepsilon}_{e,o}\sigma\tilde{T}_{e,o}^4$ can be considered as the effective irradiation from the outside environment (atmosphere and ground). The total heat transfer to the inside glass surface can be written similarly. The specific equations to be used in the calculations for the convective heat transfer coefficients $\bar{h}_o$ and $\bar{h}_i$ as well as the effective irradiation $\tilde{\varepsilon}_{e,o}\sigma\tilde{T}_{e,o}^4$ on the outside and $\tilde{\varepsilon}_{e,i}\sigma\tilde{T}_{e,i}^4$ on the inside glass panes will be given later.

The total (conduction or convection plus radiation) heat transfer rate across the air gap between the window panes can be written as

$$q_{gap} = \tilde{k}_{e,i}(T_{g2,o} - T_{g1,i})/\Delta_1 + \tilde{\varepsilon}_{21}\sigma\tilde{T}_{g2,o}^4 - \tilde{\varepsilon}_{12}\sigma T_{g1,i}^4 \tag{12}$$

where $\tilde{k}_{e,1}$ is the effective conductivity of the vertical fluid layer in the gap and $\tilde{\varepsilon}_{21}$ and $\tilde{\varepsilon}_{12}$ are the radiation exchange factors between the glass panes. The factor $\tilde{\varepsilon}_{12}$ is defined as

$$\tilde{\varepsilon}_{12} = \int_0^\infty \int_0^{\pi/2} \frac{\varepsilon_{1\lambda}(\theta)\alpha_{2\lambda}(\theta)E_{b\lambda}(T_1)\sin2\theta d\theta d\lambda}{\{1-[1-\varepsilon_{1\lambda}(\theta)][1-\alpha_{2\lambda}(\theta)]\}} \Big/ \sigma T_1^4 \tag{13}$$

where $\varepsilon_{1\lambda}(\theta)$ and $\alpha_{2\lambda}(\theta)$ are the spectral directional emissivity and absorptivity of glass pane 1 and 2, respectively; $E_{b\lambda}$ is the spectral black body flux. Since the window panes may be coated with radiation selective films, the averaging indicated in Eq. (13) is necessary to account properly for both spectral and directional effects. The factor $\tilde{\varepsilon}_{21}$ is defined in analogous manner by reversing subscripts. The detailed evaluations and discussion of these factors as they pertain to two window panes can be found elsehwere [5].

If it is assumed that the solar radiation flux which is absorbed by the window pane 1 (F_{a1}) is uniformly distributed throughout the pane, the heat transfer rate by conduction at the outer surface of glass pane 1 is given by

$$q_{g1} = \frac{k_{g1}(T_{g1,o} - T_{g1,i})}{\Delta_1} \left[1 + \frac{F_{a1}}{k_{g1}(T_{g1,i} - T_{g1,o})/\Delta_1} \right] \tag{14}$$

The second term in the brackets of this equation accounts for modification of the conductive flux due to absorption of solar radiation by the pane. Of course, at night when $F_{a1} = 0$ the transport of heat through the glass is by conduction alone. The heat transfer rate at the outer surface of glass pane 2 can be expressed similarly. It is recognized that Eq. (14) is only approximate, but a more detailed and rigorous analysis [9] is not warranted here.

The analysis presented above is applicable for two-glass-pane window systems; extension to three or more panes is straightforward and is not developed here. The results of calculations presented in the paper are based on semi-empirical convective heat transfer correlations and other parameters which have intrinsic inaccuracies. The specific correlations and parameters used are discussed below. Accordingly, it would be unrealistic to conclude that the model would give highly accurate results for absolute solar transmission and heat transfer through the window. More importantly, the results permit comparison of merits of the different window systems, and with time, relative improvements of one system over another can be determined.

Auxiliary Equations and Parameters

The convective heat transfer coefficient from a glass pane exposed to outside winds is given by McAdams [10] as,

$$\bar{h}_o = 5.7 + 2.8W \tag{15}$$

where $\bar{h}_o$ is in SI system of units and W is the wind speed in m/s. There is considerable uncertainty in determining this coefficient, but data and more reliable correlations are not available.

Free convection heat transfer from vertical surfaces has been studied extensively and a number of correlations are available in the literature [11]. The particular equations used here are typical:

$$(\bar{h}H/k) = \begin{cases} 0.55(Gr_H Pr)^{\frac{1}{4}}, & 1700 < Gr_H Pr < 10^8 \\ 0.13(Gr_H Pr)^{\frac{1}{4}}, & Gr_H Pr > 10^8 \end{cases} \tag{16}$$

where Gr_H is the Grashoff number based on the height of the window H as a characteristic dimension.

The free convection heat transfer in a vertical gap of gas enclosed between two parallel walls has been reviewed by Krings and Olink [12]. The particular correlation employed in the calculations is

$$\tilde{k}_e/k = f[Gr_\delta Pr(\delta/H)^{1/3}] \tag{17}$$

where the function f is given by Linke [13] in graphical form for a range of the arguments. This equation has been experimentally verified for windows and recommended in the literature [3, 12].

The effective emitted flux of the environment $\tilde{\varepsilon}_{e,o}\sigma\tilde{T}_{e,o}^4$ was calculated by assuming that the window "sees" half of the sky's dome and half of the ground (ground, surrounding structures, etc.) such that

$$\tilde{\varepsilon}_{e,o}\sigma\tilde{T}_{e,o}^4 = \tfrac{1}{2}\sigma(\varepsilon_a T_a^4 + \varepsilon_{gr} T_{gr}^4) \tag{18}$$

The emittance of the atmosphere ε_a was calculated from an empirical equation given by Pailey et al. [14] which determines ε_a in terms of the partial pressure (relative humidity) of air, cloud cover and air temperature.

APPLICATIONS AND RESULTS

Parameters

As an application of the analysis presented we have carried out numerical calculations and include here some results. From the excessive number of independent parameter possibilities a range of parameters expected in practical applications was selected for use in sample results.

The window design parameters were taken as follows unless otherwise indicated: H = 1.0 m, $\Delta_1 = \Delta_2 = 0.3$ cm, $\delta = 1.0$ cm. The radiation characteristics of the window in the solar and thermal part of the spectrum were determined using the analysis reported elsewhere [5]. The window panes were assumed to be of clear float glass. The spectral absorption coefficient and index of refraction were obtained from data provided by the manufacturers (Libbey-Owens-Ford Co. and PPG Industries, Inc.). A number of different materials suitable for radiation selective thin films have been identified [15, 16, 6, 2]. As a typical material In_2O_3 was selected because the spectral complex index of refraction data were available. Presented in Figure 2 are the directional transmittances of the window system averaged over the solar spectrum using as a weighting function the spectral distribution of solar radiation incident on the surface of the earth for a typical air mass value of two [17]. The hemispherical radiation characteristics of the window in the solar and thermal part of the spectrum are listed in Table 1.

Hourly measured insolation and meteorological data for Sterling, Virginia (a Washington, D. C. suburb) and Phoenix, Arizona in 1964 were used in the calculations as representative of two different climatic regions. The calculations were performed for both a winter (January) and a summer (July) month. Ground reflectances (ρ_{gr}) of 0.2 and 0.8 were used in the calculations. The lower value is representative of bare and the higher of snow covered ground [17]. The ground emittance (ε_{gr}) was taken as 0.95, and the room air temperature was held at constant values of 20°C during the winter and 22°C during the summer throughout the simulation period.

Averaging of Meteorological Data

A comparison of the solar, thermal and total heat transfer through a window based on hourly meteorological data and on standard month day (SMD) is shown in Figure 3. The SMD was obtained by averaging the hourly meteorological data for a specific hour over the entire month giving 24 average values of each meteorological parameter [18]. The figure indicates large differences between two sets of results. The primary reason is that insolation on January 1, 1964 in Sterling, Virginia was very low because of cloudy skies (100%).

Figure 4 illustrates the results for the month of January based on hour-by-hour calculations and clearly indicates large day-to-day variations in heat transfer throughout the month. This is due to the large variations in the meteorological conditions and insolation during the winter season. However, the figure and Table 2 also show that the results obtained for the entire month are in reasonably good agreement with those based on the SMD. Hence, because the hour-by-hour calculations for a period of a month or longer are rather time

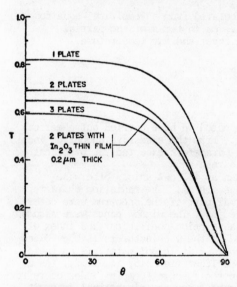

Figure 2. Solar Directional Transmittance of Different Windows: Clear Float Glass, $\Delta_1 = \Delta_2 = \Delta_3 = 0.3$ cm

Table 1. Total Hemispherical Solar Reflectance (R), Transmittance (T), and Absorptance (A) and Thermal Reflectivity (ρ), Emissivity (ε) of Uncoated Glass and Exchange Factor for the Gap ($\tilde{\varepsilon}_{12}$) for Clear Float Glass: $\Delta_1 = \Delta_2 = 0.3$ cm, In_2O_3 Thin Film Coating, Air Mass of 2.0.

System	Solar			Thermal		
	R	T	A	ρ	ε	$\tilde{\varepsilon}_{12}$
1 Plate	0.1421	0.7434	0.1145	0.1133	0.8867	---
2 Plates	0.2122	0.5968	0.1910	0.1133	0.8867	0.8085
2 Plates (0.1 μm thick film)	0.2377	0.5645	0.1978	0.1133	0.8867	0.1367
2 Plates (0.2 μm thick film)	0.2368	0.5597	0.2035	0.1133	0.8867	0.1367
3 Plates	0.2564	0.4901	0.2535	0.1133	0.8867	0.8085

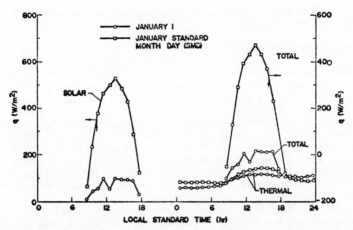

Figure 3. Diurnal Variation of Heat Transfer Through a Single Pane Window Based on Hourly and Standard Month Day Meteorological Data: Sterling, Virginia, January 1964, South Facing Window, ρ_{gr} = 0.8

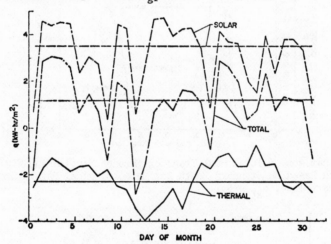

Figure 4. Comparison of Heat Transfer Through a Single Pane Window for the Month of January Based on Hourly and Standard Month Day Meteorological Data: Sterling, Virginia, January 1964, South Facing Window, ρ_{gr} = 0.8

Table 2. Comparison of Solar, Thermal and Total Heat Transfer (in kW-hr/m²) Through Single Pane (SP) and Double Pane (DP) Windows Based on Hourly and SMD Data: Sterling, Virginia, January 1964, South Facing Window, $\Delta_1 = \Delta_2 =$ 0.32 cm, δ = 1.27 cm, ρ_{g2} = 0.8.

	Solar		Thermal		Total	
	SP	DP	SP	DP	SP	DP
Hourly Data	98.0	77.8	-63.6	-25.9	35.1	51.9
SMD Data	108.2	85.6	-67.0	-26.3	41.2	59.3

consuming, the heat transfer results to be presented in the remaining of the paper are based on the SMD data.

Diurnal and Window Orientation Effects

The diurnal nature of air temperature variation and insolation based on SMD for January in Sterling, Virginia is given in Figure 5. The data indicate that the air temperature peaks later in the afternoon than the insolation. The total heat transfer through the window follows a similar trend. The results show the diurnal nature of the phenomena and make it clear that during periods of highly variable weather, typical of winter months, the SMD, daily, or monthly averaged meteorological data [18] should not be expected to estimate accurately the heat transfer through the window. The heat transfer results also indicate the importance of the window orientation. During the month of January there is a net loss (-55.4 kW-hr/m^2) of heat from a window facing north but a net gain (7.05 kW-hr/m^2) of heat through a window facing south.

A comparison of total heat transfer through single, double, and "advanced" double pane windows (with a thin film coating) is given in Figure 6. The results

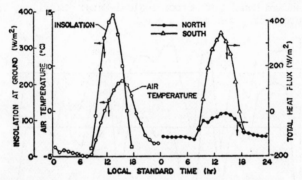

Figure 5. Comparison of Total Heat Transfer Through Single Pane Windows Facing North and South Based on Standard Month Day Meteorological Data: Sterling, Virginia, January 1964, ρ_{gr} = 0.2.

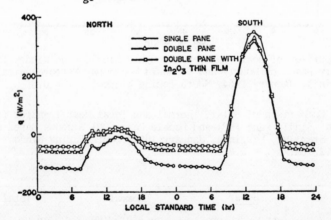

Figure 6. Comparison of Total Heat Transfer Through SP, DP and Advanced DP Windows Based on SMD Meteorological Data for January 1964 at Sterling, Virginia: Window Facing South, ρ_{gr} = 0.2.

show that a double pane (DP) window can decrease significantly the heat loss
from the building to the environment over that for a conventional single pane
(SP) design. No attempt has been made to optimize the width of the gap between
the inner and outer panes. Results reported by others [3, 12] indicate that the
gap between the glass panes should be greater than 1.0 cm wide, but the optimum
value depends on the room temperature, meteorological conditions, gas filler
(if any) between the panes, etc. Further significant reduction in window heat
loss is possible with advanced DP window systems. During the daylight hours
there is even a decrease in the transmission of solar radiation (see Figure 2)
into the room through DP windows as compared with a SP window.

The diurnal variation of the overall heat transfer coefficient, U, defined
by the equation

$$q_{thermal} = U(T_a - T_r) \qquad (19)$$

is illustrated in Figure 7. The trends for a winter month in Sterling, Virginia
and those for a summer month in Phoenix, Arizona are completely different. Since
the convective heat transfer across the gap is dominant, the thin film on a
conventional air filled DP window is not too effective in reducing the heat
transfer. However, as will be shown later, as soon as convection or conduction
in the gap is reduced, the thin film coating on the outer pane facing the room
becomes very effective in reducing the heat transfer.

Comparison of Uncoated and Coated Window Systems

A comparison of SP and DP window systems for winter and summer at Sterling,
Virginia and Phoenix, Arizona is given in Table 3. The results show that for
the winter month the thermal heat loss from the room to the environment is reduced
by more than 50%, whereas for the summer month the heat gain from the environment
to the building is decreased by about 40%. The decreased solar radiation transmission through a DP window during the summer is advantageous but not during the
winter months.

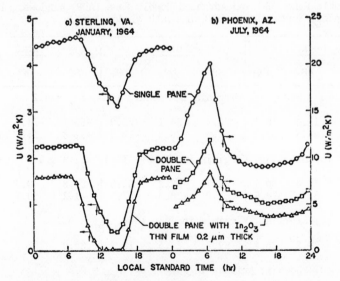

Figure 7. Comparison of Overall Heat Transfer Coefficients Based on SMD Meteorological Data: Window Facing South, ρ_{gr} = 0.2.

Inspection of the results presented in Table 3 reveals that the "advanced" double pane (ADP) window coated with a thin film and air in the gap reduces the heat loss during a winter month by about 30% over an ordinary DP window. The heat gain through an ADP window during a summer month is over 25% smaller than through a DP one.

Effect of Gap Gas Filler

The results presented for DP windows were for the case when the gap is filled with air. But air has a relatively high thermal conductivity compared to other gases, and it is desirable to examine the effect of the gas in the gap on the heat transfer through the window [2, 12]. Of course, the gas should have low thermal conductivity, be chemically inert, available, inexpensive and non-hazardous. Inert gases such as Krypton and Xenon have low conductivities but they are relatively expensive. Refrigerants (Freons) have also low conductivities but they may not be acceptable environmentally. In order to determine the sensitivity of the heat transfer through the window on the gas in the gap, air, Krypton and vacuum were considered. Some results are presented in Table 4.

The results show that from the heat transfer point of view the most desirable window design consists of an evacuated gap and a radiation selective thin film coating. Note that even for the north-facing window in the winter there is a net gain of energy through a thin film coated and evacuated DP window. This finding indicates that when the heat transfer by conduction-convection across the gap has been reduced, the thin film coating is very effective in reducing the heat transfer by radition. The design can be further improved, of course, at the expense of increased cost by using thin film coatings on both sides of the gap. The results given in the table also show the thin film coating to be as effective in reducing the heat transfer through the DP window as the evacuation of the gap.

Table 3. Comparison of Monthly Heat Transfer (in kW-hr/m^2) Through Single Pane (SP), Double Pane (DP) and Advanced Double Pane (ADP) Windows: In$_2$O$_3$ Thin Film 0.2 μm Thick, North Window (NW) or South Window (SW).

	Solar			Thermal			Total		
	SP	DP	ADP	SP	DP	ADP	SP	DP	ADP
(a) Sterling, Virginia in January 1964, $\rho_{gr} = 0.2$									
NW	15.1	11.1	10.4	-70.2	-34.1	-23.9	-55.1	-23.0	-13.5
SW	79.7	64.3	59.7	-67.9	-29.6	-19.3	-11.8	34.7	40.4
(b) Sterling, Virginia in January 1964, $\rho_{gr} = 0.8$									
NW	43.6	32.4	30.1	-69.4	-32.2	-22.0	-25.8	0.2	8.1
SW	108.2	85.6	79.5	-67.0	-26.3	-17.3	41.2	59.3	62.2
(c) Sterling, Virginia in July 1964, $\rho_{gr} = 0.2$									
SW	40.4	28.2	26.4	39.7	24.5	18.2	80.0	52.7	44.6
(d) Phoenix, Arizona in July 1964, $\rho_{gr} = 0.2$									
SW	31.8	22.2	20.7	75.2	44.0	31.7	107.0	66.2	52.4

Table 4. Comparison of Monthly Solar Transmission and Heat Transfer Through DP Windows of Different Design (in kW-hr/m^2): SMD Meteorological Data for January 1964 in Sterling, Virginia, ρ_g = 0.2, Window Facing North.

	Solar	Thermal	Total
No Film - Air	11.1	-34.1	-23.0
No Film - Krypton	11.1	-28.1	-17.0
No Film - Vacuum	11.1	-24.8	-13.7
With In$_2$O$_3$ Film - Air	10.4	-23.9	-13.5
With In$_2$O$_3$ Film - Krypton	10.4	-12.8	- 2.4
With In$_2$O$_3$ Film - Vacuum	10.4	- 6.2	4.2

CONCLUSIONS

A realistic analysis has been presented to predict the transmission of solar radiation and thermal heat transfer through architectural window systems. The emphasis here is on design considerations affecting heat transfer and quantitative calculations of potential energy savings rather than economics or cost-effectiveness. The thermal performance of a number of conventional and advanced window system design has been evaluated using actual U.S. Weather Bureau meteorological data.

The results indicate that double glazed windows are quite effective in reducing the heat loss from a building during the winter and decreasing the heat gain during the summer over a single glazed window. A selective thin film coating deposited on the outside pane facing the room reduces the heat transfer even more. A combination of design improvements such as reduction of conduction-convection across the gap by low conductivity gas filler and use of a thin film coating on the glass appears to result in significant reductions (in some cases as much as a factor of 10) in the total heat transfer through architectural windows for winter meteorological conditions.

As a result of these promising performance predictions, further study of the cost effectiveness of advanced architectural window design concepts including thin film coatings, number and spacing of glass panes, and low thermal conductivity gas fillers, etc., is warranted. There appears to be a high potential for significant payoff through energy savings by improved thermal designs of window systems.

ACKNOWLEDGMENTS

This work was initiated while the first author (R.V.) held a U.S. Senior Scientist Award from the Alexander von Humbolt Foundation. He also gratefully acknowledges the hospitality extended to him by the Technical University of Munich and helpful discussions with Prof. E. R. F. Winter. The authors wish to thank Mrs. Debbie Drake for editorial assistance and expert typing of the manuscript.

REFERENCES

1. Solar Radiation Considerations in Building Planning and Design, Proceedings of a Working Conference, National Academy of Sciences, Washington, D.C. (1976).

2. S. M. Berman and S. D. Silverstein, Editors, "Energy Conservation in Window Systems," American Physical Society (1975).

3. W. Prüfling, "Analyse des Energie-Durchgangs durch Fenster Verschiedener Bauarten," Diplomarbeit, Technische Universität München, München (1975).

4. J. A. Duffie and W. A. Beckman, Solar Thermal Processes, John Wiley and Sons, New York (1974).

5. R. Viskanta, D. L. Siebers and R. P. Taylor, "Radiation Characteristics of Multiple-Plate Glass System," submitted for publication to Inter. J. Heat Mass Transfer.

6. R. P. Taylor and R. Viskanta, "Spectral and Directional Radiation Characteristics of Thin-Film Coated Isothermal Transparent Plates," Wärme-und Stoffübertragung 8, 219-227 (1975).

7. N. Robinson, Editor, Solar Radiation, Elsevier Publishing Co., Amsterdam (1966).

8. W. Wang and G. A. Domoto, "The Radiative Effect of Aerosols in the Earth Atmosphere," J. Appl. Meteor. 5, 521-534 (1974).

9. R. Viskanta and E. E. Anderson, "Heat Transfer in Semitransparent Solids," in Advances in Heat Transfer, Volume 11, Academic Press, New York (1975), pp. 317-441.

10. W. C. McAdams, Heat Transmission, Third Edition, McGraw-Hill, New York (1954), p. 249.

11. H. Gröber, S. Erk and U. Grigull, Fundamentals of Heat Transfer, McGraw-Hill Book Co., New York (1961), pp. 320-323.

12. A. Krings and J. Th. Olink, "Wärmeübertragung durch Doppel-und Mehrfach-scheiben mit eingeschlossener Gassicht," Glastechn. Ber. 30, 175-182 (1957).

13. W. Linke, "Die Wärmeübertragung durch Thermopanefenster," Kaltechn. 8, 378-384 (1956).

14. P. P. Pailey, E. O. Macagno and J. F. Kennedy, "Winter-Regime Surface Heat Loss from Heated Stream," Institute of Hydraulic Research, The University of Iowa, IIHR Report No. 155, Iowa City, Iowa (1974).

15. H. Schröder, "Grossflächenbelegung von Glas zur Änderung der Strahlungs-durchlässigkeit," Glastechn. Berichte 39, 156-163 (1966).

16. R. Perssen, "Wärmeabsorbierende und wärmereflektierende Gläser," VDI Zeit. 110 (1), 9-15 (1968).

17. Handbook of Geophysics, Revised Edition, The McMillan Co., New York (1960), p. 16-19, and p. 2-17.

18. D. L. Siebers and R. Viskanta, "Comparison of Long Term Flat-Plate Solar Collector Performance Calculations Based on Averaged Meteorological Data," Solar Energy (in press).

AN IMPROVED MODEL FOR CALCULATION OF HEAT TRANSFER DUE TO SOLAR RADIATION THROUGH WINDOWS

L. I. KISS AND I. BENKÖ

Budapest Technical University
H-1111 Budapest, Hungary

ABSTRACT

The aim of work presented here was to construct a more reliable calculation procedure for the determination of input heat
rate,applying the basic concepts of heat transfer consequently
for single and double windows of different glass-types.

Experimental investigations were carried out on a small-
-size and on a bigger model,and also in a building.Spectral
characteristics of different glasses were measured by spectro-
-photometers.Direct and diffuse radiation was determined by
thermally compensated radiometers.Glasses in investigation were
of different types: reflective,absorbing/coloured/,opaque and
common ones.

Different forms of heat transfer through windows were considered in detail/conduction in glass with non-zero absorbance
or with a reflecting layer on one side,natural and forced convection regimes inside and outside,conduction and convection between
panes,self-radiation of glass depending on its spectral characteristics etc./.After an order-of-magnitude analysis a so called
"two-line,two-band" model was created.The system of equations
was solved numerically,and the resulting radiative and convective heat fluxes were tabulated for a series of various thermal
conditions.A set of nomograms was also worked out for practical
design purposes.

The results of calculations were checked by our measurements
and they were compared with values given by the so called "shading coefficient" method.Few results for comparison of different
window structures are given.

NOMENCLATURE

α coefficient of convective heat transfer, $W/m^2.K$

δ thickness of air layer between panes, m

ε emittance, dimensionless

λ heat conductivity, $W/m.K$; wavelength, μm

R reflectance, dimensionless

t temperature, centigrade

T absolute temperature, K

A absorptance, dimensionless

D transmittance, dimensionless

i incident angle of direct radiation, degree

I intensity of total radiation/diffuse and direct/, W/m^2

q surface heat flux, W/m^2

w velocity of air, m/s

σ Stefan-Boltzmann constant, $5,67.10^{-8}$ $W/m^2.K^4$

Nu Nusselt number, $\alpha.L/\lambda$, dimensionless

Gr Grashoff number, $g.\beta.L^3.\Delta t/\nu^2$, dimensionless

Pr Prandtl number, ν/a , dimensionless

Subscripts

e equivalent

1,2 the first and second glass respectively

12 reduced emittance for heat exchange between pane 1 and 2

out outer

H related to the height of wall

in inner, inlet

-,+ radiation property from back or front side of glass

λ value for monochromatic radiation

INTRODUCTION

Light-weight structures with large glass surfaces need continuous air conditioning during the whole year and a relatively
long cooling period.The economical optimum between investment
and running costs can be highly affected by the selection of various radiation-protective window types because of the different
heat loads on cooling system.Designing the cooling system the
maximum heat gain rate must be established.Maximal value of total heat gain is depending on meteorological,geographical
factors and also on the difficult heat transfer processes
through windows.

Many investigations were made by several researchers in the
field of determination solar heat gains,for example [1], [2],[3],
[4].In spite of the great number of the literature sources we
had to start our own research work because of the different ambient conditions,glass types and window structures.Simultaneous
experimental and theoretical work resulted in a relatively simple
calculation procedure,derived from the basic heat transfer prin-
ciples.The calculation model was used for comparison of window
structures with various types of glasses between different ambient and room conditions.

EXPERIMENTS

Experimental research was carried out in three fields: mo-
del-measurements for determination of heat gain,investigation of
spectral transmission characteristics of different glass panes
and "on site" measurements in buildings.

Two models,a small size one/300mmX300mm/ and a bigger
/1000mmX1500mm/ were constructed /Fig. 1 and 2/.The heat was measured not by calorimetry,but by heat flux gauges /six in the big
model/,fixed to the back wall.The temperature of the blackened
cooper back wall was kept constant by water from a thermostat.
One of the radiometers in the centre of the model had a surface
with reduced absorbance.Applying the so-called "two-radiometer"
method we could separate the convective and radiative parts of
the total heat flux reaching the back wall /Fig. 3/.

The temperatures of the inner and outer glass surfaces and
also of the frame were measured by thermocouples.

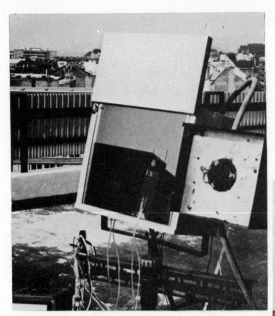

Fig. 1 Test apparatus for
    small-size glass samples.

Fig. 2 Test apparatus for measu-
    rement of heat transfer
    processes in complete
    window structures.

Fig. 3 Location of heat flux
    gauges.

Direct and diffuse radiation intensities were measured by
thermally compensated radiometers, one of them measuring total
hemisphere, the other measuring direct radiation.

Spectral transmission curves of various glass plates
were determined by two spetrophotometers, one of them working in
the visible, the other in the IR region.

Measurements in buildings included full day continuous registration of inside and outside climatic factors, radiation
heat load etc.

Evaluation of experiments was based on the following analysis.

HEAT TRANSFER THROUGH DOUBLE-GLAZED WINDOWS

There are several types of heat-protecting window structures.
We put the stress on the so called "thermopane" type factory
sealed double windows. In these structures the inner pane is generally made of common glass, the outer one can be of common, absorbing or reflecting glass.

The heat transfer through a double-glazing window even in
steady state is complicated by the simultaneous existence of conduction, convection and radiation, and exceptionally by the difficult radiative properties of glass.

Radiation properties

Directional dependence of radiation characteristics were taken from literature sources [1], [2], [4]. In Fig. 4 transmittance,
absorptance and reflectance are presented for two typical glass
types: a.-for common and b.-for reflective ones [4]. An important feature is that between an angle of $\pm$ 45° of incident rays
the changes in directional characteristics practically are not
too significant. Total radiative heat load on the outer surface
of a window consists direct and diffuse components. Diffuse component is a sum of sky scattered and ground reflected radiation.
In Budapest the pollution of air is high, solar radiation intensities are lower than those predicted by Moon for the given
air mass, and the diffuse component is high [13]. The data of Hungarian Meteorological Service were proven by our measurements,
the intensity of diffuse radiation was often well over that for
direct component.

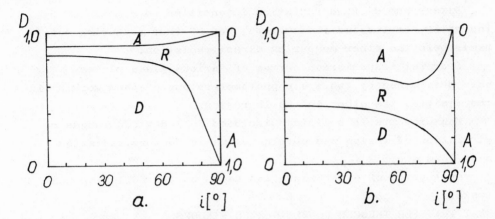

Fig. 4 Directional dependence of radiation properties of glass.
        a. Common glass pane.   b. Heat-protective glass pane.

Radiation property data for diffuse radiation of for total
hemispherical radiation of different glasses are relatively poor
in literature. So we started our analyses from data for normal radiation, supposing directional dependence to be similar fot those
in Fig. 4 .

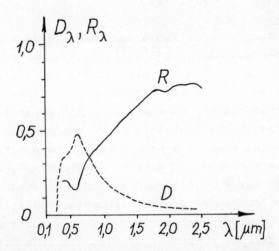

Fig. 5 Monochromatic transmittance and absorptance for
        "Thermaflex" glass pane.

Spectral dependence of radiation properties were investigated deeply in literature [1],[2],[3],[4],[6],[7],[10] etc., in many cases manufacturers also give these parameters, [11],[12].

A typical pair of curves for "Thermaflex" reflective glass can be seen in Fig. 5 .Spectral transmittance curves for a few commercial glass panes are shown in Fig. 6 . Common feature of

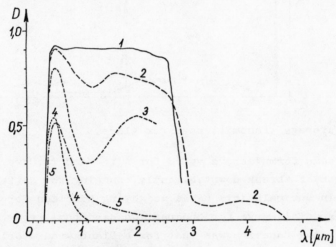

Fig. 6 Spectral transmittance of different glasses.
1. Highly transparent "Pyrex".
2. Common glass, 6 mm thick.
3. Athermic Clair, 4,8 mm thick.
4. Athermic, 5,5 mm thick.
5. Stopray.

this curves is the "break-down" point between 2,5 - 3,0 micrometers.

It makes possible to divide the spectrum into two parts, and to characterize glass by two average value of radiation properties. For example the transmittance /Fig. 7 / :

$$D_{SOLAR} = \frac{\int_0^\lambda I_{\lambda,SOLAR} . D_\lambda . d\lambda}{\int_0^\lambda I_{\lambda,SOLAR} . d\lambda} \quad , \qquad /1/$$

$$D_{THERMAL} = \frac{\int_\lambda^\infty I_{\lambda,BLACK\ BODY} . D_\lambda . d\lambda}{\int_\lambda^\infty I_{\lambda,BLACK\ BODY} . d\lambda} \quad . \qquad /2/$$

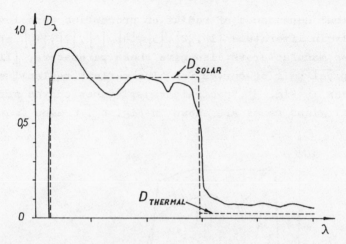

<u>Fig. 7</u> Average transmittances for glass.

The same formulae are valid for A , R and $\mathcal{E}$ .Here λ is the
wavelength of "break-down",slightly changing for different
glasses.In our calculations as weighting function for D_{SOLAR} in
lack of special curve for Budapest we used that from [2].In /2/
the weighting function was that for a Planck-radiator.

As an example,the average transmittance for common glass
plates of different thickness are shown in Fig. 8 ,depending on

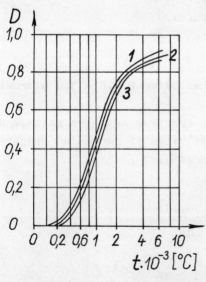

<u>Fig. 8</u> Average transmittance of
        common glass depending
        on the temperature of
        black body radiator.
        1. 1 mm thick.
        2. 2 mm thick.
        3. 5 mm thick.

the temperature of radiator.Calculated values were compared to

those by measurements in [6], the agreement was within the accuracy of graphical representation.

Conduction in glass

We made a serie of calculations for determination temperature profiles in glass, supposing a uniform absorptance in volu-
me.A typical temperature distribution without numerical values
is presented in Fig. 9 .Evaluating maximal temperature differences,their order-of-magnitude was a few Kelvins.During practical
calculations we neglected them.

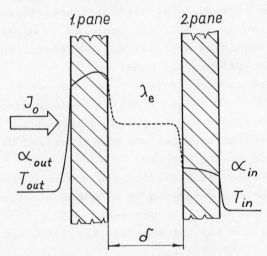

Fig. 9   Temperature profiles in glass panes.
         1st pane absorbing, 2nd common.

Convection in window structures

In our work we investigated all the combinations of forced
and natural convection in- and outside.The convective heat transfer coefficients and equivalent conductivity of air layer between panes were calculated on the base of widely known correlations of dimensionless quantites [7],[8],[9].Different types of
empirical equations were used,but final result were not highly
affected.

Basically the following equations were used.
Forced convection:

$$\alpha = 6{,}1 + 4{,}2.w \qquad \text{if} \quad w < 5 \text{ m/s} \quad , \qquad /3/$$
$$\alpha = 7{,}5.w^{0,8} \qquad \text{if} \quad w > 5 \text{ m/s} \quad .$$

Natural convection:

$$Nu_H = C \ /Gr_H.Pr/^n \qquad , \qquad\qquad\qquad /5/$$

$\underline{C}$ varies between 0,135 and 0,5, $\underline{n}$ between 0 and 1/3, depending
on the value of /Gr.Pr/ .

Equivalent conductivity of air layer :

$$\frac{\lambda_e}{\lambda} = 1 + \frac{0,0236 \ /Gr.Pr/^{1,39}}{/Gr.Pr/ + 1,01.10^4} \qquad . \qquad\qquad /6/$$

<u>Calculation model</u>

The different components of difficult heat transfer processes were calculated in detail, were compared to experimental data,
and an order-of-magnitude analysis was performed. After these
the following assumptions were made:

1. The glass plates are considered to be surfaces without thickness, i.e. conduction within glass is neglected, the surface
   temperatures of a pane are supposed to be equal.

2. Radiation properties of glass are characterized by two average
   values according to /1/ and /2/.

3. Radiation properties of glass for the sum of direct and diffuse
   radiation can be characterized by single factors, if the angle
   of incidence is between $\pm$ 45 degrees.

4. The glass panes and air between them have zero heat capacity,
   the process is steady state.

5. The temperatures of surrounding objects and outside air temperature are equal, and also the room air temperatures and room
   wall temperatures are equal.

On the base of above mentioned assumptions a so called "two-line,
two-band" model was created /Fig.10/.

The arrangement in Fig. 10 supposes the first pane to have
different emissivity from front and back side. This supposition
is necessary if first glass have a reflective layer on back side.
In this case the difference between forward and backward emittance is significant.

The heat balance of glass panes, after neglection of multiple
reflections of solar radiation:

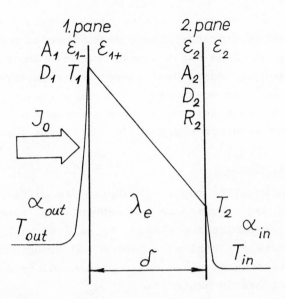

<u>Fig. 10</u> Two-line, two-band model of window structures.

$$/A_1 + D_1 R_2 A_1/I_o = \alpha_o/T_1 - T_o/ + \frac{\lambda_e}{\delta} /T_1 - T_2/ + \mathcal{E}_{1-} \, \mathcal{6}/T_1^4 - T_o^4/ +$$
$$+ \, \mathcal{E}_{12} \, \mathcal{6}/T_1^4 - T_2^4/ \quad , \qquad /7/$$

$$\frac{\lambda_e}{\delta} /T_1 - T_2/ + \, D_1 A_2 I_o + \mathcal{E}_{12} \, \mathcal{6}/T_1^4 - T_2^4/ = \alpha_i/T_2 - T_i/ + \mathcal{E}_2 \, \mathcal{6}/T_2^4 - T_1^4/ \; . \;\; /8/$$

Here A_1 , D_1 , A_2 , D_2 and R_2 correspond to solar spectrum; $\mathcal{E}_1$, $\mathcal{E}_2$ and $\mathcal{E}_{12}$ are calculated using average emissivities of glass, considering differences between forward and backward properties if necessary.

The total heat flux from the inner surface:

$$q = \underbrace{I_o D_1 D_2 \, + \, \mathcal{E}_2 \, \mathcal{6}/T_2^4 - T_i^4/}_{\text{radiative terms}} + \underbrace{\alpha_i/T_2 - T_i/}_{\text{convection}} \quad . \qquad /9/$$

A computer program was worked out to solve the set of equations. For the first cycle of computation estimate values of glass temperatures should be given. A version for Hewlett-Packard desktop computers was also worked out.

The system of equations $[7]$, $[8]$, $[9]$ was used for evaluation of measurements as well as a calculation model for design purposes.

Having a new window structures we started from literature data, and from spectral transmittance measurements. In most of cases, the second glass pane was a common one with known characteristics. So the number of independent radiation properties was reduced from 7 to 3-4. Inserting measurement data into /7/,/8/,/9/, from one experiment we could determine three independent radiation properties. This procedure was repeated for different experimental conditions.

RESULTS AND CONCLUSIONS

The above calculation model was applied to different Hungarian and foreign made thermopane and conventional windows, with several absorbing or reflecting glasses as well as with glass panes of non-smooth surface. A series of tables and nomograms for practical design purposes were worked out supposing different convective heat transfer regimes in and outside.

We made comparison between different radiation protective windows as a preparation for economical decisions application them in new buildings. We applied this method also for improvement of heat protection in metallurgical industry. In these cases the average radiation characteristics show a significant difference from those for solar radiation.

The structure of computer program makes possible to apply it for single pane window or to extend the validity for triple-glazed windows too.

As illustration we present here a comparison of three different window types at the same ambient conditions.

W/m²	I_o	q_{RD}	q_{RE}	q_{RAD}	q_{CONV}	q_{IN}
I. STOPRAY	700	144	58	202	36	238
II. HUNGAROPANE	700	517	38	555	22	577
III. SPEKTRA "B"	700	295	72	367	48	415

Natural convection on both side; $t_{IN} = 20^\circ C$, $t_{OUT} = 27^\circ C$. The data of table are schematically shown in Fig. 11 .

The three window structures represent three different types of windows. Hungaropane is a common thermopane one, Spectra "B" has absorbing character and Stopray is reflective type.

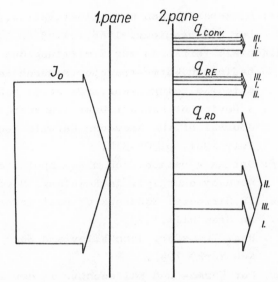

<u>Fig. 11</u>   Schematic energy flow diagram of heat transfer
              through various window structures.

SUMMARY

A calculation model for determination solar heat gains were
developed out on the base of basic heat transfer concepts and experimental data.Results given by our procedure were compared to
our measurements and to "shading-coefficient" theory.

The agreement between experimental results and our calculations were within a $\pm5\%$ range, while the use of shading coefficient
theory gave deviations $\pm2o\%$ .The agreement was the best if inner
and outer air temperatures were close to each other and radiation
intensities were high.In case of greater temperature differences
and low intensities,the deviations reached 25% from experimental
values.The two-line,two-band model fits well to the character of
difficult heat transfer process through windows.

LITERATURE

[1]   Odd Lyng   :Värmetransport genom fönster.Literature study
                 with bibliography.Rapport fran Byggforskningen,
                 Stockholm, 15. 1965.

[2]   J.YELLOT   :Transmission and absorption of solar radiation
                 in glass.Proceedings of the International Semi-
                 nar on Solar and Aeolian Energy. 1961. Plenum
                 Press. New York.

[3] F.Bruckmeyer: Glasoberflächen und Klimatisation.
                 Der Bauingenieur. 1968.vol.43.No.1.p.26-27.

[4] C.Snatzke,H.Kunzel: Verfahren zur Ermittlung des strahlungs-
                 bedingten.Wärme-transports durch Fenster.
                 Klima+Kälteingenieur,1974.No.5. p.207-216.

[5] R.Gardon   : A Review of Radiant Heat Transfer in Glass.
                 Journal of the American Ceramic Society,
                 July 1961. p.305-312.

[6] V.B.Weinberg: Optika v ustanovkach dlya ispolzovaniya sol-
                 nechnoy energiyi./in Russian/. Moscow 1959.

[7] W.M.Rohsenow,J.P.Hartnett: Handbook of Heat Transfer.
                 Mc Graw Hill.1973.

[8] M.Jakob   : Heat Transfer. John Wiley and Sons Inc.
                 New York, 1949.

[9] J.S.Cammerer: Der Wärme- und Kälteschutz in der Industrie.
                 Springer, 1962.

[10] Eichler,F. : Bauphysikalische entwurfslehre.
                 Verlag für Bauwesen. Berlin, 1971.

[11] Saint-Gobain: Product catalogue.

[12] Glaverbel   : Product catalogue.

[13] Szabó Gy.,Tárkányi Zs.: Insolation Data for Building Design.
                 1971.Budapest. /in Hungarian/.

AUTHOR INDEX

SUBJECT INDEX